MATERIALS FOR CIVIL AND CONSTRUCTION ENGINEERS

FOURTH EDITION IN SI UNITS

MICHAEL S. MAMLOUK

JOHN P. ZANIEWSKI

Pearson

Boston   Columbus   Indianapolis   New York   San Francisco   Hoboken
Amsterdam   Cape Town   Dubai   London   Madrid   Milan   Munich   Paris   Montreal   Toronto
Delhi   Mexico City   São Paulo   Sydney   Hong Kong   Seoul   Singapore   Taipei   Tokyo

Vice President and Editorial Director, ECS: *Marcia J. Horton*
Executive Editor: *Holly Stark*
Editorial Assistant: *Amanda Brands*
Acquistions Editor, Global Edition: *Abhijit Baroi*
Executive Marketing Manager: *Tim Galligan*
Director of Marketing: *Christy Lesko*
Product Marketing Manager: *Bram van Kempen*
Field Marketing Manager: *Demetrius Hall*
Marketing Assistant: *Jon Bryant*
Team Lead Program and Product Management: *Scott Disanno*
Program Manager: *Erin Ault*
Project Editor, Global Edition: *K.K. Neelakantan*

Global HE Director of Vendor Sourcing and Procurement: *Diane Hynes*
Director of Operations: *Nick Sklitsis*
Operations Specialist: *Maura Zaldivar-Garcia*
Senior Manufacturing Controller, Global Edition: *Jerry Kataria*
Creative Director: *Blair Brown*
Art Director: *Janet Slowik*
Media Production Manager, Global Edition: *Vikram Kumar*
Cover Design: *Lumina Datamatics*
Manager, Rights and Permissions: *Rachel Youdelman*
Full-Service Project Management: *SPi Global*

Pearson Education Limited
Edinburgh Gate
Harlow
Essex CM20 2JE
England

and Associated Companies throughout the world

Visit us on the World Wide Web at:
www.pearsonglobaleditions.com

British Library Cataloguing-in-Publication Data
A catalogue record for this book is available from the British Library

10 9 8 7 6 5 4 3 2 1

ISBN 10: 1-292-15440-3
ISBN 13: 978-1-292-15440-4

Typeset by SPi Global
Printed and bound in Malaysia.

CONTENTS

ONE
Materials Engineering Concepts 21

4   Contents

TWO
Nature of Materials   76

THREE

Steel 109

6 **Contents**

FOUR
Aluminum   168

FIVE
Aggregates   193

SIX

Portland Cement, Mixing Water, and Admixtures 243

SEVEN

Portland Cement Concrete 287

EIGHT

Masonry 369

NINE

Asphalt Binders and Asphalt Mixtures 385

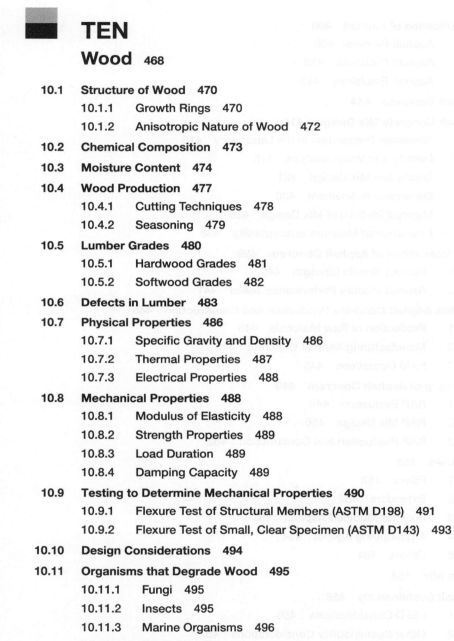

TEN

Wood 468

ELEVEN
Composites 522

Appendix

PREFACE

A basic function of civil and construction engineering is to provide and maintain the infrastructure needs of society. The infrastructure includes buildings, water treatment and distribution systems, waste water removal and processing, dams, and highway and airport bridges and pavements. Although some civil and construction engineers are involved in the planning process, most are concerned with the design, construction, and maintenance of facilities. The common denominator among these responsibilities is the need to understand the behavior and performance of materials. Although not all civil and construction engineers need to be material specialists, a basic understanding of the material selection process, and the behavior of materials, is a fundamental requirement for all civil and construction engineers performing design, construction, and maintenance.

Material requirements in civil engineering and construction facilities are different from material requirements in other engineering disciplines. Frequently, civil engineering structures require tons of materials with relatively low replications of specific designs. Generally, the materials used in civil engineering have relatively low unit costs. In many cases, civil engineering structures are formed or fabricated in the field under adverse conditions. Finally, many civil engineering structures are directly exposed to detrimental effects of the environment.

The subject of engineering materials has advanced greatly in the past few decades. As a result, many of the conventional materials have either been replaced by more efficient materials or modified to improve their performance. Civil and construction engineers have to be aware of these advances and be able to select the most cost-effective material or use the appropriate modifier for the specific application at hand.

This text is organized into three parts: (1) introduction to materials engineering, (2) characteristics of materials used in civil and construction engineering, and (3) laboratory methods for the evaluation of materials.

The introduction to materials engineering includes information on the basic mechanistic properties of materials, environmental influences, and basic material classes. In addition, one of the responsibilities of civil and construction engineers is the inspection and quality control of materials in the construction process. This requires an understanding of material variability and testing procedures. The atomic structure of materials is covered in order to provide basic understanding of material behavior and to relate the molecular structure to the engineering response.

The second section, which represents a large portion of the book, presents the characteristics of the primary material types used in civil and construction engineering: steel, aluminum, concrete, masonry, asphalt, wood, and composites. Since the discussion of concrete and asphalt materials requires a basic knowledge of aggregates, there is a chapter on aggregates. Moreover, since composites are gaining wide acceptance among engineers and are replacing many of the conventional materials, there is a chapter introducing composites.

The discussion of each type of material includes information on the following:

- Basic structure of the materials
- Material production process
- Mechanistic behavior of the material and other properties
- Environmental influences
- Construction considerations
- Special topics related to the material discussed in each chapter

Finally, each chapter includes an overview of various test procedures to introduce the test methods used with each material. However, the detailed description of the test procedures is left to the appropriate standards organizations such as the American Society for Testing and Materials (ASTM) and the American Association of State Highway and Transportation Officials (AASHTO). These ASTM and AASHTO standards are usually available in college libraries, and students are encouraged to use them. Also, there are sample problems in most chapters, as well as selected questions and problems at the end of each chapter. Answering these questions and problems will lead to a better understanding of the subject matter.

There are volumes of information available for each of these materials. It is not possible, or desirable, to cover these materials exhaustively in an introductory single text. Instead, this book limits the information to an introductory level, concentrates on current practices, and extracts information that is relevant to the general education of civil and construction engineers.

The content of the book is intended to be covered in one academic semester, although quarter system courses can definitely use it. The instructor of the course can also change the emphasis of some topics to match the specific curriculum of the department. Furthermore, since the course usually includes a laboratory portion, a number of laboratory test methods are described. The number of laboratory tests in the book is more than what is needed in a typical semester in order to provide more flexibility to the instructor to use the available equipment. Laboratory tests should be coordinated with the topics covered in the lectures so that the students get the most benefit from the laboratory experience.

The first edition of this textbook served the needs of many universities and colleges. Therefore, the second edition was more of a refinement and updating of the book, with some notable additions. Several edits were made to the steel chapter to improve the description of heat treatments, phase diagram, and the heat-treating effects of welding. Also, a section on stainless steel was added, and current information on the structural uses of steel was provided. The cement and concrete chapters have been augmented with sections on hydration-control admixtures, recycled wash water, silica fume, self-consolidating concrete, and flowable fill. When the first edition was published, the Superpave mix design method was just being introduced to the industry. Now Superpave is a well-established method that has been field tested and revised to better meet the needs of the paving community. This development required a complete revision to the asphalt chapter to accommodate the current methods and procedures for both Performance Grading of asphalt binders and the Superpave mix design method. The chapter on wood was revised to provide information on recent manufactured wood products that became available in the past several years. Also, since fiber-reinforced polymer composites have been more commonly used in retrofitting old and partially damaged structures, several examples were added in the chapter on composites. In the laboratory manual, an experiment on dry-rodded unit weight of aggregate that is used in portland cement concrete (PCC) proportioning was added, and the experiment on creep of asphalt concrete was deleted for lack of use.

What's New in This Edition

The primary focus of the updates presented in this edition was on the sustainability of materials used in civil and construction engineering. The information on sustainability in Chapter 1 was updated and expanded to include recent information on sustainability. In addition, a section was added to Chapters 3 through 11 describing the sustainability considerations of each material. The problem set for each chapter was updated and increased to provide some fresh Exercises and to cover other topics discussed in the chapter. References were updated and increased in all chapters to provide students with additional reading on current issues related to different materials. Many figures were added and others were updated throughout the book to provide visual illustrations to students. Other specific updates to the chapters include:

- Chapter 1 now includes a more detailed section on viscoelastic material behavior and a new sample problem.
- Chapter 3 was updated with recent information about the production of steel.
- A sample problem was added to Chapter 5 about the water absorbed by aggregate in order to highlight the fact that absorbed water is not used to hydrate the cement or improve the workability of plastic concrete.
- Two new sample problems were added to Chapter 6 on the acceptable criteria of mixing water and to clarify the effect of water reducer on the properties of concrete.
- Chapter 7 was augmented with a discussion of concrete mixing water and a new sample problem. A section on pervious concrete was added to reflect the current practice on some parking lots and pedestrian walkways.
- Chapter 9 was updated with reference to the multiple stress creep recovery test, and the information about the immersion compression test was replaced with the tensile strength ratio method to reflect current practices. The selection of the binder was refined to incorporate the effect of load and speed. The section on the diameteral tensile resilient modulus was removed for lack of use. The sample problem on the diameteral tensile resilient modulus was also removed and replaced with a sample problem on the freeze-thaw test and the tensile strength ratio.

- Chapter 10 was updated to include more information about wood deterioration and preservation. The first two sample problems were edited to provide more accurate solutions since the shrinkage values used in wood are related to the green dimensions at or above the fiber saturation point (FSP), not the dry dimensions. The third sample problem was expanded to demonstrate how to determine the modulus of elasticity using the third-point bending test.
- Chapter 11 was updated to reflect information about the effective length of fibers and the ductility of fiber-reinforced polymers (FRP). The discussion was expanded with several new figures to incorporate fibers, fabrics, laminates, and composites used in civil engineering applications. The first sample problem was expanded to apply other concepts covered in the chapter.
- The laboratory manual in the appendix was updated to include two new experiments on creep in polymers and the effect of fiber orientation on the elastic modulus of fiber reinforced composites. The experiment on the tensile properties of composites was updated. This would allow more options to the instructor to choose from in assigning lab experiments to students.

Acknowledgments

The authors would like to acknowledge the contributions of many people who assisted with the development of this new edition. First, the authors wish to thank the reviewers and recognize the fact that most of their suggestions have been incorporated into the fourth edition, in particular Dr. Dimitrios Goulias of University of Maryland, Tyler Witthuhn of the National Concrete Masonry Association, Mr. Philip Line of American Wood Council, Dr. Baoshan Huang of University of Tennessee, and Dr. Steve Krause of Arizona State University. Appreciation is also extended to Drs. Narayanan Neithalath, Shane Underwood, Barzin Mobasher, and Kamil Kaloush of Arizona State University for their valuable technical contributions. The photos of FRP materials contributed by Dr. Hota GangaRao of the Constructed Facilities Center at West Virginia University are appreciated. Appreciation also goes to Dr. Javed Bari, formerly with the Arizona Department of Transportation for his contribution in preparing the slides and to Dr. Mena Souliman of the University of Texas at Tyler for his contribution in the preparation of the solution manual.

Acknowledgments for the Global Edition

Pearson would like to thank and acknowledge Weena Lokuge of the University of Southern Queensland and Tayfun Altug Soylev of Gezbe Technical University for contributing to the Global Edition, and Pang Sze Dai of the National University of Singapore, Prakash Nanthagopalan of the Indian Institute of Technology Bombay, and Supratic Gupta of the Indian Institute of Technology Delhi for reviewing the Global Edition.

About the Authors

Michael S. Mamlouk is a Professor of Civil, Environmental, and Sustainable Engineering at Arizona State University. He has many years of experience in teaching courses of civil engineering materials and other related subjects at both the undergraduate and graduate levels. He has been actively involved in teaching materials and pavement design courses to practicing engineers. Dr. Mamlouk has directed many research projects and is the author of numerous publications in the fields of pavement and materials. He is a professional engineer in the state of Arizona. Dr. Mamlouk is a fellow of the American Society of Civil Engineers and a member of several other professional societies.

John P. Zaniewski is the Asphalt Technology Professor in the Civil and Environmental Engineering Department of West Virginia University. Dr. Zaniewski earned teaching awards at both WVU and Arizona State University. In addition to materials, Dr. Zaniewski teaches graduate and undergraduate courses in pavement materials, design and management, and construction engineering and management. Dr. Zaniewski has been the principal investigator on numerous research projects for state, federal, and international sponsors. He is a member of several professional societies and has been a registered engineer in three states. He is the director of the WV Local Technology Assistance Program and has been actively involved in adult education related to pavement design and materials.

CHAPTER

MATERIALS ENGINEERING CONCEPTS

Materials engineers are responsible for the selection, specification, and quality control of materials to be used in a job. These materials must meet certain classes of criteria or materials properties (Ashby and Jones, 2011). These classes of criteria include

- economic factors
- mechanical properties
- nonmechanical properties
- production/construction considerations
- aesthetic properties

In addition to this traditional list of criteria, civil engineers must be concerned with environmental quality. In 1997, the ASCE *Code of Ethics* was modified to include "sustainable development" as an ethics issue. Sustainable development basically recognizes the fact that our designs should be sensitive to the ability of future generations to meet their needs. There is a strong tie between the materials selected for design and sustainable development.

When engineers select the material for a specific application, they must consider the various criteria and make compromises. Both the client and the purpose of the facility or structure dictate, to a certain extent, the emphasis that will be placed on the different criteria.

Civil and construction engineers must be familiar with materials used in the construction of a wide range of structures. Materials most frequently used include steel, aggregate, concrete, masonry, asphalt, and wood. Materials used to a lesser extent include aluminum, glass, plastics, and fiber-reinforced composites. Geotechnical engineers make a reasonable case for including soil as the most widely used engineering material, since it provides the basic support for all civil engineering structures. However, the properties of soils will not be discussed in this text because soil properties are generally the topic of a separate course in civil and construction engineering curriculums.

Recent advances in the technology of civil engineering materials have resulted in the development of better quality, more economical, and safer materials. These materials are commonly referred to as high-performance materials. Because more is known about the molecular structure of materials and because of the continuous research efforts by scientists and engineers, new materials such as polymers, adhesives, composites, geotextiles, coatings, cold-formed metals, and various synthetic products are competing with traditional civil engineering materials. In addition, improvements have been made to existing materials by changing their molecular structures or including additives to improve quality, economy, and performance. For example, superplasticizers have made a breakthrough in the concrete industry, allowing the production of much stronger concrete. Joints made of elastomeric materials have improved the safety of high-rise structures in earthquake-active areas. Lightweight synthetic aggregates have decreased the weight of concrete structures, allowing small cross-sectional areas of components. Polymers have been mixed with asphalt, allowing pavements to last longer under the effect of vehicle loads and environmental conditions.

The field of fiber composite materials has developed rapidly in the past 30 years. Many recent civil engineering projects have used fiber-reinforced polymer composites. These advanced composites compete with traditional materials due to their higher strength-to-weight ratio and their ability to overcome such shortcomings as corrosion. For example, fiber-reinforced concrete has much greater toughness than conventional portland cement concrete. Composites can replace reinforcing steel in concrete structures. In fact, composites have allowed the construction of structures that could not have been built in the past.

The nature and behavior of civil engineering materials are as complicated as those of materials used in any other field of engineering. Due to the high quantity of materials used in civil engineering projects, the civil engineer frequently works with locally available materials that are not as highly refined as the materials used in other engineering fields. As a result, civil engineering materials frequently have highly variable properties and characteristics.

This chapter reviews the manner in which the properties of materials affect their selection and performance in civil engineering applications. In addition, this chapter reviews some basic definitions and concepts of engineering mechanics required for understanding material behavior. The variable nature of material properties is also discussed so that the engineer will understand the concepts of precision and accuracy, sampling, quality assurance, and quality control. Finally, instruments used for measuring material response are described.

1.1   Economic Factors

The economics of the material selection process are affected by much more than just the cost of the material. Factors that should be considered in the selection of the material include

- availability and cost of raw materials
- manufacturing costs

- transportation
- placing
- maintenance

The materials used for civil engineering structures have changed over time. Early structures were constructed of stone and wood. These materials were in ready supply and could be cut and shaped with available tools. Later, cast iron was used, because mills were capable of crudely refining iron ore. As the industrial revolution took hold, quality steel could be produced in the quantities required for large structures. In addition, portland cement, developed in the mid-1800s, provided civil engineers with a durable inexpensive material with broad applications.

Due to the efficient transportation system in the United States, availability is not as much of an issue as it once was in the selection of a material. However, transportation can significantly add to the cost of the materials at the job site. For example, in many locations in the United States, quality aggregates for concrete and asphalt are in short supply. The closest aggregate source to Houston, Texas, is 150 km from the city. This haul distance approximately doubles the cost of the aggregates in the city, and hence puts concrete at a disadvantage compared with steel.

The type of material selected for a job can greatly affect the ease of construction and the construction costs and time. For example, the structural members of a steel-frame building can be fabricated in a shop, transported to the job site, lifted into place with a crane, and bolted or welded together. In contrast, for a reinforced concrete building, the forms must be built; reinforcing steel placed; concrete mixed, placed, and allowed to cure; and the forms removed. Constructing the concrete frame building can be more complicated and time consuming than constructing steel structures. To overcome this shortcoming, precast concrete units commonly have been used, especially for bridge construction.

All materials deteriorate over time and with use. This deterioration affects both the maintenance cost and the useful life of the structure. The rate of deterioration varies among materials. Thus, in analyzing the economic selection of a material, the life cycle cost should be evaluated in addition to the initial costs of the structure.

1.2 Mechanical Properties

The mechanical behavior of materials is the response of the material to external loads. All materials deform in response to loads; however, the specific response of a material depends on its properties, the magnitude and type of load, and the geometry of the element. Whether the material "fails" under the load conditions depends on the failure criterion. Catastrophic failure of a structural member, resulting in the collapse of the structure, is an obvious material failure. However, in some cases, the failure is more subtle, but with equally severe consequences. For example, pavement may fail due to excessive roughness at the surface, even though the stress levels are well within the capabilities of the material. A building may have to be closed due to excessive vibrations by wind or other live loads, although it could be structurally sound. These are examples of *functional* failures.

1.2.1 ▨ Loading Conditions

One of the considerations in the design of a project is the type of loading that the structure will be subjected to during its design life. The two basic types of loads are static and dynamic. Each type affects the material differently, and frequently the interactions between the load types are important. Civil engineers encounter both when designing a structure.

Static loading implies a sustained loading of the structure over a period of time. Generally, static loads are slowly applied such that no shock or vibration is generated in the structure. Once applied, the static load may remain in place or be removed slowly. Loads that remain in place for an extended period of time are called *sustained* (dead) loads. In civil engineering, much of the load the materials must carry is due to the weight of the structure and equipment in the structure.

Loads that generate a shock or vibration in the structure are *dynamic* loads. Dynamic loads can be classified as *periodic, random,* or *transient,* as shown in Figure 1.1 (Richart et al., 1970). A periodic load, such as a harmonic or sinusoidal load, repeats itself with time. For example, rotating equipment in a building can produce a vibratory load. In a random load, the load pattern never repeats, such as that produced by earthquakes. Transient load, on the other hand, is an impulse load that is applied over a short time interval, after which the vibrations decay until the

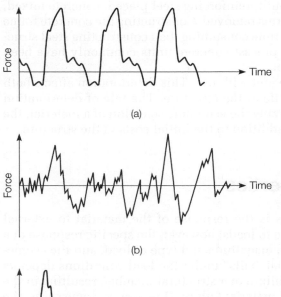

(a)

(b)

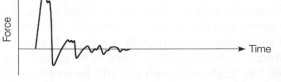

(c)

FIGURE 1.1 Types of dynamic loads: (a) periodic, (b) random, and (c) transient.

system returns to a rest condition. For example, bridges must be designed to withstand the transient loads of trucks.

1.2.2 ■ Stress–Strain Relations

Materials deform in response to loads or forces. In 1678, Robert Hooke published the first findings that documented a linear relationship between the amount of force applied to a member and its deformation. The amount of deformation is proportional to the properties of the material and its dimensions. The effect of the dimensions can be normalized. Dividing the force by the cross-sectional area of the specimen normalizes the effect of the loaded area. The force per unit area is defined as the stress σ in the specimen (i.e., σ = force/area). Dividing the deformation by the original length is defined as strain ε of the specimen (i.e., ε = change in length/original length). Much useful information about the material can be determined by plotting the stress–strain diagram.

Figure 1.2 shows typical uniaxial tensile or compressive stress–strain curves for several engineering materials. Figure 1.2(a) shows a linear stress–strain relationship up to the point where the material fails. Glass and chalk are typical of materials exhibiting this tensile behavior. Figure 1.2(b) shows the behavior of steel in tension. Here, a linear relationship is obtained up to a certain point (proportional limit), after which the material deforms without much increase in stress. On the other hand, aluminum alloys in tension exhibit a linear stress–strain relationship up to the proportional limit, after which a nonlinear relation follows, as illustrated in Figure 1.2(c). Figure 1.2(d) shows a nonlinear relation throughout the whole range. Concrete and other materials exhibit this relationship, although the first portion of the curve for concrete is very close to being linear. Soft rubber in tension differs from most materials in such a way that it shows an almost linear stress–strain relationship followed by a reverse curve, as shown in Figure 1.2(e).

1.2.3 ■ Elastic Behavior

If a material exhibits true elastic behavior, it must have an instantaneous response (deformation) to load, and the material must return to its original shape when the load is removed. Many materials, including most metals, exhibit elastic behavior, at

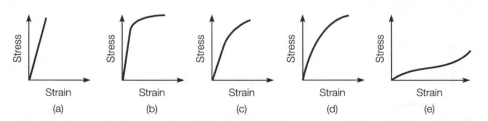

FIGURE 1.2 Typical uniaxial stress–strain diagrams for some engineering materials: (a) glass and chalk, (b) steel, (c) aluminum alloys, (d) concrete, and (e) soft rubber.

least at low stress levels. As will be discussed in Chapter 2, elastic deformation does not change the arrangement of atoms within the material, but rather it stretches the bonds between atoms. When the load is removed, the atomic bonds return to their original position.

Young observed that different elastic materials have different proportional constants between stress and strain. For a homogeneous, isotropic, and linear elastic material, the proportional constant between normal stress and normal strain of an axially loaded member is the *modulus of elasticity* or *Young's modulus, E,* and is equal to

$$E = \frac{\sigma}{\varepsilon} \tag{1.1}$$

where σ is the normal stress and ε is the normal strain.

In the axial tension test, as the material is elongated, there is a reduction of the cross section in the lateral direction. In the axial compression test, the opposite is true. The ratio of the lateral strain, ε_l, to the axial strain, ε_a, is *Poisson's ratio,*

$$v = \frac{-\varepsilon_l}{\varepsilon_a} \tag{1.2}$$

Since the axial and lateral strains will always have different signs, the negative sign is used in Equation 1.2 to make the ratio positive. Poisson's ratio has a theoretical range of 0.0 to 0.5, where 0.0 is for a compressible material in which the axial and lateral directions are not affected by each other. The 0.5 value is for a material that does not change its volume when the load is applied. Most solids have Poisson's ratios between 0.10 and 0.45.

Although Young's modulus and Poisson's ratio were defined for the uniaxial stress condition, they are important when describing the three-dimensional stress–strain relationships, as well. If a homogeneous, isotropic cubical element with linear elastic response is subjected to normal stresses σ_x, σ_y, and σ_z in the three orthogonal directions (as shown in Figure 1.3), the normal strains ε_x, ε_y, and ε_z can be computed by the *generalized Hooke's law,*

$$\varepsilon_x = \frac{\sigma_x - v(\sigma_y + \sigma_z)}{E}$$

$$\varepsilon_y = \frac{\sigma_y - v(\sigma_z + \sigma_x)}{E}$$

$$\varepsilon_z = \frac{\sigma_z - v(\sigma_x + \sigma_y)}{E} \tag{1.3}$$

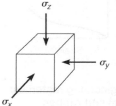

FIGURE 1.3   Normal stresses applied on a cubical element.

Sample Problem 1.1

A cube made of an alloy with dimensions of 50 mm $\times$ 50 mm $\times$ 50 mm is placed into a pressure chamber and subjected to a pressure of 90 MPa. If the modulus of elasticity of the alloy is 100 GPa and Poisson's ratio is 0.28, what will be the length of each side of the cube, assuming that the material remains within the elastic region?

Solution

$$\varepsilon_x = [\sigma_x - \nu(\sigma_y + \sigma_z)]/E = [-90 - 0.28 \times (-90 - 90)]/100000$$
$$= -0.000396 \text{ m/m}$$
$$\varepsilon_y = \varepsilon_z = -0.000396 \text{ m/m}$$
$$\Delta x = \Delta y = \Delta z = -0.000396 \times 50 = -0.0198 \text{ mm}$$
$$L_{new} = 50 - 0.0198 = 49.9802 \text{ mm}$$

Linearity and elasticity should not be confused. A *linear material*'s stress–strain relationship follows a straight line. An *elastic material* returns to its original shape when the load is removed and reacts instantaneously to changes in load. For example, Figure 1.4(a) represents a linear elastic behavior, while Figure 1.4(b) represents a nonlinear elastic behavior.

For materials that do not display any linear behavior, such as concrete and soils, determining a Young's modulus or elastic modulus can be problematical. There are several options for arbitrarily defining the modulus for these materials. Figure 1.5 shows four options: the initial tangent, tangent, secant, and chord moduli. The *initial tangent modulus* is the slope of the tangent of the stress–strain curve at the origin. The *tangent modulus* is the slope of the tangent at a point on the stress–strain curve. The *secant modulus* is the slope of a chord drawn between the origin and an arbitrary point on the stress–strain curve. The *chord modulus* is the slope of a chord drawn between two points on the stress–strain curve. The selection of which modulus to use for a nonlinear material depends on the stress or strain level at which the material typically is used. Also, when determining the tangent, secant, or chord modulus, the stress or strain levels must be defined.

Table 1.1 shows typical modulus and Poisson's ratio values for some materials at room temperature. Note that some materials have a range of modulus values rather

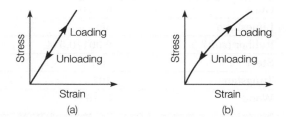

FIGURE 1.4   Elastic behavior: (a) linear and (b) nonlinear.

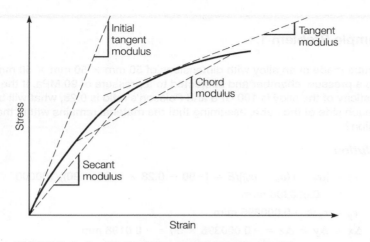

FIGURE 1.5   Methods for approximating modulus.

than a distinct value. Several factors affect the modulus, such as curing level and proportions of components of concrete or the direction of loading relative to the grain of wood.

1.2.4 ■ Elastoplastic Behavior

For some materials, as the stress applied on the specimen is increased, the strain will proportionally increase up to a point; after this point, the strain will increase with little additional stress. In this case, the material exhibits linear elastic behavior

TABLE 1.1   Typical Modulus and Poisson's Ratio Values (Room Temperature)

Material	Modulus GPa	Poisson's Ratio
Aluminum	69–75	0.33
Brick	10–17	0.23–0.40
Cast iron	75–169	0.17
Concrete	14–40	0.11–0.21
Copper	110	0.35
Epoxy	3–140	0.35–0.43
Glass	62–70	0.25
Limestone	58	0.2–0.3
Rubber (soft)	0.001–0.014	0.49
Steel	200	0.27
Tungsten	407	0.28
Wood	6–15	0.29–0.45

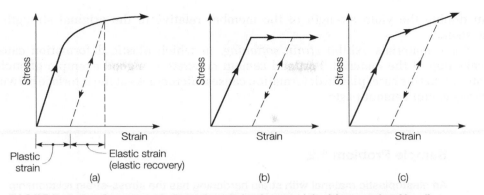

FIGURE 1.6 Stress–strain behavior of plastic materials: (a) example of loading and unloading, (b) elastic–perfectly plastic, and (c) elasto–plastic with strain hardening.

followed by plastic response. The stress level at which the behavior changes from elastic to plastic is the *elastic limit*. When the load is removed from the specimen, some of the deformation will be recovered and some of the deformation will remain as seen in Figure 1.6(a). As discussed in Chapter 2, plastic behavior indicates permanent deformation of the specimen so that it does not return to its original shape when the load is removed. This indicates that when the load is applied, the atomic bonds stretch, creating an elastic response; then the atoms actually slip relative to each other. When the load is removed, the atomic slip does not recover; only the atomic stretch is recovered (Callister, 2006).

Several models are used to represent the behavior of materials that exhibit both elastic and plastic responses. Figure 1.6(b) shows a linear elastic–perfectly plastic response in which the material exhibits a linear elastic response upon loading, followed by a completely plastic response. If such material is unloaded after it has plasticly deformed, it will rebound in a linear elastic manner and will follow a straight line parallel to the elastic portion, while some permanent deformation will remain. If the material is loaded again, it will have a linear elastic response followed by plastic response at the same level of stress at which the material was unloaded (Popov, 1968).

Figure 1.6(c) shows an elastoplastic response in which the first portion is an elastic response followed by a combined elastic and plastic response. If the load is removed after the plastic deformation, the stress–strain relationship will follow a straight line parallel to the elastic portion; consequently, some of the strain in the material will be removed, and the remainder of the strain will be permanent. Upon reloading, the material again behaves in a linear elastic manner up to the stress level that was attained in the previous stress cycle. After that point the material will follow the original stress–strain curve. Thus, the stress required to cause plastic deformation actually increases. This process is called *strain hardening* or *work hardening*. Strain hardening is beneficial in some cases, since it allows more stress to be applied without permanent deformation. In the production of cold-formed steel framing members, the permanent deformation used in the production process can double the yield strength of the member relative to the original strength of the steel.

Some materials exhibit *strain softening,* in which plastic deformation causes weakening of the material. Portland cement concrete is a good example of such a material. In this case, plastic deformation causes microcracks at the interface between aggregate and cement paste.

Sample Problem 1.2

An elastoplastic material with strain hardening has the stress–strain relationship shown in Figure 1.6(c). The modulus of elasticity is 175 GPa, yield strength is 480 MPa, and the slope of the strain-hardening portion of the stress–strain diagram is 20.7 GPa.

  a. Calculate the strain that corresponds to a stress of 550 MPa.
  b. If the 550-MPa stress is removed, calculate the permanent strain.

Solution

  (a) $\varepsilon = (480/175 \times 10^3) + [(550 - 480)/20.7 \times 10^3] = 0.0061$ m/m

  (b) $\varepsilon_{permanent} = 0.0061 - [550/(175 \times 10^3)] = 0.0061 - 0.0031$
        $= 0.0030$ m/m

Materials that do not undergo plastic deformation prior to failure, such as concrete, are said to be *brittle,* whereas materials that display appreciable plastic deformation, such as mild steel, are *ductile.* Generally, ductile materials are preferred for construction. When a brittle material fails, the structure can collapse in a catastrophic manner. On the other hand, overloading a ductile material will result in distortions of the structure, but the structure will not necessarily collapse. Thus, the ductile material provides the designer with a margin of safety.

Figure 1.7(a) demonstrates three concepts of the stress–strain behavior of elastoplastic materials. The lowest point shown on the diagram is the *proportional limit,* defined as the transition point between linear and nonlinear behavior. The second point is the *elastic limit,* which is the transition between elastic and plastic behavior. However, most materials do not display an abrupt change in behavior from elastic to plastic. Rather, there is a gradual, almost imperceptible transition between the behaviors, making it difficult to locate an exact transition point (Polowski and Ripling, 2005). For this reason, arbitrary methods such as the *offset* and the *extension* methods, are used to identify the elastic limit, thereby defining the *yield stress (yield strength).* In the offset method, a specified offset is measured on the abscissa, and a line with a slope equal to the initial tangent modulus is drawn through this

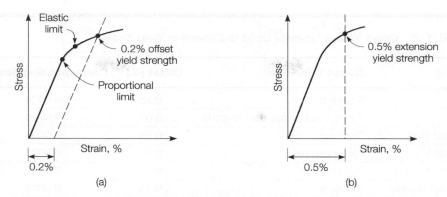

FIGURE 1.7   Methods for estimating yield stress: (a) offset method and (b) extension method.

point. The point where this line intersects the stress–strain curve is the *offset yield stress* of the material, as seen in Figure 1.7(a). Different offsets are used for different materials (Table 1.2). The *extension yield stress* is located where a vertical projection, at a specified strain level, intersects the stress–strain curve. Figure 1.7(b) shows the yield stress corresponding to 0.5% extension.

Sample Problem 1.3

A rod made of aluminum alloy, with a gauge length of 100 mm, diameter of 10 mm, and yield strength of 150 MPa, was subjected to a tensile load of 5.85 kN. If the gauge length was changed to 100.1 mm and the diameter was changed to 9.9967 mm, calculate the modulus of elasticity and Poisson's ratio.

Solution

$$\sigma = P/A = (5850 \text{ N})/[\pi(5 \times 10^{-3} \text{ m})^2] = 74.5 \times 10^6 \text{ Pa} = 74.5 \text{ MPa}$$

Since the applied stress is well below the yield strength, the material is within the elastic region.

$$\varepsilon_a = \Delta L/L = (100.1 - 100)/100 = 0.001$$
$$E = \sigma/\varepsilon_a = 74.5/0.001 = 74{,}500 \text{ MPa} = 74.5 \text{ GPa}$$
$$\varepsilon_l = \text{change in diameter/diameter} = (9.9967 - 10)/10 = -0.00033$$
$$\nu = -\varepsilon_l/\varepsilon_a = 0.00033/0.001 = 0.33$$

TABLE 1.2   Offset Values Typically Used to Determine Yield Stress

Material	Stress Condition	Offset (%)	Corresponding Strain
Steel	Tension	0.20	0.0020
Wood	Compression parallel to grain	0.05	0.0005
Gray cast iron	Tension	0.05	0.0005
Concrete	Compression	0.02	0.0002
Aluminum alloys	Tension	0.20	0.0020
Brass and bronze	Tension	0.35	0.0035

1.2.5 ▪ Viscoelastic Behavior

The previous discussion assumed that the strain was an immediate response to stress. This is an assumption for elastic and elastoplastic materials. However, no material has this property under all conditions. In some cases, materials exhibit both viscous and elastic responses, which are known as *viscoelastic.* Typical viscoelastic materials used in construction applications are asphalt and plastics.

Time-Dependent Response   Viscoelastic materials have a delayed response to load application. For example, Figure 1.8(a) shows a sinusoidal axial load applied on a viscoelastic material, such as asphalt concrete, versus time. Figure 1.8(b) shows the

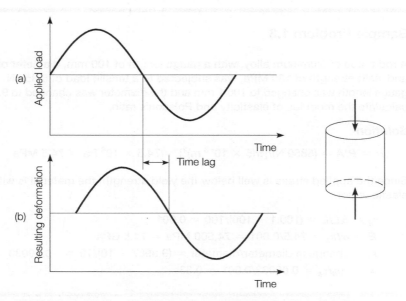

FIGURE 1.8   Load-deformation response of a viscoelastic material.

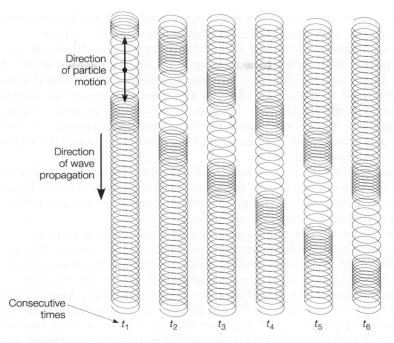

Direction
of particle
motion

Direction
of wave
propagation

Consecutive
times

t_1 t_2 t_3 t_4 t_5 t_6

FIGURE 1.9   Delay of propagation of compression and dilation
waves in a Slinky®.

resulting deformation versus time, where the deformation lags the load—that is, the
maximum deformation of the sample occurs after the maximum load is applied. The
amount of time delayed of the deformation depends on the material characteristics
and the temperature.

The delay in the response of viscoelastic materials can be simulated by the movement of the Slinky® toy in the hand of a child, as illustrated in Figure 1.9. As the
child moves her hand up and down, waves of compression and dilation are developed in the Slinky. However, the development of the waves in the Slinky does not
happen exactly at the same time as the movements of the child's hand. For example,
a compression wave could be propagating in one part of the Slinky at the same time
when the child is moving her hand upward and *vice versa*. This occurs because of
the delay in response relative to the action. Typical viscoelastic civil engineering
materials, such as asphalt, have the same behavior, although they are not as flexible
as a Slinky.

There are several mechanisms associated with time-dependent deformation, such
as *creep* and *viscous flow*. There is no clear distinction between these terms. Creep is
generally associated with long-term deformations and can occur in metals, ionic and
covalent crystals, and amorphous materials. On the other hand, viscous flow is associated only with amorphous materials and can occur under short-term load duration. For example, concrete, a material with predominantly covalent crystals, can
creep over a period of decades. Asphalt concrete pavements, an amorphous-binder

material, can have ruts caused by the accumulated effect of viscous flows resulting from traffic loads with a load duration of only a fraction of a second.

Creep of metals is a concern at elevated temperatures. Steel can creep at temperatures greater than 30% of the melting point on the absolute scale. This can be a concern in the design of boilers and nuclear reactor containment vessels. Creep is also considered in the design of wood and advanced composite structural members. Wood elements loaded for a few days can carry higher stresses than elements designed to carry "permanent" loads. On the other hand, creep of concrete is associated with microcracking at the interface of the cement paste and the aggregate particles (Mehta and Monteiro, 2013).

The viscous flow models are similar in nature to Hooke's law. In linearly viscous materials, the rate of deformation is proportional to the stress level. These materials are not compressible and do not recover when the load is removed. Materials with these characteristics are *Newtonian fluids.*

Figure 1.10(a) shows a typical creep test in which a constant compressive stress is applied to an asphalt concrete specimen. In this case, an elastic strain will develop, followed by time-dependent strain or creep. If the specimen is unloaded, a part of the strain will recover instantaneously, while the remaining strain will recover, either completely or partially, over a period of time. Another phenomenon typical of time-dependent materials is relaxation, or dissipation of stresses with time. For example, if an asphalt concrete specimen is placed in a loading machine and subjected to a constant strain, the stress within the specimen will initially be high, then gradually dissipate due to relaxation as shown in Figure 1.10(b). Relaxation is an important concern in the selection of steel for a prestressed concrete design.

In viscoelasticity, there are two approaches used to describe how stresses, strains, and time are interrelated. One approach is to postulate mathematical

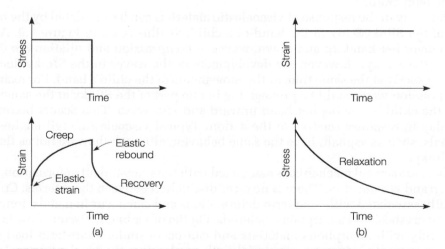

FIGURE 1.10 Behavior of time-dependent materials: (a) creep and (b) relaxation.

relations between these parameters based on material functions obtained from laboratory tests. The other approach is based on combining a number of discrete *rheological elements* to form *rheological models,* which describe the material response.

Rheological Models   Rheological models are used to model mechanically the time-dependent behavior of materials. There are many different modes of material deformation, particularly in polymer materials. These materials cannot be described as simply elastic, viscous, etc. However, these materials can be modeled by a combination of simple physical elements. The simple physical elements have characteristics that can be easily visualized. Rheology uses three basic elements, combined in either series or parallel to form models that define complex material behaviors. The three basic rheological elements, Hookean, Newtonian, and St. Venant, are shown in Figure 1.11 (Polowski and Ripling, 2005).

The *Hookean* element, as in Figure 1.11(a), has the characteristics of a linear *spring.* The deformation δ is proportional to force F by a constant M:

$$F = M\delta \tag{1.4}$$

This represents a perfectly linear elastic material. The response to a force is instantaneous and the deformation is completely recovered when the force is removed. Thus, the Hookean element represents a perfectly linear elastic material.

A *Newtonian* element models a perfectly viscous material and is modeled as a *dashpot* or shock absorber as seen in Figure 1.11(b). The deformation for a given level of force is proportional to the amount of time the force is applied. Hence, the rate of deformation, for a constant force, is a constant β:

$$F = \beta\dot{\delta} \tag{1.5}$$

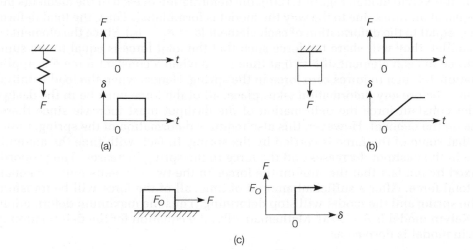

FIGURE 1.11   Basic elements used in rheology: (a) Hookean, (b) Newtonian, and (c) St. Venant.

The dot above the δ defines this as the rate of deformation with respect to time. If $\delta = 0$ at time $t = 0$ when a constant force F is applied, the deformation at time t is

$$\delta = \frac{Ft}{\beta} \tag{1.6}$$

When the force is removed, the specimen retains the deformed shape. There is no recovery of any of the deformation.

The *St. Venant* element, as seen in Figure 1.11(c), has the characteristics of a sliding block that resists movement by friction. When the force F applied to the element is less than the critical force F_O, there is no movement. If the force is increased to overcome the static friction, the element will slide and continue to slide as long as the force is applied. This element is unrealistic, since any sustained force sufficient to cause movement would cause the block to accelerate. Hence, the St. Venant element is always used in combination with the other basic elements.

The basic elements are usually combined in parallel or series to model material response. Figure 1.12 shows the three primary two-component models: the Maxwell, Kelvin, and Prandtl models. The Maxwell and Kelvin models have a spring and dashpot in series and parallel, respectively. The Prandtl model uses a spring and St. Venant elements in series.

In the *Maxwell* model [Figure 1.12(a)], the total deformation is the sum of the deformations of the individual elements. The force in each of the elements must be equal to the total force ($F = F_1 = F_2$). Thus, the equation for the total deformation at any time after a constant load is applied is simply:

$$\delta = \delta_1 + \delta_2 = \frac{F}{M} + \frac{Ft}{\beta} \tag{1.7}$$

In the *Kelvin* model, Figure 1.12(b), the deformation of each of the elements must be equal at all times due to the way the model is formulated. Thus, the total deformation is equal to the deformation of each element ($\delta = \delta_1 = \delta_2$). Since the elements are in parallel, they will share the force such that the total force is equal to the sum of the force in each element. If $\delta = 0$ at time $t = 0$ when a constant force F is applied, Equation 1.4 then requires zero force in the spring. Hence, when the load is initially applied, before any deformation takes place, all of the force must be in the dashpot. Under constant force the deformation of the dashpot must increase since there is force on the element. However, this also requires deformation of the spring, indicating that some of the force is carried by the spring. In fact, with time the amount of force in the dashpot decreases and the force in the spring increases. The proportion is fixed by the fact that the sum on the forces in the two elements must be equal to the total force. After a sufficient amount of time, all of the force will be transferred to the spring and the model will stop deforming. Thus the maximum deformation of the Kelvin model is $\delta = F/M$. Mathematically, the equation for the deformation in a Kelvin model is derived as:

$$F = F_1 + F_2 = M\delta + \beta\dot{\delta} \tag{1.8}$$

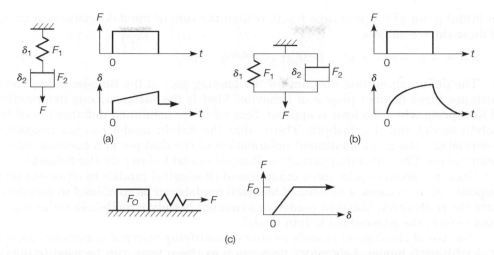

FIGURE 1.12 Two-element rheological models: (a) Maxwell, (b) Kelvin, and (c) Prandtl.

Integrating Equation 1.8, using the limits that $\delta = 0$ at $t = 0$, and solving for the deformation δ at time t results in

$$\delta = (F/M)(1 - e^{-Mt/\beta}) \tag{1.9}$$

The *Prandtl* model [Figure 1.12(c)] consists of St. Venant and Hookean bodies in series. The Prandtl model represents a material with an elastic–perfectly plastic response. If a small load is applied, the material responds elastically until it reaches the yield point, after which the material exhibits plastic deformation.

Neither the Maxwell nor Kelvin model adequately describes the behavior of some common engineering materials, such as asphalt concrete. However, the Maxwell and the Kelvin models can be put together in series, producing the *Burgers* model, which can be used to describe simplistically the behavior of asphalt concrete. As shown in Figure 1.13, the Burgers model is generally drawn as a spring in series with a Kelvin model in series with a dashpot. The total deformation at time t, with

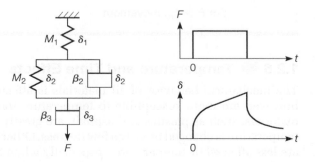

FIGURE 1.13 Burgers model of viscoelastic materials.

an initial point of $\delta = 0$ at time $t = 0$, is then the sum of the deformations at time t of these three elements.

$$\delta = \delta_1 + \delta_2 + \delta_3 = F/M_1 + (F/M_2)(1 - e^{-M_2 t/\beta_2}) + Ft/\beta_3 \qquad (1.10)$$

The deformation-time diagram for the loading part of the Burgers model demonstrates three distinct phases of behavior. First is the instantaneous deformation of the spring when the load is applied. Second is the combined deformation of the Kelvin model and the dashpot. Third, after the Kelvin model reaches maximum deformation, there is a continued deformation of the dashpot at a constant rate of deformation. The unloading part of the Burgers model follows similar behavior.

Some materials require more complicated rheological models to represent their response. In such cases, a number of Maxwell models can be combined in parallel to form the generalized Maxwell model, or a number of Kelvin models in series can be used to form the generalized Kelvin model.

The use of rheological models requires quantifying material parameters associated with each model. Laboratory tests, such as creep tests, can be used to obtain deformation–time curves from which material parameters can be determined.

Although the rheological models are useful in describing the time-dependent response of materials, they can be used only to represent uniaxial responses. The three-dimensional behavior of materials and the Poisson's effect cannot be represented by these models.

Sample Problem 1.4

Derive the response relation for the following model assuming that the force F is constant and instantaneously applied.

FIGURE SP1.4

Solution

For $F \leq F_o$: $\delta = F/M$

For $F > F_o$: movement

1.2.6 ■ Temperature and Time Effects

The mechanical behavior of all materials is affected by temperature. Some materials, however, are more susceptible to temperature than others. For example, viscoelastic materials, such as plastics and asphalt, are greatly affected by temperature, even if the temperature is changed by only a few degrees. Other materials, such as metals or concrete, are less affected by temperatures, especially when they are near ambient temperature.

Ferrous metals, including steel, demonstrate a change from ductile to brittle behavior as the temperature drops below the *transition temperature*. This change

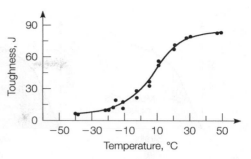

FIGURE 1.14 Fracture toughness of steel under impact testing.

from ductile to brittle behavior greatly reduces the toughness of the material. While this could be determined by evaluating the stress–strain diagram at different temperatures, it is more common to evaluate the toughness of a material with an impact test that measures the energy required to fracture a specimen. Figure 1.14 shows how the energy required to fracture a mild steel changes with temperature (Flinn and Trojan, 1995). The test results seen in Figure 1.14 were achieved by applying impact forces on bar specimens with a "defect" (a simple V notch) machined into the specimens (ASTM E23). During World War II, many Liberty ships sank because the steel used in the ships met specifications at ambient temperature, but became brittle in the cold waters of the North Atlantic.

In addition to temperature, some materials, such as viscoelastic materials, are affected by the load duration. The longer the load is applied, the larger is the amount of deformation or creep. In fact, increasing the load duration and increasing the temperature cause similar material responses. Therefore, temperature and time can be interchanged. This concept is very useful in running some tests. For example, a creep test on an asphalt concrete specimen can be performed with short load durations by increasing the temperature of the material. A *time–temperature shift factor* is then used to adjust the results for lower temperatures.

Viscoelastic materials are affected not only by the duration of the load but also by the rate of load application. If the load is applied at a fast rate, the material is stiffer than if the load is applied at a slow rate. For example, if a heavy truck moves at a high speed on an asphalt pavement, no permanent deformation may be observed. However, if the same truck is parked on an asphalt pavement on a hot day, some permanent deformations on the pavement surface may be observed.

1.2.7 ▉ Work and Energy

When a material is tested, the testing machine is actually generating a force in order to move or deform the specimen. Since work is force times distance, the area under a force–displacement curve is the work done on the specimen. When the force is divided by the cross-sectional area of the specimen to compute the stress, and the deformation is divided by the length of the specimen to compute the strain, the force–displacement diagram becomes a stress–strain diagram. However, the area under the stress–strain diagram no longer has the units of work. By manipulating the units of the stress–strain diagram, we can see that the area under the stress–strain

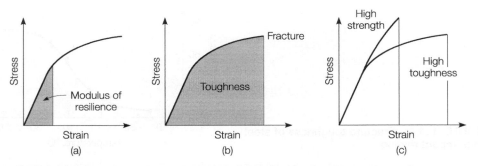

FIGURE 1.15   Areas under stress–strain curves: (a) modulus of resilience, (b) toughness, and (c) high-strength and high-toughness materials.

diagram equals the work per unit volume of material required to deform or fracture the material. This is a useful concept, for it tells us the energy that is required to deform or fracture the material. Such information is used for selecting materials to use where energy must be absorbed by the member. The area under the elastic portion of the curve is the *modulus of resilience* [Figure 1.15(a)]. The amount of energy required to fracture a specimen is a measure of the *toughness* of the material, as in Figure 1.15(b). As shown in Figure 1.15(c), a high-strength material is not necessarily a tough material. For instance, as discussed in Chapter 3, increasing the carbon content of steel increases the yield strength but reduces ductility. Therefore, the strength is increased, but the toughness may be reduced.

1.2.8 ▇ Failure and Safety

Failure occurs when a member or structure ceases to perform the function for which it was designed. Failure of a structure can take several modes, including fracture, fatigue, general yielding, buckling, and excessive deformation. *Fracture* is a common failure mode. A brittle material typically fractures suddenly when the static stress reaches the strength of the material, where the strength is defined as the maximum stress the material can carry. On the other hand, a ductile material may fracture due to excessive plastic deformation.

Many structures, such as bridges, are subjected to repeated loadings, creating stresses that are less than the strength of the material. Repeated stresses can cause a material to fail or *fatigue,* at a stress well below the strength of the material. The number of applications a material can withstand depends on the stress level relative to the strength of the material. As shown in Figure 1.16, as the stress level decreases, the number of applications before failure increases. Ferrous metals have an apparent *endurance limit,* or stress level, below which fatigue does not occur. The endurance limit for steels is generally in the range of one-quarter to one-half the ultimate strength (Flinn and Trojan, 1995). Another example of a structure that may fail due to fatigue is pavement. Although the stresses applied by traffic are typically much less than the strength of the material, repeated loadings may eventually lead to a loss of the structural integrity of the pavement surface layer, causing fatigue cracks as shown in Figure 1.17.

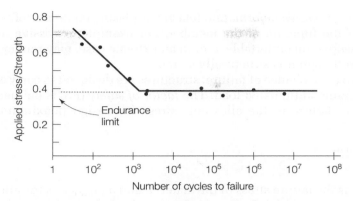

FIGURE 1.16   An example of endurance limit under repeated loading.

Another mode of failure is *general yielding*. This failure happens in ductile materials, and it spreads throughout the whole structure, which results in a total collapse.

Long and slender members subjected to axial compression may fail due to *buckling*. Although the member is intended to carry axial compressive loads, a small lateral force might be applied, which causes deflection and eventually might cause failure.

FIGURE 1.17   Fatigue failure of asphalt pavement due to repeated traffic loading.

Sometimes *excessive deformation* (elastic or plastic) could be defined as failure, depending on the function of the member. For example, excessive deflections of floors make people uncomfortable and, in an extreme case, may render the building unusable even though it is structurally sound.

To minimize the chance of failure, structures are designed to carry a load greater than the maximum anticipated load. The *factor of safety* (*FS*) is defined as the ratio of the stress at failure to the allowable stress for design (maximum anticipated stress):

$$FS = \frac{\sigma_{\text{failure}}}{\sigma_{\text{allowable}}} \qquad (1.11)$$

where σ_{failure} is the failure stress of the material and $\sigma_{\text{allowable}}$ is the allowable stress for design. Typically, a high factor of safety requires a large structural cross section and consequently a higher cost. The proper value of the factor of safety varies from one structure to another and depends on several factors, including the

- cost of unpredictable failure in lives, dollars, and time,
- variability in material properties,
- degree of accuracy in considering all possible loads applied to the structure, such as earthquakes,
- possible misuse of the structure, such as improperly hanging an object from a truss roof,
- degree of accuracy of considering the proper response of materials during design, such as assuming elastic response although the material might not be perfectly elastic.

1.3   Nonmechanical Properties

Nonmechanical properties refer to characteristics of the material, other than load response, that affect selection, use, and performance. There are several types of properties that are of interest to engineers, but those of the greatest concern to civil engineers are density, thermal properties, and surface characteristics.

1.3.1   Density and Unit Weight

In many structures, the dead weight of the materials in the structure significantly contributes to the total design stress. If the weight of the materials can be reduced, the size of the structural members can also be reduced. Thus, the weight of the materials is an important design consideration. In addition, in the design of asphalt and concrete mixes, the weight–volume relationship of the aggregates and binders must be used to select the mix proportions.

There are three general terms used to describe the mass, weight, and volume relationship of materials. *Density* is the mass per unit volume of material. *Unit weight* is the weight per unit volume of material. By manipulation of units, it can be shown that

$$\gamma = \rho g \qquad (1.12)$$

where
$\gamma = $ unit weight
$\rho = $ density
$g = $ acceleration of gravity

Specific gravity is the ratio of the mass of a substance relative to the mass of an equal volume of water at a specified temperature. The density of water is 1 Mg/m^3 at 4°C. According to the definition, specific gravity is equivalent to the density of a material divided by the density of water. Since the density of water in the metric system has a numerical value of 1, the numerical value of density and specific gravity are equal. This fact is often used in the literature where density and specific gravity terms are used interchangeably.

For solid materials, such as metals, the unit weight, density, and specific gravity have definite numerical values. For other materials such as wood and aggregates, voids in the materials require definitions for a variety of densities and specific gravities. As shown in Figure 1.18(a) and (b), the bulk volume aggregates will occupy depends on the compaction state of the material. In addition, the density of the material will change depending on how the volume of individual particles is measured. Several types of particle volume can be used, such as the total volume enclosed within the boundaries of the individual particles, volume not accessible to water or asphalt, and volume of solids, as seen in Figure 1.18(c), (d), and (e), respectively. These are important factors in the mix designs of portland cement concrete and asphalt concrete.

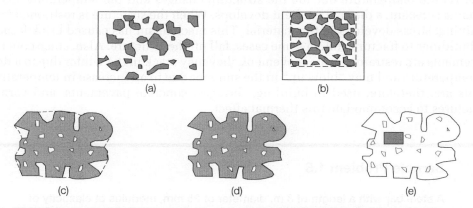

(a)   (b)

(c)   (d)   (e)

FIGURE 1.18 Definitions of volume used for determining density: (a) loose, (b) compacted, (c) total particle volume, (d) volume not accessible to water, and (e) volume of solids.

1.3.2 ▓ Thermal Expansion

Practically all materials expand as temperature increases and contract as temperature falls. The amount of expansion per unit length due to one unit of temperature increase is a material constant and is expressed as the *coefficient of thermal expansion*

$$\alpha_L = \frac{\delta L/dT}{L} \tag{1.13}$$

$$\alpha_V = \frac{\delta V/dT}{V} \tag{1.14}$$

where
  α_L = linear coefficient of thermal expansion
  α_V = volumetric coefficient of thermal expansion
  δL = change in the length of the specimen
  δT = change in temperature
   L = original length of the specimen
  δV = change in the volume of the specimen
   V = original volume of the specimen

For isotropic materials, $\alpha_V = 3\alpha_L$.

The coefficient of thermal expansion is very important in the design of structures. Generally, structures are composed of many materials that are bound together. If the coefficients of thermal expansion are different, the materials will strain at different rates. The material with the lesser expansion will restrict the straining of other materials. This constraining effect will cause stresses in the materials that can lead directly to fracture.

Stresses can also be developed as a result of a thermal gradient in the structure. As the temperature outside the structure changes and the temperature inside remains constant, a thermal gradient develops. When the structure is restrained from straining, stress develops in the material. This mechanism has caused brick facades on buildings to fracture and, in some cases, fall off the structure. Also, since concrete pavements are restrained from movement, they may crack in the winter due to a drop in temperature and may "blow up" in the summer due to an increase in temperature. Joints are, therefore, used in buildings, bridges, concrete pavements, and various structures to accommodate this thermal effect.

Sample Problem 1.5

A steel bar with a length of 3 m, diameter of 25 mm, modulus of elasticity of 207 GPa, and linear coefficient of thermal expansion of 0.000009 m/m/°C is fixed at both ends when the ambient temperature is 40°C. If the ambient temperature is decreased to 15°C, what internal stress will develop due to this temperature change? Is this stress tension or compression? Why?

Solution

If the bar was fixed at one end and free at the other end, the bar would have contracted and no stresses would have developed. In that case, the change in length can be calculated by using Equation 1.13 as follows:

$$\delta L = \alpha_L \times \delta T \times L = 0.000009 \times (-25) \times 3 = -0.000675 \text{ m}$$
$$\varepsilon = \delta L/L = -0.000675/3 = -0.000225 \text{ m/m}$$

Since the bar is fixed at both ends, the length of the bar will not change. Therefore, a tensile stress will develop in the bar as follows:

$$\sigma = \varepsilon E = 0.000225 \times 207000 = 46.575 \text{ MPa}$$

The stress will be tension; in effect, the length of the bar at 15°C without restraint would be 2.999325 m and the stress would be zero. Restraining the bar into a longer condition requires a tensile force.

1.3.3 ▉ Surface Characteristics

The surface properties of materials of interest to civil engineers include corrosion and degradation, the ability of the material to resist abrasion and wear, and surface texture.

Corrosion and Degradation   Nearly all materials deteriorate over their service lives. The mechanisms contributing to the deterioration of a material differ depending on the characteristics of the material and the environment. Crystalline materials, such as metals and ceramics, deteriorate through a *corrosion* process in which there is a loss of material, either by dissolution or by the formation of nonmetallic scale or film. Polymers, such as asphalt, deteriorate by *degradation* of the material, including the effects of solvent and ultraviolet radiation on the material.

The protection of materials from environmental degradation is an important design concern, especially when the implications of deterioration and degradation on the life and maintenance costs of the structure are considered. The selection of a material should consider both how the material will react with the environment and the cost of preventing the resulting degradation.

Abrasion and Wear Resistance   Since most structures in civil engineering are static, the abrasion or wear resistance is of less importance than in other fields of engineering. For example, mechanical engineers must be concerned with the wear of parts in the design of machinery. This is not to say that wear resistance can be totally ignored in civil engineering. Pavements must be designed to resist the wear and polishing from vehicle tires in order to provide adequate skid resistance for braking and turning. Resistance to abrasion and wear is, therefore, an important property of aggregates used in pavements.

Surface Texture   The surface texture of some materials and structures is of importance to civil engineers. For example, smooth texture of aggregate particles is needed in portland cement concrete to improve workability during mixing and placing. In contrast, rough texture of aggregate particles is needed in asphalt concrete mixtures to provide a stable pavement layer that resists deformation under the action of load. Also, a certain level of surface texture is needed in the pavement surface to provide adequate friction resistance and prevent skidding of vehicles when the pavement is wet.

1.4   Production and Construction

Even if a material is well suited to a specific application, production and construction considerations may block the selection of the material. Production considerations include the availability of the material and the ability to fabricate the material into the desired shapes and required specifications. Construction considerations address all the factors that relate to the ability to fabricate and erect the structure on site. One of the primary factors is the availability of a trained work force. For example, in some cities, high-strength concrete is used for skyscrapers, whereas in other cities, steel is the material of choice. Clearly, either concrete or steel can be used for high-rise buildings. Regional preferences for one material develop as engineers in the region become comfortable and confident in designing with one of the materials and constructors respond with a trained work force and specialized equipment.

1.5   Aesthetic Characteristics

The aesthetic characteristics of a material refer to the appearance of the material. Generally, these characteristics are the responsibility of the architect. However, the civil engineer is responsible for working with the architect to ensure that the aesthetic characteristics of the facility are compatible with the structural requirements. During the construction of many public projects, a certain percentage of the capital budget typically goes toward artistic input. The collaboration between the civil engineer and the architect is greatly encouraged, and the result can increase the value of the structure (see Figure 1.19).

In many cases, the mix of artistic and technical design skills makes the project acceptable to the community. In fact, political views are often more difficult to deal with than technical design problems. Thus, engineers should understand that there are many factors beyond the technical needs that must be considered when selecting materials and designing public projects.

FIGURE 1.19   An example of artist-engineer collaboration in an engineering project: Air Force Academy, Colorado Springs, Colorado.

 1.6   Sustainable Design

Sustainable design is the philosophy of designing physical objects, the built environment and services to comply with the principles of economic, social, and ecological sustainability. Specifically, civil and construction engineers should be concerned with design and construction decisions which are sensitive to both the current and future needs of society. Some owners are mandating the use of a building assessment system for design, construction, and long-term performance. Since 2003, the General Services Administration, GSA, in the United States, has required a sustainablility rating of all building projects (Fowler and Rauch, 2006). With the emphasis on sustainability, several rating systems have been developed including:

- BREEAM (Building Research Establishment's Environmental Assessment Method) developed in the United Kingdom.
- CASBEE (Comprehensive Assessment System for Building Environmental Efficiency) developed in Japan.
- GBTool an internationally developed system that was initially developed in 1998 and has undergone multiple revisions.

- Green Globes™ developed in Cananda based on the BREEAM method. A US version has been developed by the Green Building Initiative.
- LEED® (Leadership in Energy and Environmental Design).
- Living Building Challenge.

The LEED system is the most widely used with thousands of users (Wang et al., 2012) so it is the only one reviewed in this text.

The materials used for a civil engineering project are a fundamental consideration for determining the sustainability of a project. While sustainable design is a philosophy, implementation of the concept requires direct and measureable actions.

To this end, the Green Building Council developed the Leadership in Environment and Energy Design, LEED, building rating system (LEED, 2013).

The LEED rating system is used to evaluate the sustainability of a project. Elements of the project are rated in several areas. Based on the cumulative rating, the project is awarded a status of Certified, Silver, Gold, or Platinum. For new construction and major renovations, the rating areas include:

- Sustainable sites
- Water efficiency
- Energy and atmosphere
- Materials and resources
- Indoor environmental quality
- Innovation in design
- Regional priority

There is an obvious connection between this book and the Materials and Resources area of LEED. The items considered in the Materials and Resource area include:

- Storage and collection of recyclables
- Building reuse–maintain existing walls, floors and roof, interior walls, and nonstructural elements
- Construction waste management
- Materials reuse
- Recycled content
- Regional materials
- Rapidly renewable materials
- Certified wood

If an owner wishes to pursue a specific level of LEED certification, the engineer is charged with the responsibility of selecting materials that meet the functional requirements of the building and provide the desired LEED rating. For example, by selecting regional materials, the LEED rating is increased and there may be little or no impact on the cost of the facility.

In addition to the direct tie to the materials and resources, there are other areas of the LEED certification that are affected by material selection. Selection of engineered wood products which are fabricated with glues rated for low off gassing of volatile organic compounds, VOCs, improves the LEED score with respect to Indoor Environmental Quality. Selection of a porous pavement surface for parking lots improves the LEED score for site development.

It should be noted that the LEED rating system is for the certification of buildings, not the materials that are used for construction. Manufactures can seek third party certification of their products to use as evidence that these products support a claim for LEED credits. Environmental and health reviews and certifications are available from a variety of organizations such as Scientific Certification Systems, Forest Stewardship Council, Building Green Inc., Green Seal, Green Guard, Building Green, etc (USGBC, 2015).

1.7   Material Variability

It is essential to understand that engineering materials are inherently variable. For example, steel properties vary depending on chemical composition and method of manufacture. Concrete properties change depending on type and amount of cement, type of aggregate, air content, slump, method of curing, etc. The properties of asphalt concrete vary depending on the binder amount and type, aggregate properties and gradation, amount of compaction, and age. Wood properties vary depending on the tree species, method of cut, and moisture content. Some materials are more homogeneous than others, depending on the nature of the material and the method of manufacturing. For example, the variability of the yield strength of one type of steel is less than the variability of the compressive strength of one batch of concrete. Therefore, variability is an important parameter in defining the quality of civil engineering materials.

When materials from a particular lot are tested, the observed variability is the cumulative effect of three types of variance: the inherent variability of the material, variance caused by the sampling method, and variance associated with the way the tests was conducted. Just as materials have an inherent variability, sampling procedures and test methods can produce variable results. Frequently, statisticians call variance associated with sampling and testing *error*. However, this does not imply the sampling or testing was performed incorrectly. When an incorrect procedure is identified, it is called a *blunder*. The goal of a sampling and testing program is to minimize sampling and testing variance so the true statistical features of the material can be identified.

The concepts of precision and accuracy are fundamental to the understanding of variability. *Precision* refers to the variability of repeat measurements under carefully controlled conditions. *Accuracy* is the conformity of results to the true value or the absence of bias. *Bias* is a tendency of an estimate to deviate in one direction from the true value. In other words, bias is a systematic error between a test value and the true value. A simple analogy to the relationship between precision and accuracy is the target shown in Figure 1.20. When all shots are concentrated at one location away from the center, that indicates good precision and poor accuracy (biased) [Figure 1.20(a)]. When shots are scattered around the center, that indicates poor precision and good accuracy [Figure 1.20(b)]. Finally, good precision and good accuracy are obtained if all shots are concentrated close to the center [Figure 1.20(c)] (Burati and Hughes, 1990). Many standardized test methods, such as those of the

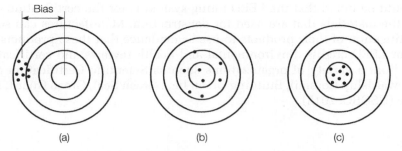

FIGURE 1.20 Exactness of measurements: (a) precise but not accurate, (b) accurate but not precise, and (c) precise and accurate. (Federal Highway Administration)

American Society for Testing and Materials (ASTM) and the American Association of State Highway and Transportation Officials (AASHTO), contain precision and bias statements. These statements provide the limits of acceptable test results variability. Laboratories are usually required to demonstrate testing competence and can be certified by the American Material Reference Laboratory (AMRL).

1.7.1 ▨ Sampling

Typically, *samples* are taken from a *lot* or *population,* since it is not practical or possible to test the entire lot. By testing sufficient samples, it is possible to estimate the properties of the entire lot. In order for the samples to be valid they must be *randomly* selected. Random sampling requires that all elements of the population have an equal chance for selection. Another important concept in sampling is that the sample must be *representative* of the entire lot. For example, when sampling a stockpile of aggregate, it is important to collect samples from the top, middle, and bottom of the pile and to combine them, since different locations within the pile are likely to have different aggregate sizes. The sample size needed to quantify the characteristics of a population depends on the variability of the material properties and the confidence level required in the evaluation.

Statistical parameters describe the material properties. The mean and the standard deviation are two commonly used statistics. The *arithmetic mean* is simply the average of test results of all specimens tested. It is a measure of the central tendency of the population. The *standard deviation* is a measure of the dispersion or spread of the results. The equations for the mean $\bar{x}$ and standard deviation s of a sample are

$$\bar{x} = \frac{\sum_{i=1}^{n} x_i}{n} \tag{1.15}$$

$$s = \left(\frac{\sum_{i=1}^{n} (x_i - \bar{x})^2}{n-1} \right)^{1/2} \tag{1.16}$$

where n is the sample size. The mean and standard deviation of random samples are estimates of the mean and standard deviation of the population, respectively. The standard deviation is an inverse measure of precision; a low standard deviation indicates good precision.

1.7.2 ▨ Normal Distribution

The normal distribution is a symmetrical function around the mean, as shown in Figure 1.21. The normal distribution describes many populations that occur in nature, research, and industry, including material properties. The area under the curve between any two values represents the probability of occurrence of an event of interest. Expressing the results in terms of mean and standard deviation, it is possible to determine the probabilities of an occurrence of an event. For example, the probability of occurrence of an event between the mean and ± 1 standard deviation is 68.3%, between the mean and ± 2 standard deviations is 95.5%, and between the mean and ± 3 standard deviations is 99.7%. If a materials engineer tests 20 specimens of concrete and determines the average as 22 MPa and the standard deviation as 3 MPa, the statistics will show that 95.5% of the time the data will be in the range of 22 $\pm$ (2 $\times$ 3), or 16 to 28 MPa. If one wishes to estimate how close the mean of the data is to the mean of the population then the standard deviation of the mean, $\sigma/(n)^{0.5}$ would be used. For the above example the standard deviation of the mean would be $3/(20)^{0.5} = 0.67$ and 95.5% of the time the population mean would be in the range of 22 $\pm$ 2 $\times$ 0.67 or 20.66 to 23.34.

1.7.3 ▨ Control Charts

Control charts have been used in manufacturing industry and construction applications to verify that a process is in control. It is important to note that control charts do not get or keep a process under control; they provide only a visual warning mechanism to identify when a contractor or material supplier should look for possible problems with the process. Control charts have many benefits (Burati and Hughes, 1990), such as

- detect trouble early
- decrease variability
- establish process capability
- reduce price adjustment cost

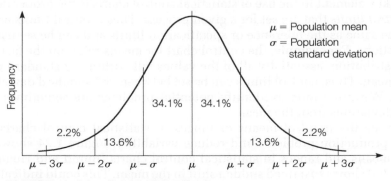

FIGURE 1.21 A normal distribution.

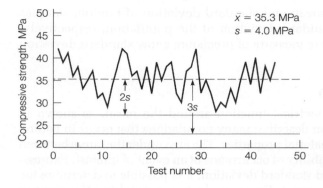

$\bar{x} = 35.3$ MPa
$s = 4.0$ MPa

FIGURE 1.22   Control chart of compressive strength of concrete specimens.

- decrease inspection frequency
- provide a basis for altering specification limits
- provide a permanent record of quality
- provide basis for acceptance
- instill quality awareness

There are many types of control charts, the simplest of which plots individual results in chronological order. For example, Figure 1.22 shows a control chart of the compressive strength of concrete specimens tested at a ready-mix plant. The control chart can also show the specification tolerance limits so that the operator can identify when the test results are out of the specification requirements. Although this type of control chart is useful, it is based on a sample size of one, and it therefore fails to consider variability within the sample.

Statistical control charts can be developed, such as the control chart for means (X-bar chart) and the control chart for the ranges (R chart) in which the means or the ranges of the test results are chronologically plotted. Figure 1.23(a) shows a control chart for the moving average of each three consecutive compressive strength tests. For example, the first point represents the mean of the first three tests, the second point represents the mean of tests two through four, and so on. Figure 1.23(b) shows a control chart for the moving range of each three consecutive compressive strength tests. The key element in the use of statistical control charts is the proper designation of the *control limits* that are set for a given process. These control limits are not necessarily the same as the tolerance or specification limits and can be set using probability functions. For example, the control chart for means relies on the fact that, for a normal distribution, essentially all of the values fall within ± 3 standard deviations from the mean. Thus, control limits can be set between ± 3 standard deviations from the mean. Warning limits to identify potential problems are sometimes set at ± 2 standard deviations from the mean.

Observing the trend of means or ranges in statistical control charts can help eliminate production problems and reduce variability. Figure 1.24 shows possible trends of means and ranges in statistical control charts (Burati and Hughes, 1990). Figure 1.24(a) shows sustained sudden shift in the mean. This could indicate a change

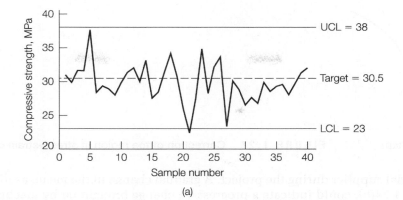

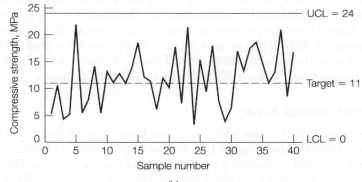

FIGURE 1.23 Statistical control charts: (a) X-bar chart and (b) R chart. UCL indicates upper control limit; LCL indicates lower control limit.

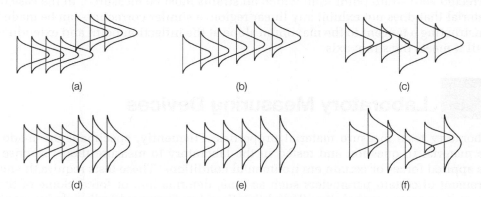

FIGURE 1.24 Possible trends of means and ranges in statistical control charts: (a) sudden change in mean, (b) gradual change in mean, (c) irregular change in mean, (d) sudden change in range, (e) gradual change in range, and (f) irregular change in mean and range.

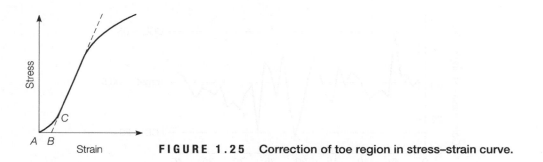

FIGURE 1.25 Correction of toe region in stress–strain curve.

of a material supplier during the project. A gradual change in the mean, as illustrated in Figure 1.24(b), could indicate a progressive change brought on by machine wear. An irregular shift in the mean, as shown in Figure 1.24(c), may indicate that the operator is making continuous, but unnecessary, adjustments to the process settings. Figure 1.24(d) shows a sudden change in range, which could also indicate a change of a material supplier during the project. Figure 1.24(e) shows a gradual increase in the range, which may indicate machine wear. Finally, Figure 1.24(f) shows an irregular shift in both mean and range, which indicates a flawed process.

1.7.4 Experimental Error

When specimens are tested in the laboratory, inaccuracy could occur due to machine or human errors. For example, Figure 1.25 shows a stress–strain curve in which a toe region (AC) that does not represent a property of the material exists. This toe region is an artifact caused by taking up slack and alignment or seating of the specimen. In order to obtain correct values of such parameters as modulus, strain, and offset yield point, this artifact must be compensated for in order to give the corrected zero point on the strain axis. This is accomplished by extending the linear portion of the curve backward until it meets the strain axis at point B. In this case, point B is the corrected zero strain point from which all strains must be measured. In the case of a material that does not exhibit any linear region, a similar correction can be made by constructing a tangent to the maximum slope at the inflection point and extending it until it meets the strain axis.

1.8 Laboratory Measuring Devices

Laboratory tests measure material properties. Frequently, specimens are made of the material in question and tested in the laboratory to measure their response to the applied forces or certain environmental conditions. These tests require the measurement of certain parameters such as time, deformation, or force. Some of these parameters are measured directly, while others are measured indirectly by relating parameters to each other. Length and deformation can be measured directly using simple devices such as rulers, dial gauges, and calipers. In other cases, indirect measurements are made by measuring electric voltage and relating it to deformation, force, stress, or strain. Examples of such devices include linear variable differential transformers (LVDTs), strain gauges, and load cells. Noncontact deformation measuring devices using lasers and various optical devices are also available. Electronic measuring devices can easily be connected to chart recorders, digital readout devices, or computers, where the results can be easily displayed and processed.

Each device has a certain *sensitivity,* which is the smallest value that can be read on the device's scale. Sensitivity should not be mistaken for accuracy or precision. Magnification can be designed into a gauge to increase its sensitivity, but wear, friction, noise, drift, and other factors may introduce errors that limit the accuracy and precision.

Measurement accuracy cannot exceed the sensitivity of the measuring device. For example, if a stopwatch with a sensitivity of 0.01 second is used to measure time, the smallest time interval that can be recorded is also 0.01 second. The selection of the measuring device and its sensitivity depends on the required accuracy of measurement. The required accuracy, on the other hand, depends on the significance and use of the measurement. For example, when expressing distance of travel from one city to another, an accuracy of 1 km or even 10 km may be meaningful. In contrast, manufacturing a computer microchip may require an accuracy of one-millionth of a meter or better. In engineering tests, the accuracy of measurement must be determined in advance to ensure proper use of such measurements and, at the same time, to avoid unnecessary effort and expense during testing. Many standardized test methods, such as those of ASTM and AASHTO, state the sensitivity of the measuring devices used in a given experiment. In any case, care must be taken to ensure proper calibration, connections, use, and interpretation of the test results of various measuring devices.

Next we will briefly describe measuring devices commonly used in material testing, such as dial gauges, LVDTs, strain gauges, proving rings, and load cells.

1.8.1 ■ Dial Gauge

Dial gauges are used in many laboratory tests to measure deformation. The dial gauge is attached at two points, between which the relative movement is measured. Most of the dial gauges include two scales with two different pointers, as depicted in Figure 1.26(a). The smallest division of the large scale determines the sensitivity of the device and is usually recorded on the face of the gauge. One division of the small pointer corresponds to one full rotation of the large pointer. The full range of the small pointer determines the range of measurement of the dial gauge. Dial gauges used in civil engineering material testing frequently have sensitivities ranging from 0.1 to 0.002 mm. The gauge can be "zeroed" by rotating the large scale in order to start the reading at the current pointer position. Digital dial gauges [Figure 1.26(b)] are also available, which allow for faster readings and simpler interpretation.

Dial gauges can be attached to frames or holders with different configurations to measure the deformation of a certain gauge length or the relative movement between two points. For example, the *extensometer* shown in Figure 1.27 is used to measure the deformation of the gauge length of a metal bar during the tensile test. Note that because of the extensometer configuration shown in the figure, the deformation of the bar is one-half of the reading indicated by the dial gauge.

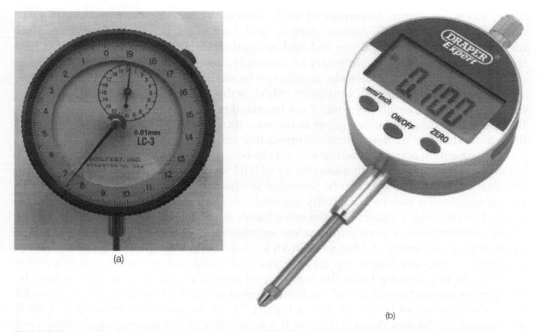

(a)

(b)

FIGURE 1.26   Dial gauges: (a) analog and (b) digital. (Courtesy of Draper Tools Ltd.)

FIGURE 1.27   Extensometer with a dial gauge.

1.8.2 ■ Linear Variable Differential Transformer (LVDT)

The linear variable differential transformer (transducer), or LVDT, is an electronic device commonly used in laboratory experiments to measure small movements or deformations of specimens. The LVDT consists of a nonmagnetic shell and a magnetic core. The shell contains one primary and two secondary electric coils, as illustrated in Figure 1.28. An electric voltage is input to the LVDT and an output voltage is obtained. When the core is in the null position at the center of the shell, the output voltage is zero. When the core is moved slightly in one direction, an output voltage is obtained. Displacing the core in the opposite direction produces an output voltage with the opposite sign. The relationship between the core position and the output voltage is linear within a certain range determined by the manufacturer. If this relation is known, the displacement can be determined by measuring the output voltage using a voltmeter or a readout device. LVDTs can measure both static and dynamic movements.

LVDTs vary widely in sensitivity and range. The sensitivity of commercially available LVDTs ranges from 0.003 to 0.25 V/mm of displacement per volt of excitation. Normal excitation supplied to the primary coil is 3 Vac with a frequency ranging from 50 Hz to 10 kHz. If 3 V is used, the most sensitive LVDTs provide an output of 18.9 mV/mm (Dally and Riley, 2005). In general, very sensitive LVDTs have small linear ranges, whereas LVDTs with greater linear ranges are less sensitive. The sensitivity and the linear range needed depend on the accuracy and the amount of displacement required for the measurement.

Before use, the LVDT must be calibrated to determine the relation between the output voltage and the displacement. A calibration device consisting of a micrometer, a voltmeter, and a holder is used to calibrate the LVDT, as shown in Figure 1.29.

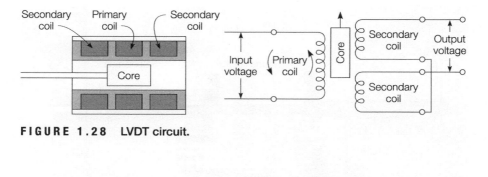

FIGURE 1.28   LVDT circuit.

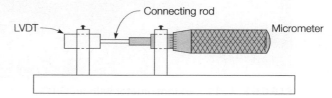

FIGURE 1.29   LVDT calibration device.

The shell and the core of the LVDT can be either separate or attached in a spring-loaded arrangement (Figure 1.30). When the former type is used, a nonmagnetic threaded connecting rod is attached to the core used to attach the LVDT to the measured object. In either case, to measure the relative movement between two points, the core is attached to one point and the shell is attached to the other point. When the distance between the two points changes, the core position changes relative to the shell, proportionally altering the output voltage. Figure 1.31 shows an extensometer with an LVDT that can be used to measure the deformation of a metal rod during the tensile test.

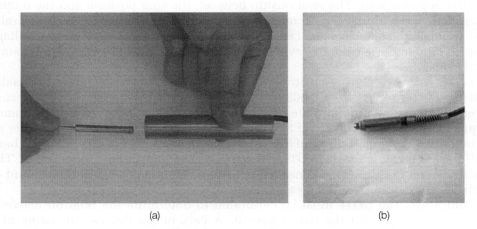

(a)   (b)

FIGURE 1.30 Types of LVDT: (a) nonspring loaded (core and connecting rod taken out of the shell), and (b) spring loaded (core and connecting rod inside the shell).

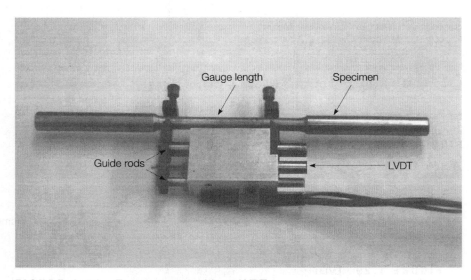

FIGURE 1.31 Extensometer with an LVDT.

1.8.3 ■ Strain Gauge

Strain gauges are used to measure small deformations within a certain gauge length. There are several types of strain gauges, but the most dominant type is the electrical strain gauge, which consists of a foil or wire bonded to a thin base of plastic or paper (Figure 1.32). An electric current is passed through the element (foil or wire). As the element is strained, its electrical resistance changes proportionally. The strain gauge is bonded via an adhesive to the surface on which the strain measurement is desired. As the surface deforms, the strain gauge also deforms and, consequently, the resistance changes. Since the amount of resistance change is very small, an ordinary ohmmeter cannot be used. Therefore, special electric circuits, such as the Wheatstone bridge, are used to detect the change in resistance (Dally and Riley, 2005).

Strain gauges are manufactured with different sizes, but the most convenient strain gauges have a gauge length of about 5 to 15 mm. Larger strain gauges can also be made and used in some applications.

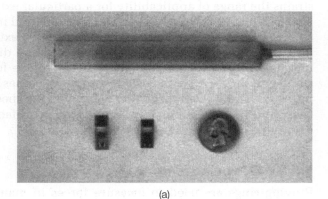

(a)

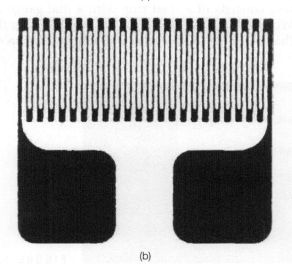

FIGURE 1.32 Strain gauges: (a) strain gauges with different sizes and (b) typical foil strain gauge.

(b)

A wire gauge consists of a length of very fine wire (about 0.025 mm diameter) that is looped into a pattern. A foil gauge is made by etching a pattern on a very thin metal foil (about 0.0025 mm thick). Foils or wires are made in a great variety of shapes, sizes, and types and are bonded to a plastic or paper base. When the strain gauge is bonded to the object, it is cemented firmly with the foil or wire side out. Foil–plastic gauges are more commonly used than wire gauges.

When using strain gauges, it is very important to have a tight bond between the gauge and the member. The surface must be carefully cleaned and prepared, and the adhesive must be properly applied and cured. The adhesive must be compatible with the material being tested.

1.8.4 ■ Noncontact Deformation Measurement Technique

Traditional deformation measuring techniques rely on the use of LVDTs and strain gauges both of which must be specialized for a specific specimen configuration, testing temperature, rate of loading, and range of strain. Each of these variables constrains the range of applicability for a particular extensometer. The use of a noncontact deformation measuring technique is preferred in many situations. Methods such as Laser Doppler Velocimetry (LDV) and laser extensometer, digital laser speckle technique, rotating drum of high-speed camera, diffraction-grating technique, and image correlation technique are currently in use for noncontact deformation measurement. Figure 1.33 shows a test setup that uses a laser extensometer for finding the stress–strain relationship of aluminum. The specimen is illuminated with a laser beam and the reflections from the specimen surface are received and processed to determine the changes in the gage length.

1.8.5 ■ Proving Ring

Proving rings are used to measure forces in many laboratory tests. The proving ring consists of a steel ring with a dial gauge attached, as shown in Figure 1.34. When a force is applied on the proving ring, the ring deforms, as measured with the dial gauge. If the relation between the force and dial gauge reading is known,

FIGURE 1.33   Noncontact laser extensometer.

FIGURE 1.34 Proving ring. (Courtesy of Humboldt Mfg. Co.)

the proving ring can be used to measure the applied force. Therefore, the proving ring comes with a calibration relationship, in either a linear equation or a table, which allows the user to determine the magnitude of the force based on measuring the deformation of the proving ring. To avoid damage, it is important not to apply a force on the proving ring higher than the capacity specified by the manufacturer. Moreover, periodic calibration of the proving ring is advisable to insure proper measurements.

1.8.6 ▨ Load Cell

The load cell is an electronic force-measuring device used for many laboratory tests. In this device, strain gauges are attached to a member within the load cell, which is subjected to either axial loading or bending. An electric voltage is input to the load cell and an output voltage is obtained. If the relation between the force and the output voltage is known, the force can easily be determined by measuring the output voltage.

Load cells are manufactured in different shapes and load capacities. Figure 1.35(a) illustrates a tensile load cell fabricated by mounting four strain gauges on the central region of a tension specimen. Figure 1.35(b) shows an S-shaped load cell in which strain gauges are bonded to the central portion and calibrated to measure the force applied on the top and bottom of the load cell. Figure 1.35(c) illustrates a diaphragm strain gauge bonded to the inside surface of an enclosure that measures the amount of pressure applied on the load cell.

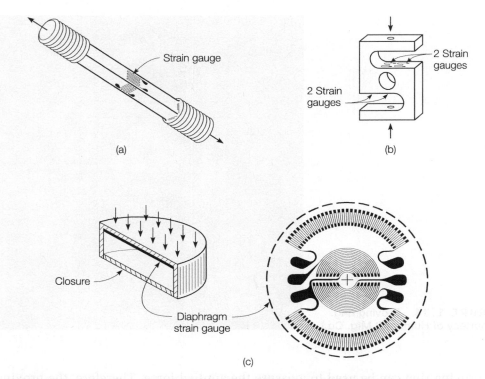

FIGURE 1.35 Load cells: (a) strain gauges on a tension rod, (b) strain gauges on an S-shaped element, and (c) diaphragm strain gauge on a closure.

Load cells must be regularly calibrated using either dead loads or a calibrated loading machine. Care must be taken not to overload the load cell. If the load applied on the load cell exceeds the capacity recommended by the manufacturer, permanent deformation can develop, ruining the load cell.

Summary

Civil and construction engineers are involved in the selection of construction materials with the mechanical properties needed for each project. In addition, the selection process must weigh other factors beyond the material's ability to carry loads. Economics, production, construction, maintenance, and aesthetics must all be considered when selecting a material.

Lately, in all fields of engineering, there has been tremendous growth in the use of new high-performance materials. For example, in the automotive industry, applications of ceramics and plastics are increasing as manufacturers strive for better performance, economy, and safety, while pushing to reduce emissions. Likewise, civil and construction engineers are continuously looking for materials with better quality and higher performance. Advanced composite materials, geotextiles, and various synthetic products are currently competing with traditional civil engineering materials. Although traditional materials such as steel, concrete, wood, and asphalt will continue to be used for some time, improvements of these materials will proceed by changing the molecular structure of such materials and using modifiers to improve their performance. Examples of such improvements include fiber-reinforced concrete, polymer-modified concrete and asphalt, low temperature–susceptible asphalt binder, high-early-strength concrete, superplasticizers, epoxy-coated steel reinforcement, synthetic bar reinforcement, rapid-set concrete patching compounds, prefabricated drainage geocomposites, lightweight aggregates, fire-resistant building materials, and earthquake-resistant joints. Civil engineers are also recycling old materials in an effort to save materials cost, reduce energy, and improve the environment.

Questions and Problems

1.1   State three examples of a static load application and three examples of a dynamic load application.

1.2   A material has the stress–strain behavior shown in Figure P1.2. What is the material strength at rupture? What is the toughness of this material?

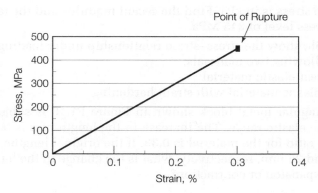

FIGURE P1.2

1.3   A tensile load of 190 kN is applied to a round metal bar with a diameter of 16 mm and a gauge length of 50 mm. Under this load, the bar elastically deforms so that the gauge length increases to 50.1349 mm and the diameter decreases to 15.99 mm. Determine the modulus of elasticity and Poisson's ratio for this metal.

1.4   A cylinder with a 150 mm diameter and 300 mm length is put under a compressive load of 665 kN. The modulus of elasticity for this specimen is 55 GPa and Poisson's ratio is 0.35. Calculate the final length and the final diameter of this specimen under this load assuming that the material remains within the linear elastic region.

1.5   A metal rod with 10 mm diameter is subjected to 9 kN tensile load. Calculate the resulting diameter of the rod after loading. Assume that the modulus of elasticity is 70 GPa, Poisson's ratio is 0.33, and the yield strength is 145 MPa.

1.6   A rectangular block of aluminum 30 mm × 60 mm × 90 mm is placed in a pressure chamber and subjected to a pressure of 100 MPa. If the modulus of elasticity is 70 GPa and Poisson's ratio is 0.333, what will be the decrease in the longest side of the block, assuming that the material remains within the linear elastic region? What will be the decrease in the volume of the block?

1.7   A plastic cube with a 100 mm × 100 mm × 100 mm. is placed in a pressure chamber and subjected to a pressure of 100 MPa. If the modulus of elasticity is 7 GPa and Poisson's ratio is 0.39, what will be the decrease in each side of the block, assuming that the material remains within the linear elastic region? What will be the decrease in the volume of the cube?

1.8   A material has a stress–strain relationship that can be approximated by the equation

$$\varepsilon = 0.3 \times 10^{-16} \times \sigma^3$$

where the stress is in kPa. Find the secant modulus and the tangent modulus for the stress level of 345 MPa.

1.9   On a graph, show the stress–strain relationship under loading and unloading for the following two materials:
a. nonlinear elastic material
b. elastoplastic material with strain hardening

1.10   The rectangular metal block shown in Figure P1.10 is subjected to tension within the elastic range. The increase in the length of a is 0.05 mm and the Poisson's ratio for the material is 0.33. If the original lengths of a and b were 50 mm and 25 mm, respectively, what is the change in the length of b? Is this change expansion or contraction?

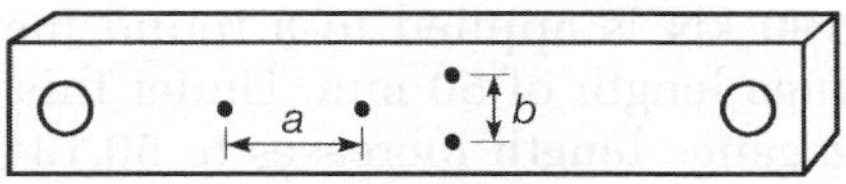

FIGURE P1.10

1.11 A cylindrical rod with a length of 380 mm and a diameter of 10 mm is to be subjected to a tensile load. The rod must not experience plastic deformation or an increase in length of more than 0.9 mm when a load of 24.5 kN is applied. Which of the four materials listed in Table P1.11 are possible candidates? Justify your answer.

TABLE P1.11

Material	Elastic Modulus, GPa	Yield Strength, MPa	Tensile Strength, MPa
Copper	110	248	289
Aluminum alloy	70	255	420
Steel	207	448	551
Brass alloy	101	345	420

1.12 A cylindrical rod with a diameter of 15.24 mm and a gauge length of 400 mm is subjected to a tensile load. If the rod is to experience neither plastic deformation nor an elongation of more than 0.45 mm when the applied load is 31000 N, which of the four metals or alloys listed below are possible candidates? Justify your choice(s).

TABLE P1.12

Material	E (GPa)	Yield Strength (MPa)	Tensile Strength (MPa)
Steel alloy 1	180	860	502
Steel alloy 2	200	400	250
Titanium alloy	110	900	730
Copper	117	220	70

1.13 The stress–strain relationship shown in Figure P1.13 was obtained during the tensile test of an aluminum alloy specimen.

Determine the following:
a. Young's modulus within the linear portion
b. Tangent modulus at a stress of 310 MPa
c. Yield stress using an offset of 0.002 strain
d. If the yield stress in part c is considered failure stress, what is the maximum working stress to be applied to this material if a factor of safety of 1.5 is used?

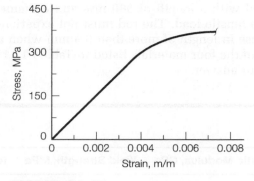

FIGURE P1.13

1.14 A tension test performed on a metal specimen to fracture produced the stress–strain relationship shown in Figure P1.14. Graphically determine the following (show units and all work):

a. Modulus of elasticity within the linear portion.
b. Yield stress at an offset strain of 0.002 m/m.
c. Yield stress at an extension strain of 0.005 m/m.
d. Secant modulus at a stress of 525 MPa.
e. Tangent modulus at a stress of 525 MPa.

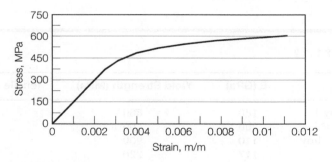

FIGURE P1.14

1.15 Use Problem 1.14 to graphically determine the following:

a. Modulus of resilience
b. Toughness

   Hint: The toughness (u_t) can be determined by calculating the area under the stress–strain curve

$$u_t = \int_0^{\varepsilon_f} \sigma \, d\varepsilon$$

where ε_f is the strain at fracture. The preceding integral can be approximated numerically by using a trapezoidal integration technique:

$$u_t = \sum_{i=1}^{n} u_i = \sum_{i=1}^{n} \frac{1}{2}(\sigma_i + \sigma_{i-1})(\varepsilon_i - \varepsilon_{i-1})$$

    c. If the specimen is loaded to 275 MPa only and the lateral strain was found to be −0.00057 m/m, what is Poisson's ratio of this metal?

    d. If the specimen is loaded to 480 MPa only and then unloaded, what is the permanent strain?

1.16   Figure P1.16 shows the stress–strain relations of metals A and B during tension tests until fracture. Determine the following for the two metals (show all calculations and units):

    a. Proportional limit

    b. Yield stress at an offset strain of 0.002 m/m.

    c. Ultimate strength

    d. Modulus of resilience

    e. Toughness

    f. Which metal is more ductile? Why?

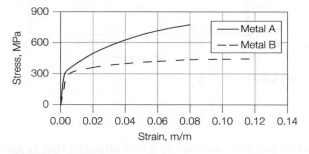

FIGURE P1.16

1.17   A brass alloy has a yield strength of 280 MPa, a tensile strength of 390 MPa, and an elastic modulus of 105 GPa. A cylindrical specimen of this alloy 12.7 mm in diameter and 250 mm long is stressed in tension and found to elongate 7.6 mm. On the basis of the information given, is it possible to compute the magnitude of the load that is necessary to produce this change in length? If so, calculate the load. If not, explain why.

1.18   Figure P1.18 shows (i) elastic–perfectly plastic and (ii) elastoplastic with strain hardening idealized responses. What stress is needed in each case to have:

    a. a strain of 0.001?

    b. a strain of 0.004?

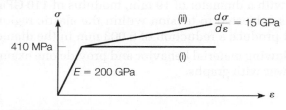

FIGURE P1.18

1.19   An elastoplastic material with strain hardening has the stress–strain relationship shown in Figure P1.19. The yield point corresponds to 600 MPa stress and 0.003 m/m strain.

    a. If a bar made of this material is subjected to a stress of 650 MPa and then released, what is the permanent strain?

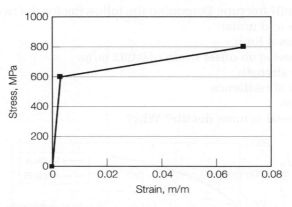

FIGURE P1.19

    b. What is the percent increase in yield strength that is gained by the strain hardening shown in part (a)?

    c. After strain hardening, if the material is subjected to a stress of 625 MPa, how much strain is obtained? Is this strain elastic, permanent, or a combination of both?

1.20   A brass alloy rod having a cross sectional area of 100 mm² and a modulus of 110 GPa is subjected to a tensile load. Plastic deformation was observed to begin at a load of 39872 N.

    a. Determine the maximum stress that can be applied without plastic deformation.

    b. If the maximum length to which a specimen may be stretched without causing plastic deformation is 67.21 mm, what is the original specimen length?

1.21   A copper rod with a diameter of 19 mm, modulus of 110 GPa, and a Poisson's ratio of 0.35 is subjected to tension within the elastic region. Determine the force that will produce a reduction of 0.003 mm in the diameter.

1.22   Define the following material behavior and provide one example of each. Support your answer with graphs.

    a. Elastic

    b. Plastic

    c. Viscoelastic

1.23 An asphalt concrete cylindrical specimen with a 100 mm diameter and a 150 mm height is subjected to an axial compressive static load of 0.7 kN for 1 hour, after which the load is removed. The test was performed at a temperature of 40°C. The height of the specimen was measured at different time intervals during loading and unloading for a total of 2 hours and produced the results shown in Table P1.23.

a. Using a spreadsheet program, plot the stress versus time and the strain versus time on two separate graphs.

b. How much is the elastic strain?

c. What is the permanent strain at the end of the experiment?

d. What is the phenomenon of the change of specimen height during static loading called? What is the phenomenon of the change of specimen height during unloading called?

TABLE P1.23

Time, minutes	Specimen Height, mm	Time, minutes	Specimen Height, mm
0	152.40	60.01	151.83
0.01	152.19	62	151.91
2	152.07	65	151.99
5	151.98	70	152.09
10	151.88	80	152.21
20	151.77	90	152.25
30	151.70	100	152.28
40	151.66	110	152.29
50	151.63	120	152.30
60	151.61		

1.24 Draw a sketch of the Maxwell model and label all components. Draw a graph showing displacement versus time when the model is subjected to a constant force for a time period t and then released. Comment on why the model responds this way.

1.25 Derive the response relation for each of the models shown in Figure P1.25 assuming that the force F is constant and instantaneously applied.

(a)   (b)

FIGURE P1.25

1.26   What are the differences between "modulus of resilience" and "toughness"? Explain graphically.

1.27   State four failure modes of materials. Describe typical examples of each mode.

1.28   A metal rod having a diameter of 10 mm is subjected to a repeated tensile load. The material of the rod has a tensile strength of 290 MPa and a fatigue failure behavior as shown in Figure 1.16. How many load repetitions can be applied to this rod before it fails if the magnitude of the load is (a) 5 kN and (b) 11 kN?

1.29   What is the factor of safety? On what basis is its value selected?

1.30   State the typical values of the specific gravity of three materials commonly used in construction.

1.31   Define the coefficient of thermal expansion. What is the relation between the linear and the volumetric coefficients of thermal expansion?

1.32   A steel rod, which is free to move, has a length of 200 mm and diameter of 20 mm at a temperature of 15°C. If the rod is heated uniformly to 115°C, determine the length and the diameter of this rod to the nearest micron at the new temperature if the linear coefficient of thermal expansion of steel is 12.5×10^{-6} m/m/°C. Is there a stress on the rod at 115°C?

1.33   In Problem 1.32, if the rod is snugly fitted against two immovable non-conducting walls at a temperature of 15°C and then heated uniformly to 115°C, what is its length at 115°C? If the modulus of elasticity of steel is 207 GPa, what is the stress induced in the bar? Is this stress tension or compression?

1.34   A 4-m-long steel plate with a rectangular cross section (10 mm $\times$ 50 mm) is resting on a frictionless surface under the sun. The plate temperature is measured to be at 40°C. The plate is then moved into a cold room and is left to rest on a frictionless surface. After several hours, the plate temperature is measured to be at 5°C. The steel has a modulus of elasticity equal to 200 GPa and a coefficient of thermal expansion equal to 1.1×10^{-5} m/m/°C.
a. What is the length of shrinkage?
b. What tension load is needed to return the length to the original value of 4 meters?
c. What is the longitudinal strain under this load?

1.35   Estimate the tensile strength required to prevent cracking in a concrete-like material that is cast into a bar, 1.25 m long, and is fully restrained at each end against axial movement. The concrete is initially cast and cured at a temperature of 40°C and subsequently cools to a temperature of −20°C. Assume that the modulus of elasticity is 35 GPa and the thermal coefficient is 9×10^{-6} m/m/°C.

1.36   State two examples in which corrosion plays an important role in selecting the material to be used in a structure.

1.37   Briefly discuss the variability of construction materials. Define the terms accuracy and precision when tests on materials are performed.

1.38   In order to evaluate the properties of a material, samples are taken and tested. What are the two most important factors that must be satisfied when the sample is collected? Show how these factors can be satisfied.

1.39   A contractor claims that the mean compressive strength for a concrete mix is 32.4 MPa and that it has a standard deviation of 2.8 MPa. If you break 16 cylinders and obtain a mean compressive strength of 30.3 MPa, would you believe the contractor's claim? Why? (*Hint:* Use statistical *t*-test.)

1.40   In a ready-mix plant, cylindrical samples are prepared and tested periodically to detect any mix problem and to ensure that the compressive strength is higher than the lower specification limit. The minimum target value was set at 34 MPa. The compressive strength data shown in Table P1.40 were collected.
   a. Calculate the mean, standard deviation, and the coefficient of variation of the data.
   b. Using a spreadsheet program, create a control chart for these data showing the lower specification limit. Is the plant production meeting the specification requirement? If not, comment on possible reasons. Comment on the data scatter.

TABLE P1.40

Sample No.	Compressive Strength (MPa)	Sample No.	Compressive Strength (MPa)
1	38.6	11	46.1
2	35.4	12	37.6
3	43.8	13	36.4
4	35.8	14	46.2
5	35.77	15	35.8
6	36.1	16	41.5
7	41.1	17	36.7
8	40.5	18	39.0
9	42.4	19	45.5
10	35.3	20	36.1

1.41   During construction of the surface layer of asphalt pavement, two samples were taken every day to measure the asphalt content in the mix to ensure that it is within the specification limits. The target value was set at 5.5% with a tolerance of ±0.4%. The asphalt content data shown in Table P1.41 were collected:
   a. Calculate the mean, standard deviation, and the coefficient of variation of the data.

b. Using a spreadsheet program, create a control chart for these data showing the target value and the upper and lower specification limits. Are the
asphalt content data within the specification limits? Comment on any
trend and possible reasons.

TABLE P1.41

Date	Asphalt Content, %	Date	Asphalt Content, %
Oct. 15	5.7	Oct. 20	5.3
Oct. 15	5.8	Oct. 20	5.6
Oct. 16	5.3	Oct. 21	5.8
Oct. 16	5.4	Oct. 21	5.1
Oct. 17	5.7	Oct. 22	5.8
Oct. 17	5.7	Oct. 22	5.9
Oct. 18	5.6	Oct. 23	5.1
Oct. 18	5.8	Oct. 23	6.2
Oct. 19	5.4	Oct. 24	5.2
Oct. 19	5.5	Oct. 24	4.8

1.42   Linear variable differential transformers (LVDTs) are used in some laboratory
experiments. What is their function? Show a sketch showing the main components of an LVDT and discuss the theory behind its use.

1.43   Briefly discuss the concept behind each of the following measuring devices:
a. LVDT
b. strain gauge
c. proving ring
d. load cell

1.44   Referring to the dial gauge shown in Figure P1.44, determine
a. Accuracy
b. Sensitivity
c. Range, given that the small pointer moves one division for each full turn
of the large pointer
d. Which of the preceding items can be improved through calibration?
e. For measuring devices, which of the following two statements is always
true?

FIGURE P1.44

    (i) Accuracy cannot exceed sensitivity
    (ii) Sensitivity cannot exceed accuracy

1.45   Measurements should be reported to the nearest 1000, 100, 10, 1, 0.1, 0.01, 0.001 units, etc., depending on their variability and intended use. Using your best judgment, how should the following measurements be reported?
    a. Deformation of a steel specimen during the tension test (mm)
    b. Tensile strength of steel (MPa)
    c. Modulus of elasticity of aluminum (MPa)
    d. Weight of aggregate in a sieve analysis test (grams)
    e. Compressive strength of portland cement concrete (MPa)
    f. Moisture content in a concrete masonry unit (percent)
    g. Asphalt content in hot-mix asphalt (percent)
    h. Specific gravity of wood
    i. Distance between New York City and Los Angeles (km)
    j. Dimensions of a computer microchip (mm)

1.46   During calibration of an LVDT, the data shown in Table P1.46 were obtained. Using a spreadsheet program, plot the relation between the micrometer reading and voltage. What is the linear range of the LVDT? Determine the calibration factor of the LVDT by obtaining the best fit line of the data within the linear range.

TABLE P1.46

Micrometer Reading, mm	Voltage, Volts	Micrometer Reading, mm	Voltage, Volts
−2.997	−10.12	0.305	0.985
−2.743	−10.12	0.559	1.99
−2.489	−10.121	0.813	3.023
−2.235	−9.134	1.067	4.035
−1.981	−8.131	1.321	5.071
−1.727	−7.111	1.575	6.098
−1.473	−6.1	1.829	7.115
−1.219	−5.108	2.083	8.143
−0.965	−4.089	2.337	9.144
−0.711	−3.097	2.591	10.157
−0.457	−2.059	2.845	10.156
−0.203	−1.053	3.099	10.156
0	0		

TABLE P1.47

Micrometer Reading, mm	Voltage, Volts
−13.22	−0.671
−12.72	−0.671
−11.47	−0.671
−8.98	−0.526
−6.48	−0.380
−3.98	−0.235
−1.49	−0.090
1.01	0.055
3.51	0.200
6.00	0.345
8.50	0.491
10.99	0.491
13.49	0.491

1.47 During calibration of an LVDT, the data shown in Table P1.47 were obtained. Using a spreadsheet program, plot the relation between the micrometer reading and voltage. What is the linear range of the LVDT? Determine the calibration factor of the LVDT by obtaining the best fit line of the data within the linear range.

1.48 Using the Internet, research the LEED rating system for new construction. What are the prerequisite requirements for LEED certification? What are the requirements for LEED certification of wood? Which LEED credit area can be awarded the greatest number of points?

1.49 Although the LEED system is the most widely used with many users, several other rating systems have been developed and used throughout the world. Summarize some of these systems.

1.9 References

Ashby, M. F. and D. R. H. Jones. *Engineering Materials 1: An Introduction to Properties, Applications and Design.* 4th ed. Butterworth-Heinemann, 2011.

Burati, J. L. and C. S. Hughes. *Highway Materials Engineering, Module I: Materials Control and Acceptance—Quality Assurance.* Publication No. FHWA-HI-90-004. Washington, DC: Federal Highway Administration, 1990.

Callister, W. D., Jr. *Materials Science and Engineering—An Introduction.* 6th ed. New York: John Wiley and Sons, 2006.

Dally. J. W. and W. F. Riley. *Experimental Stress Analysis.* 4th ed. College House Enterprises, ISBN 0-9762413-0-7, 2005.

Flinn, R. A. and P. K. Trojan. *Engineering Materials and Their Applications.* 4th ed. New York: John Wiley and Sons, 1995.

Fowler, K. M. and E. M. Rauch, Sustainable Building Rating Systems Summary, GSA Contract DE-AC05-76RL061830, Pacific Northwest National Laboratory, Report PNNL-15858, 2006.

LEED v4. LEED v4 Users Guide, http://www.usgbc.org/resources/leed-v4-user-guide, 2013.

Mehta, P. K. and P. J. M. Monteiro. *Concrete: Microstructure, Properties, and Materials.* 4th ed. McGraw-Hill Education, 2013.

Pawlowski, N. H. and E. J. Ripling. *Strength and Structure of Engineering Materials.* New York: McGraw-Hill Professional, 2005.

Popov, E. P. *Introduction to Mechanics of Solids.* Upper Saddle River, NJ: Prentice Hall, 1968.

Richart, F. E., Jr., J. R. Hall, Jr., and R. D. Woods. *Vibrations of Soils and Foundations.* Upper Saddle River, NJ: Prentice Hall, 1970.

USGBC, LEED Frequently Asked Questions, Focus: Building Materials, United States Green Building Council, web publication, http://www.usgbc.org/Docs/LEEDdocs/LEEDfaq-materials2.pdf, 2015.

Wang, N., K. M. Fowler, and R. S. Sullivan, Green Building System Certification Review, Report PNNL-20366, US Department of Energy, 2012.

CHAPTER

NATURE OF MATERIALS

To a large extent, the behavior of materials is dictated by the structure and bonding of the atoms that are the building blocks for all matter. Knowledge of the bonding and structure of materials at the molecular level allows us to understand their behavior. This chapter presents a broad overview of concepts essential to our understanding of these behaviors. The chapter reviews the basic types of bonds and then, based on the type of bond, classifies materials as metallic, ceramic, or amorphous. The general nature of each of these classes of materials is presented.

2.1   Basic Materials Concepts

Atoms are the basic building blocks of all materials. For the purpose of this text, atoms will be considered to consist of three subatomic particles: *protons, neutrons,* and *electrons*. The protons and neutrons are at the center of the atom, while the electrons travel about the nucleus in paths or shells. The *atomic number* is the number of protons in the nucleus of the atom. The *atomic mass* of an atom is the number of protons plus the number of neutrons in the center of the atom. An *element* is an atom or group of atoms with the same atomic number. *Isotopes* are elements with different numbers of neutrons in the nucleus.

2.1.1   Electron Configuration

The behavior of an atom's electrons controls, to a large extent, the characteristics of an element. An electrically neutral (or complete) atom has equal numbers of electrons and protons. However, an atom may either release or attract electrons to reach a more stable configuration. Electrons travel around the nucleus in orbital paths, or orbits, around the nucleus. The distance between electrons and the nucleus is not fixed; it is better described as a random variable with a distribution, as shown in Figure 2.1.

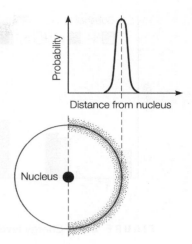

FIGURE 2.1   Orbital path of model electrons.

Each orbital path, or *shell,* can hold only a fixed number of electrons. The maximum number of electrons in a shell is equal to $2n^2$, where n is the principal quantum number of the shell. The orbital path of electrons is defined by four parameters: the principal quantum number or shell designation, the subshell designation, the number of energy states of the subshell, and the spin of the electron. Table 2.1 lists the number of electrons that can be in each shell and subshell.

The energy level of an electron depends on its shell and subshell location, as shown in Figure 2.2 (Callister, 2006). Electrons always try to fill the lowest energy location first. As shown in Figure 2.2, for a given subshell, the larger the principal quantum number, the higher is the energy level (e.g., the energy of subshell 1s is less than

TABLE 2.1   Number of Available Electron States in Some of the Electron Shells and Subshells

Principal Quantum Number	Shell Designation	Subshell	Number of States	Number of Electrons	
				Per Subshell	Per Shell
1	K	s	1	2	2
2	L	s	1	2	8
		p	3	6	
3	M	s	1	2	18
		p	3	6	
		d	5	10	
4	N	s	1	2	32
		p	3	6	
		d	5	10	
		f	7	14	

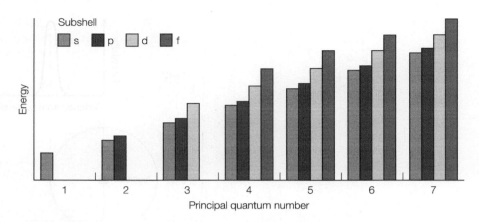

FIGURE 2.2   Energy levels of electrons in different shells and subshells.

subshell 2s). Within a shell, the energy level increases with the subshell location: Subshell f is a higher energy state than subshell d, and so on. There is an energy overlap between the shells: Subshell 4s is at a lower energy state than subshell 3d. The electron configuration, or structure of an atom, describes the manner in which the electrons are located in the shells and subshells. The convention for listing the location of shells in the atom is to list the quantum number, followed by the subshell designation, followed by the number of electrons in the subshell raised to a superscript. This sequence is repeated for each subshell that contains electrons. Table 2.2 provides some examples.

TABLE 2.2   Sample Electron Configurations

Element	Atomic Number	Electron Configuration
Hydrogen	1	$1s^1$
Helium	2	$1s^2$
Carbon	6	$1s^2 2s^2 2p^2$
Neon	10	$1s^2 2s^2 2p^6$
Sodium	11	$1s^2 2s^2 2p^6 3s^1$
Aluminum	13	$1s^2 2s^2 2p^6 3s^2 3p^1$
Silicon	14	$1s^2 2s^2 2p^6 3s^2 3p^2$
Sulfur	16	$1s^2 2s^2 2p^6 3s^2 3p^4$
Argon	18	$1s^2 2s^2 2p^6 3s^2 3p^6$
Calcium	20	$1s^2 2s^2 2p^6 3s^2 3p^6 4s^2$
Chromium	24	$1s^2 2s^2 2p^6 3s^2 3p^6 3d^5 4s^1$
Iron	26	$1s^2 2s^2 2p^6 3s^2 3p^6 3d^6 4s^2$
Copper	29	$1s^2 2s^2 2p^6 3s^2 3p^6 3d^{10} 4s^1$

The electrons in the outermost filled shell are the *valence electrons.* These are the electrons that participate in the formation of primary bonds between atoms. The eight electrons needed to fill the s and p subshells are particularly important. If these subshells are completely filled, the atoms are very stable and virtually nonreactive, as is the case for the noble gases neon and argon. In many cases, atoms will release, attract, or share electrons to complete these subshells and reach a stable configuration. Calcium and chromium are examples of atoms with electrons that fill the 4s subshell, while the 3d subshell is incomplete or empty, as would be expected from Figure 2.2. Copper demonstrates that there are exceptions to the energy rule. One would expect copper to have two electrons in the 4s subshell, leaving nine electrons for the 3d subshell. However, it has 10 electrons in the 3d subshell and only 1 in the 4s subshell. A similar disparity exists for chromium. Note that iron has two electrons in the 4s subshell; thus, it has two electrons more than it needs for a stable configuration. These two electrons are readily released to form iron molecules. Aluminum is also an exception, since it has an excess of three electrons.

2.1.2 ■ Bonding

As two atoms are brought together, both attractive and repulsive forces develop. The effects of these forces are additive, as shown in Figure 2.3, such that once the atoms are close enough to interact, they will reach a point at which the attractive and repulsive forces are balanced and an equilibrium is reached. Energy is required either to bring the atoms closer together (*compression*) or to separate them (*tension*). The distance at which the net force is zero corresponds to the minimum energy level

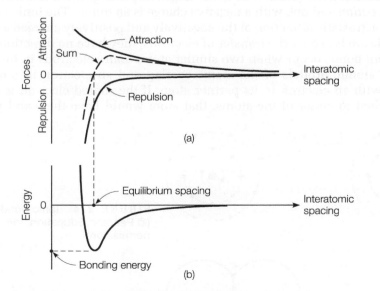

FIGURE 2.3 Attractive and repulsive (a) forces and (b) energies between atoms. (© Pearson Education, Inc. Used by permission)

and is called the *equilibrium spacing*. The minimum energy is represented by a negative sign. The largest negative value is defined as the bonding energy. The bonding energy can be computed from equations for the attractive and repulsive forces. Based on the strength of the bonds, the theoretical strength of a material can be estimated. However, this theoretical strength grossly overestimates the actual strength due to flaws in the molecular structure (Van Vlack, 1964, 1989).

The bonding energy depends on the molecular mechanism holding the atoms together. There are two basic categories of bonds: primary and secondary. *Primary bonds* form when atoms interact to change the number of electrons in their outer shells so as to achieve a stable and nonreactive electron structure similar to that of a noble gas. *Secondary bonds* are formed when the physical arrangement of the atoms in the molecule results in an imbalanced electric charge; one side is positive and the other is negative. The molecules are then bonded together through electrostatic force.

Primary Bonds   Three types of primary bonds are defined, based on the manner in which the valence electrons interact with other atoms:

1. *ionic bonds*—transfer of electrons from one elemental atom to another (Figure 2.4)
2. *covalent bonds*—sharing of electrons between specific atoms (Figure 2.5)
3. *metallic bonds*—mass sharing of electrons among several atoms (Figure 2.6)

Ionic bonds are the result of one atom releasing electrons to other atoms that accept the electrons. Each of the elements reaches a stable electron configuration of the outer s and p subshells. All of the atoms are ions, since they have an electrical charge. When an atom releases an electron, the atom becomes positively charged; the atom receiving the electron becomes negatively charged. An ion with a positive charge is a *cation* and one with a negative charge is an *anion*. The ionic bond results from the electrostatic attraction of the negatively and positively charged atoms. Since these bonds are based on the transfer of electrons, they have no directional nature.

Covalent bonds occur when two similar atoms *share* electrons in the outer subshell. The atomic orbitals of the atoms overlap and an electron in one atom can exchange with an electron in its partner atom. If the shared electron is considered to be attached to either of the atoms, that atom would have the s and p subshells

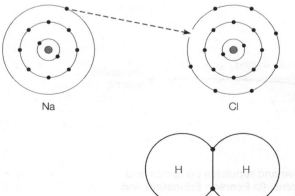

Na   Cl

FIGURE 2.4   Ionic bonding.
(© Pearson Education, Inc. Used by permission)

H   H

FIGURE 2.5   Covalent bonding.
(© Pearson Education, Inc. Used by permission)

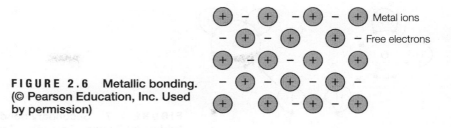

FIGURE 2.6 Metallic bonding. (© Pearson Education, Inc. Used by permission)

filled and would therefore be a stable atom. Since the orbital paths of the atoms must overlap for the covalent bond to form, these bonds are highly directional. In materials such as diamond, the covalent bonds are very strong. However, the carbon chain structure of polymers is also formed by covalent bonds and these elements display a wide range of bond energy. The number of bonds that form depends on the number of valence electrons. An atom with N electrons in the valence shell can bond with only $8N$ neighbors by sharing electrons with them. When the number of electrons N equals 7, the atoms join in pairs. When N equals 6, as in sulfur, long chains can form since the atom can bond with two neighbors. When N equals 5, a layered structure can be developed. If there are four valence electrons, three-dimensional covalent bonds can result (e.g., the structure of carbon in diamonds) (Jastrzebski, 1987). The calcium-silicate chain in portland cement concrete is based on covalent bonds.

Most interatomic bonds are partially ionic and partially covalent, and few compounds have pure covalent or ionic bonds. In general, the degree of either type of bonding depends on the relative positions of the elements in the periodic table. The wider the separation, the more ionic bonds form. Since the electrons in ionic and covalent bonds are fixed to specific atoms, these materials have good thermal and electrical insulation properties.

Metallic bonds are the result of the metallic atoms having loosely held electrons in the outer s subshell. When similar metallic atoms interact, the outer electrons are released and are free to float between the atoms. Thus, the atoms are ions that are electrically balanced by the free electrons. In other words, the free electrons disassociate themselves from the original atom and do not get attached to another atom. Metallic bonds are not directional, and the spacing of the ions depends on the physical characteristics of the atoms. The atoms tend to pack together to give simple, high density structures, like marbles shaken down in a box. The easy movement of the electrons leads to the high electrical and thermal conductivity of metals.

Secondary Bonds   Secondary bonds are much weaker than primary bonds, but they are important in the formation of links between polymer materials. These bonds result from a dipole attraction between uncharged atoms. As the electrons move about the nucleus, at any instant, the charge is distributed in a nonsymmetrical manner relative to the nucleus. Thus, at any instant, one side of the atom has a negative charge and the other side of the atom has a positive charge. Secondary bonds result from electrostatic attraction of the dipoles of the atoms. These dipole interactions occur between induced dipoles or polar molecules that have permanent dipoles; both types of interactions are classified as *van der Waals forces*. Hydrogen, however,

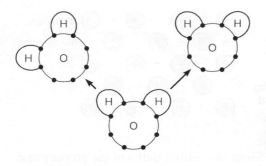

has only one proton and one electron; thus, it tends to form a polar molecule when bonded with other atoms. The electrostatic bond formed due to the hydrogen bond is generally stronger than van der Waals forces, so the hydrogen bond, shown in Figure 2.7, is a special form of the secondary bond.

2.1.3 ▓ Material Classification by Bond Type

Based on the predominant type of bond a material's atoms may form, materials are generally classified as metal, inorganic solids, and organic solids. These materials have predominantly metallic, covalent and ionic, and covalent bonds, respectively. Solids with each of these types of bonds have distinctly different characteristics. Metals and inorganic solids generally have a *crystalline* structure, a repeated pattern or arrangement of the atoms. On the other hand, organic solids usually have a random molecular structure. The following are the predominant materials used in civil engineering in each category:

    Metallic
        steel
        iron
        aluminum
    Inorganic solids
        Portland cement concrete
        bricks and cinder blocks
        glass
        aggregates (rock products)
    Organic solids
        asphalt
        plastics
        wood

2.2   Metallic Materials

The chemical definition of a metal is an element with one, two, or three valence electrons. These elements bond into a mass with metallic bonds. Due to the nature of metallic bonds, metals have a very regular and well-defined structure. Since the metallic bonds are nondirectional, the atoms are free to pack into a dense configuration. The regular three-dimensional geometric pattern of the atoms in a metal is called a *unit cell.* Repeated coalescing of unit cells forms a space lattice of the material. However, in a mass of material, a perfect structure can be achieved only through carefully controlled conditions. Generally, metallic solids are formed by cooling a mass of molten material. As the material cools, the atoms are vibrating. This can cause one atom to occupy the space of two atoms, generating a defect in the lattice structure. In addition, during the cooling process, crystals grow simultaneously from several nuclei. As the material continues to cool, these crystals grow together with a boundary forming between the grains. This produces flaws or slip planes in the structure that have an important influence on the behavior and characteristics of the material. In addition, rarely are pure elemental metals used for engineering applications. Even highly refined materials contain impurities that were not removed in the refining process. In addition, most metals do not have desirable properties in a pure state. For example, iron and aluminum used for structural applications have alloying elements that impart special characteristics to the metal. As a result, understanding the nature of metals at the molecular level requires an examination of the primary structure of the metal, the effect of cooling rates, and the impact of impurities and alloying elements.

2.2.1 ▩ Lattice Structure

As metals cool from the molten phase, the atoms are arranged into definite structures dependent on the size of the atom and the valence electrons. Certain characteristics are apparent in the three-dimensional array of points formed by the intersection of the parallel lines shown in Figure 2.8. In this configuration, the arrangement of neighboring points about any specific point is identical with the arrangement around any other internal point. This property can be described mathematically by

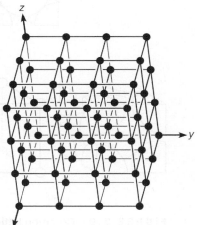

FIGURE 2.8 Parallelogram structure for crystal lattices. (© Pearson Education, Inc. Used by permission)

three unit vectors, **a**, **b**, and **c**. The location of any point, **r′**, relative to a reference point, can be defined in terms of an integer number of vector movements:

$$\mathbf{r'} = \mathbf{r} + n_1\mathbf{a} + n_2\mathbf{b} + n_3\mathbf{c} \tag{2.1}$$

A continuous repetition of Equation 2.1, using incremental integers for n_1, n_2, and n_3 will result in the parallelogram shown in Figure 2.8, where **r** is the position vector of the origin relative to the reference point. It should be noted that the angles between the axes need not be 90 degrees. Such a network of lines is called a *space lattice.* There are 14 possible space lattices in three dimensions that can be described by vectors **a**, **b**, and **c**. However, the space lattices of common engineering metals can be described by two cubic structures and one hexagonal structure, as shown in Figure 2.9.

A simple cubic lattice structure has one atom on each corner of a cube that has axes at 90 degrees and equal vector lengths. However, this is not a common structure, although it does exist in some metals. There are two important variations on the cubic structure: the *face center cubic* and the *body center cubic.* The face center cubic (FCC) structure has an atom at each corner of the cube plus an atom on each of the faces, as shown in Figure 2.9(a). The body center cubic (BCC) structure has one atom on each of the corners plus one in the center of the cube, as shown in Figure 2.9(b).

The third common metal lattice structure is the *hexagonal close pack* (HCP). As shown in Figure 2.9(c), the HCP has top and bottom layers, with atoms at each of the corners of the hexagon and one atom in the center of the top and bottom planes; in addition, there are three atoms in a center plane. The atoms in the center plane are equidistant from all neighboring atoms. See Table 2.3 for the crystal structures and the atomic radii of some metals.

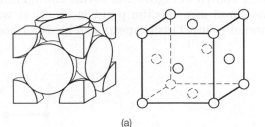

(a)

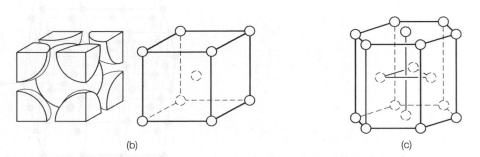

(b)   (c)

FIGURE 2.9   Common lattice structures for metals: (a) face center cubic (FCC), (b) body center cubic (BCC), and (c) hexagonal close pack (HCP).

TABLE 2.3   Lattice Structure of Metals

Metal	Crystal Structure	Atomic Radius (nm)	Metal	Crystal Structure	Atomic Radius (nm)
Aluminum	FCC	0.1413	Molybdenum	BCC	0.1363
Cadmium	HCP	0.1490	Nickel	FCC	0.1246
Chromium	BCC	0.1249	Platinum	FCC	0.1387
Cobalt	HCP	0.1253	Silver	FCC	0.1445
Copper	FCC	0.1278	Tantalum	BCC	0.1430
Gold	FCC	0.1442	Titanium	HCP	0.1445
Iron	BCC	0.1241	Tungsten	BCC	0.1371
Lead	FCC	0.1750	Zinc	HCP	0.1332

Two of the important characteristics of the crystalline structure are the coordination number and the atomic packing factor. The *coordination number* is the number of "nearest neighbors." The coordination number is 12 for FCC and HCP structures and 8 for BCC structures. This can be confirmed by examination of Figure 2.9. The *atomic packing factor* (APF) is the fraction of the volume of the unit cell that is occupied by the atoms of the structure:

$$\text{APF} = \frac{\text{volume of atoms in the unit cell}}{\text{total unit volume of the cell}} \qquad (2.2)$$

In order to compute the atomic packing factor, the *equivalent number of atoms* associated with each unit cell must be determined, along with the atomic radius of the atoms, which are given in Table 2.3. The equivalent number of atoms associated with a cell is the number of atoms in a large block of material divided by the number of unit cells in the block. However, by properly considering the fraction of atoms in Figure 2.9, we can count the number of "whole" atoms in each unit cell. The FCC structure has eight corner atoms, each of which is shared with seven other unit cells; thus, all the corner atoms contribute one atom to the atom count for the unit cell. The face atoms are shared only with one other unit cell so that each of the face atoms contributes one-half atom. Since there are six faces, the face atoms add three atoms to the count. Adding the face and corner atoms yields a total of four equivalent atoms in the FCC unit cell. The BCC structure has the equivalent of only two atoms. The HCP structure has an equivalent of six atoms. Each of the 12 corner atoms is shared between six unit cells, the face atoms are shared between two unit cells, and the three atoms on the center planes are not shared with any other unit cells, so 12/6 + 2/2 + 3 = 6.

The volume of the atoms in the unit cell can then be computed as the volume of one atom times the number of equivalent atoms. The volume of a sphere is $V = (4/3)\pi r^3$, where r is the radius. The volume of the unit cell can be determined by knowing the radius of the atoms and the fact that the atoms are in contact with each other.

Sample Problem 2.1

Show that the atomic packing factor for the FCC lattice structure (Fig. SP2.1) is 0.74.

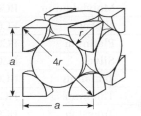

FIGURE SP2.1

Solution

Number of equivalent whole atoms in each unit cell = 4

Volume of the sphere = $\left(\dfrac{4}{3}\right)\pi r^3$

Volume of atoms in the unit cell = $4 \times \left(\dfrac{4}{3}\right)\pi r^3 = \left(\dfrac{16}{3}\right)\pi r^3$

By inspection, the diagonal of the face of an FCC unit cell = 4r

Length of each side of unit cell = $2\sqrt{2}r$

Volume of unit cell = $(2\sqrt{2}r)^3$

$$\text{APF} = \frac{\text{volume of atoms in the unit cell}}{\text{total unit volume of the cell}} = \frac{\left(\dfrac{16}{3}\right)\pi r^3}{(2\sqrt{2}r)^3} = 0.74$$

Similar geometric considerations allow for the calculation of the atomic packing factor for the BCC and HCP structures as 0.68 and 0.74, respectively.

The theoretical density of a metal is a function of the type of lattice structure and can be determined from

$$\rho = \frac{nA}{V_c N_A} \tag{2.3}$$

where

ρ = density of the material
n = number of equivalent atoms in the unit cell
A = atomic mass of the element (grams/mole)
V_c = volume of the unit cell
N_A = Avogadro's number (6.023×10^{23} atoms/mole)

Sample Problem 2.2

Calculate the radius of the aluminum atom, given that aluminum has an FCC crystal structure, a density of 2.70 Mg/m³, and an atomic mass of 26.98 g/mole. Note that the APF for the FCC lattice structure is 0.74.

Solution

$$\rho = \frac{nA}{V_c N_A}$$

For an FCC lattice structure, $n = 4$, so

$$V_c = \frac{4 \times 26.98}{2.70 \times 10^6 \times 6.023 \times 10^{23}} = 6.636 \times 10^{-29} \text{ m}^3$$

$$\text{APF} = 0.74 = \frac{4 \times \left(\frac{4}{3}\right)\pi r^3}{6.636 \times 10^{-29}}$$

$$r^3 = 0.293 \times 10^{-29}$$

$$r = 0.143 \times 10^{-9} \text{ m} = 0.143 \text{ nm}$$

Knowledge of the type of lattice structure is important when determining the mechanical behavior of a metal. Under elastic behavior, the bonds of the atoms are stretched, but when the load is removed, the atoms return to their original position. On the other hand, plastic deformation, by definition, is a permanent distortion of the materials; therefore, plastic deformation must be associated with a change in the atomic arrangement of the metal. Plastic deformation is the result of planes of atoms slipping over each other due to the action of shear stress. Naturally, the slip will occur on the planes that are the most susceptible to distortion. Since the basic bonding mechanism of the various metals is similar, differences in the theoretical strength of a material are attributed to the differences in the number and orientation of the slip planes that result from the different lattice structures.

2.2.2 ▨ Lattice Defects

Even under carefully controlled conditions, it is very difficult to grow perfect crystalline structures. Generally, pure crystalline structures are limited to approximately 1 μm in diameter. These pure materials have a strength and modulus of elasticity approaching the maximum theoretical values based on the bonding characteristics. However, the strength and deformation of all practical materials are limited by defects. There are several causes for the development of defects in the crystal structure. These can be classified as follows:

1. point defects or missing atoms
2. line defects or rows of missing atoms, commonly called an *edge dislocation*
3. area defects or grain boundaries
4. volume defects or cavities in the material

In the case of point defects, single atoms can be missing in the lattice structure because the atoms are vibrating as they transition from liquid to solid. As a result, one atom may vibrate in the area where two atoms should be in the lattice. Vacancies have little effect on the properties of the material.

In considering the differences between manufactured and theoretical material behavior, an understanding of line defects becomes important. A typical line defect is shown in Figure 2.10, where a line of missing atoms extends back into the illustration (Van Vlack, 1989; Guy and Hren, 1974). The atoms above the dislocation are in compression and those below the dislocation are in tension. As a result, the atoms are not at their natural spacing, so the bonds of these atoms are not at the point of minimum energy, as shown in Figure 2.2. Thus, when a shear stress is applied to this location, there will be a tendency for the atoms to slip in a progressive manner from position (a) to (b) to (c), as shown in Figure 2.11 (Flinn and Trojan, 1995; Budinski and Budinski, 2010).

Volume defects are flaws in the manufactured material; they will not be discussed further. Area defects are discussed in the following sections.

2.2.3 ▨ Grain Structure

The structure of metals has been described in terms of the unit cell or the repeated crystalline structure. However, equally important to the behavior of the material are the size and arrangement of the grains in the material. The grain structure (microscopic structure) of the material should be distinguished from the atomic structure.

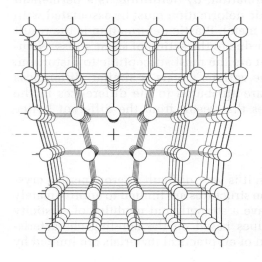

FIGURE 2.10 Atomic packing at a line defect. (© Pearson Education, Inc. Used by permission)

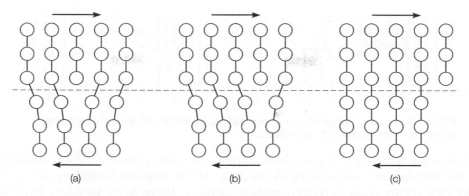

FIGURE 2.11 Plastic deformation involving movement of atoms along a slip plane.

Figure 2.12, for example, is an optical photomicrograph of a low-carbon steel. This photomicrograph was obtained using a scanning electron microscope with a magnification of 500 times. Note that this microscopic scale is much different than the atomic scale of Figure 2.9, which has a magnification in the order of 10 million times.

In metals manufacturing, the material is heated to a liquid, impurities are removed from the stock material, and alloying agents are added. An *alloy* is simply the addition of one or more elements to a host metal; most commercial applications of metals have multiple alloying elements. As the material cools from the liquid state, crystals form. Under normal cooling conditions, multiple nuclei will form, producing multiple crystals. As these crystals grow, they will eventually contact each other, forming boundaries. For a given material, the size of the grains depends primarily on the rate of cooling. Under rapid cooling, multiple nuclei are formed, resulting in small grains with extensive boundaries.

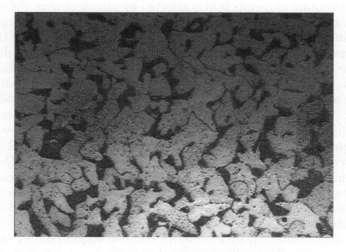

FIGURE 2.12 Optical photomicrograph of low-carbon steel (magnification: 500x).

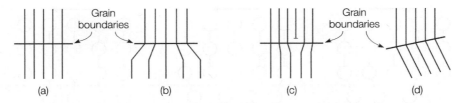

FIGURE 2.13   Types of grain boundaries: (a) coherent, (b) coherent strain, (c) semicoherent, and (d) incoherent.

There are four types of grain boundaries: coherent, coherent strain, semicoherent, and incoherent, as shown in Figure 2.13. At the coherent boundary, the lattices of the two grains align perfectly, and, in essence, there is no physical boundary. Coherent strain boundaries have a different spacing of atoms on each side of the boundary, which may be the effect of an alloying agent. As with the line dislocation, there will be a strain in the atoms on each side of the boundary due to the difference in the spacing of the atoms. The semicoherent boundary has a different number of atoms on each side of the boundary; therefore, not all of the boundaries can match up. Again, this is similar to line dislocation, with strain occurring on each side of the boundary. The incoherent boundary has a different orientation of the crystals on each side of the boundary; thus, the atoms do not match up in a natural manner.

Grain boundaries have an important effect on the behavior of a material. Although the bonds across the grain boundary do not have the strength of the pure crystal structure, the grain boundaries are at a higher energy state than the atoms away from the grain boundary. As a result, when slip occurs along one of the slip planes in the crystal, it is blocked from crossing the grain boundary and can be diverted to run along the grain boundary. This increases the length of the slip path, thus requiring more energy to deform or fracture the material. Therefore, reducing grain size increases the strength of a material.

The grain structure of metals is also affected by plastic strains. Fabrication methods frequently involve plastic straining to produce a desired shape (e.g., wire is produced by forcing a metal through successively smaller dies to reduce the dimension of the metal from the shape produced in the mill to the desired wire diameter.)

Heat treatments are used to refine grain structure. There are two basic heat treatment methods: *annealing* and *hardening.* Both processes involve heating the material to a point at which the existing grain boundaries will break down and reform upon cooling. The differences between the processes include the temperature to which the material is heated, the amount of time it is held at the elevated temperature, and the rate of cooling. In hardening, rapid cooling is achieved by immersing the material in a liquid. In annealing, the material is slowly cooled. The slowest rate of cooling is achieved by leaving the material in the furnace and gradually reducing the temperature. This is an expensive process, since it ties up the furnace. More commonly, the material is cooled in air. Heat treatments of steel and aluminum are discussed in Chapters 3 and 4, respectively.

The grain size is affected by the rate of cooling. Rapid cooling limits the time available for grain growth, resulting in small grains, whereas slow cooling results in large grains.

2.2.4 ■ Alloys

The engineering characteristics of most metallic elements make them unsuitable for use in a pure form. In most cases, the properties of the materials can be significantly improved with the addition of alloying agents. An alloying agent is simply an element or chemical that is compounded or in solution in the crystalline structure of a metal. Steel, composed primarily of iron and carbon, is perhaps the most common alloy. The way the alloying agent fits into the crystalline structure is extremely important. The alloying atoms can either fit into the voids between the atoms (*interstitial* atoms) or can replace the atoms in the lattice structure (*substitutional* atoms).

Because metals have a close-packed structure, the radius of the interstitial atom must be less than 0.6 of the radius of the host element (Derucher et al., 1998). Also, the solubility limit of interstitial atoms—less than 6%—is relatively low. Interstitial atoms can be larger than the size of the void in the lattice structure, but this will result in a strain of the structure and, therefore, will limit solubility.

If the characteristics of two metal elements are sufficiently similar, the metals can have complete miscibility; that is, there is no solubility limit. The atoms of the elements are completely interchangeable. The similarity criteria are defined by the Hume–Rothery rules (Shackelford, 2014). Under these rules, the elements must have the following characteristics:

1. less than 15% difference in the atomic radius
2. the same crystal structure
3. similar electronegatives (the ability of the atom to attract an electron)
4. the same valence

Violation of any of the Hume–Rothery rules reduces the solubility of the atoms. Atoms that are either too large or too small will result in a strain of the lattice structure.

The arrangement of the alloy atoms in the structure can be either random or ordered. In a random arrangement, there is no pattern to the placement of the alloy atoms. An ordered arrangement can develop if the alloy element has a preference for a certain location in the lattice structure. For example, in a gold–copper alloy, an FCC structure, the copper preferably occupies the face positions and the gold preferably occupies the corner positions.

Frequently, more than one alloying agent is used to modify the characteristics of a metal. Steel is a good example of a multiple element alloy. By definition, steel contains iron with carbon; however, steel frequently includes other alloying elements such as chromium, copper, nickel, phosphorous, etc.

2.2.5 ■ Phase Diagrams

To produce alloy metals, the components are heated to a molten state, mixed, and then cooled. The temperature at which the material transitions between a liquid and a solid is a function of the percentages of the components. The liquid and solid states of a material are called *phases,* and a phase diagram displays the relationship between the percentages of the elements and the transition temperatures.

Soluble Materials　The simplest type of phase diagram is for two elements that are completely soluble in both the liquid and solid phases. Solid solutions occur when the elements in the alloys remain dispersed throughout the matrix of the material in the solid state. A two-element or *binary* phase diagram is shown in Figure 2.14(a) for two completely soluble elements. In this diagram, temperature is plotted on the vertical axis and the percent weight of each element is plotted on the horizontal axis. In this case, the top axis is used for element A and the bottom axis is used for element B. The percentage of element B increases linearly across the axis, while the percentage of element A starts at 100% on the left and decreases to 0% on the right. Since this is a binary phase diagram, the sum of the percentage of elements A and B must equal 100%. In Figure 2.14(a), there are three areas. The areas at the top and bottom of the diagram have a single phase of liquid and solid material, respectively. Between the two single-phase areas, there is a two-phase area where the material is both liquid and solid. The line between the liquid and two-phase areas is the *liquidus,* and the line between the two-phase area and the solid area is the *solidus.* For a given composition of elements A and B, the liquidus defines the temperature at which, upon cooling, the first solid crystals form. The solidus defines the temperature at which all material has crystallized. It should be noted that, for a pure element, the transition between liquid and solid occurs at a single temperature. This is indicated on the phase diagram by the convergence of the liquidus and solidus on the left and right sides of Figure 2.14(a), where there are pure elements A and B, respectively.

　　A specific composition of elements at a specific temperature is defined as the *state point,* as shown in Figure 2.14(b). If the state point is above the liquidus, all the material is liquid and the composition of the liquid is the same as the total composition of the material. Similarly, if the state point is below the solidus, all the material is solid and the composition of the solid is the same as for the material. In the two-phase region between the liquidus and the solidus, the percent of material that is in either the liquid or solid phase varies with the temperature. In addition, the composition of the liquid and solid phases in this region changes with temperature. The compositions of the liquid and solid can be determined directly from the phase diagram by using the lever rule. First a *tie line* is established by connecting the liquidus and solidus with a horizontal line that passes through the state point, as shown in Figure 2.14(b). A vertical projection from the intersection of the tie line and the liquidus defines the composition of the liquid phase. A vertical projection from the intersection of the tie line and the solidus defines the composition of the solid phase. For the example in Figure 2.14(b), the alloy is composed of 50% material A and 50% material B. For the defined state point, 79% of the liquid material is element A and 21% is element B, and 31% of the solid phase material is element A and 69% of the solid material is element B. In addition, the percent of the material in the liquid and solid phases can be determined from the phase diagram. From mass balance, the total material must equal the sum of the masses of the components; that is,

$$m_t = m_l + m_s \tag{2.4}$$

where

　　m_t = mass of total material
　　m_l = mass of the total material that is in the liquid phase
　　m_s = mass of the total material that is in the solid phase

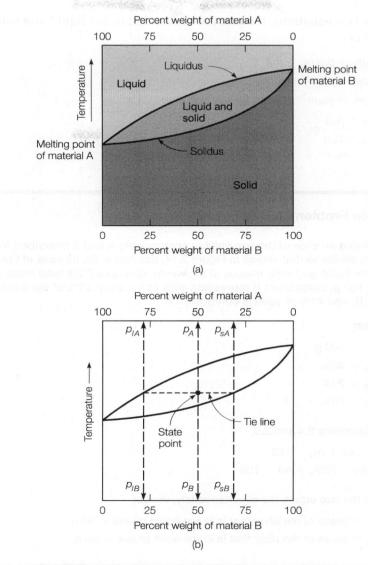

FIGURE 2.14 Binary phase diagram, two soluble elements.

This mass balance also applies to each of the component materials; that is,

$$p_B m_t = p_{lB} m_l + p_{sB} m_s \tag{2.5}$$

where

p_{lB} = percent of the liquid phase that is composed of material B
p_{sB} = percent of the solid phase that is composed of material B
p_B = percent of the material that is component B

From these two equations, the amount of material in the liquid and solid phases can be derived as

$$p_B m_t = p_{lB} m_l + p_{sB}(m_t - m_l)$$

$$p_B m_t = p_{lB} m_l + p_{sB} m_t - p_{sB} m_l$$

$$p_B m_t - p_{sB} m_t = p_{lB} m_l - p_{sB} m_l$$

$$m_l = \frac{(p_B - p_{sB})}{(p_{lB} - p_{sB})} m_t$$

$$m_s = m_t - m_l \tag{2.6}$$

Sample Problem 2.3

Considering an alloy of the two soluble components A and B described by a phase diagram similar to that shown in Figure 2.14, determine the masses of the alloy that are in the liquid and solid phases at a given temperature if the total mass of the alloy is 100 g, component B represents 40% of the alloy, 20% of the liquid is component B, and 70% of solid is component B.

Solution

$$m_t = 100 \text{ g}$$
$$p_B = 40\%$$
$$p_{lB} = 20\%$$
$$p_{sB} = 70\%$$

From Equations 2.4 and 2.5,

$$m_l + m_s = 100$$
$$20m_l + 70m_s = 40 \times 100$$

Solving the two equations simultaneously, we get

$$m_l = \text{mass of the alloy that is in the liquid phase} = 60 \text{ g}$$
$$m_s = \text{mass of the alloy that is in the solid phase} = 40 \text{ g}$$

Insoluble Materials   The discussion thus far has dealt with two completely soluble materials. It is equally important to understand the phase diagram for immiscible materials, that is, for components that are so dissimilar that their solubility in each other is nearly negligible in the solid phase. Figure 2.15 shows the phase diagram for this situation. The intersections of the liquidus with the right and left axes are the melting points for each of the components. As the materials are blended together, the liquidus forms a V shape. The point of the V defines the combination of the components that will change from liquid to solid without the formation of two phases.

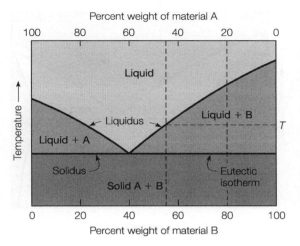

FIGURE 2.15 Binary phase
diagram, insoluble solids.

This point defines the eutectic temperature and the eutectic composition for the components. The solidus is horizontal and passes through the eutectic temperature. There are two areas on the graph with two phases. The area to the left of the eutectic composition will have solid component A, and the liquid phase will be a mixture of the A and B components (vice versa for the area to the right of the eutectic composition).

The sudden phase transformation at the eutectic temperature means that the grains do not have time to grow as the material cools. Thus, the eutectic material will have a fine grain structure.

As the material cools at a temperature other than the eutectic temperature, one of the components becomes solid as the temperature drops below the liquidus. As a result, the amount of this component in the liquid continuously decreases as the material cools in the two-phase area. In fact, the composition of the liquid will follow the liquidus as the temperature decreases. When the eutectic temperature is reached, the remainder of the liquid becomes solid and has a fine grain structure. Starting with a composition of 80% B and 20% A, as in Figure 2.15, when the temperature is lowered to T, the liquid will contain 55% B and 45% A. The amount of each component in the liquid and solid phases can be determined by mass balance (Equations 2.4 and 2.5). At T, this will yield 44.4% liquid and 55.6% solid; all of the solid will be component B. Similarly, the composition of the solid at the eutectic temperature can be computed to be 66.7% solid B and 33.3% eutectic mixture. Since these materials are insoluble in the solid state, the eutectic mixture is composed of an intimate mixture of fine crystals of A and B, usually in a plate-like structure.

Partially Soluble Materials   In between purely soluble and insoluble materials are the materials that are partially soluble. In other words, there is a solubility limit between the components of A and B. If the percent of component B is less than or equal to the solubility limit, on cooling all of the B atoms will be in solution with the A component. If the percent of the B component is above the solubility limit, the atoms

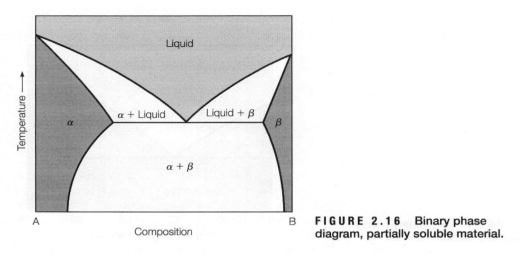

FIGURE 2.16 Binary phase diagram, partially soluble material.

in excess of the amount that will go into solution will form separate grains of the component B. The result is shown on the phase diagram in Figure 2.16. Note that the only difference between this phase diagram and the one shown in Figure 2.15 is the presence of the solid solution regions on each side of the graph. The composition analysis of the two-phase region is the same as that described for Figure 2.15.

Eutectoid Reaction Up to now, the phase diagram has been used to describe the transition between liquid and solid phases of materials. However, the lattice structure of some elements (e.g., iron) is a function of the temperature of the solid. As a result of the lattice transformation, the microstructure of the solid material changes as a function of temperature, as shown by the phase diagram in Figure 2.17. When this occurs,

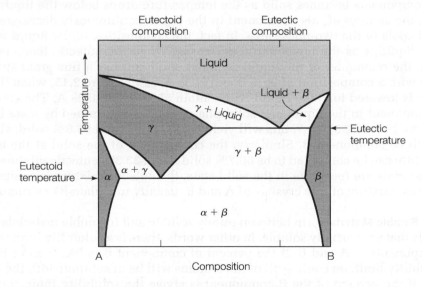

FIGURE 2.17 Phase diagram for a eutectoid reaction.

a eutectoid reaction occurs on the phase diagram; it has characteristics similar to the *eutectic* reaction but is for a lattice structure transformation of the material rather than for a liquid–solid transformation. The rules for the analysis of the components of the material are the same as those discussed for the eutectic material. As with the phase transformation at the eutectic temperature, the transformation of the lattice structure at the eutectoid temperature will result in fine-grained materials.

2.2.6 ▓ Combined Effects

The topics of lattice structure, grain size, heat treatments, and alloying are closely interrelated. The behavior of a metal is dictated by the combination of each of these factors. Clearly, the properties of a metal depend on the elemental make up, the refining and production process, and the types and extent of alloys used in the metal. These topics are too complex for detailed treatment in this text. A practicing engineer in this area must devote considerable study to material characteristics and the impact of alloying and heat treatments. Due to the importance of steel in civil engineering, the phase diagram and heat treatment of steel is presented in Chapter 3.

2.3   Inorganic Solids

Inorganic solids include all materials composed of nonmetallic elements or a combination of metallic and nonmetallic elements. This class of materials is sometimes referred to as ceramic materials. By definition, ceramic elements have five, six, and seven valence electrons. Ceramic materials are formed by a combination of ionic and covalent bonds. In the generic sense, ceramics encompass a broad range of materials, including glass, pottery, inorganic cements, and various oxides. Fired clay products, including bricks and pottery, are some of the oldest ceramic products made by humans. In terms of tonnage, portland cement concrete is the most widely used manufactured material. In the 1980s, the search for highly durable products with unique strength and thermal properties initiated the rapid development of advanced ceramics, such as zirconias, aluminum oxides, silicon carbides, and silicon nitrates. These materials have high strength, stiffness, and wear resistance and are corrosion resistant. The high-performance or engineered ceramics have found many applications in machine and tool design. Although the availability of high-performance ceramics has grown rapidly, the civil and construction engineering applications of these materials have been generally limited due to the cost of the sophisticated ceramics and their lack of fracture toughness that is needed for structural design.

Five classes of ceramic materials have been defined (Ashby and Jones, 1986):

1. glasses—based on silica
2. vitreous ceramics—clay products used for pottery, bricks, etc.
3. high-performance ceramics—highly refined inorganic solids used for specialty applications in which properties, not available from other materials, compensate for the high cost

4. cement and concrete—a multiphase material widely used in civil engineering applications
5. rocks and minerals

Ceramics can also be classified by the predominant type of their atomic bonding. Materials composed of a combination of nonmetallic and metallic elements have predominantly ionic bonds. Materials composed of two nonmetals have predominantly covalent bonds. The type of bond dictates the crystal structure of the compound. As with metals, inorganic solids have a well-defined, although more complicated, unit cell structure. This structure is repeated for the formation of the crystals.

In a simple ionic compound, the nonmetallic element will form either a face centered cubic or hexagonal close-packed structure, as shown in Figure 2.18. Octahedral holes fit in the middle of the octahedron that is formed by connecting all the face atoms of the FCC unit cell. Tetrahedral holes are in the middle of the tetrahedron formed by connecting a corner atom with the adjacent face atoms. In the FCC structure, there is one octahedral hole and six tetrahedral holes. The number of atoms and the way they fit into the holes in the close-packed structure is determined by the number of valence electrons in the metallic and nonmetallic elements. The atoms pack to maximize density, with the constraint that like ions are not nearest neighbors (Ashby and Jones, 1986).

Simple covalent bonds form materials that are highly durable and strong. Diamond is the preeminent example of an elemental high-strength material. Diamond is widely used in industrial applications needing wear resistance, such as cutting tools. However, the most important of the predominately covalent bond materials used by civil engineers are the silicate compounds of portland cement concrete. Silicon atoms link with four oxygen atoms to form a stable tetrahedron (Figure 2.19) that is the basic building block for all silicates. When combined with metal oxides, MO, with a ratio of $MOSiO_2$ of 2 or greater, the resulting silicate is made up of separate SiO_4 monomers linked by the MO molecules. Figure 2.20 shows the calcium silicate crystal structure. The primary reaction compounds of portland cement are tricalcium silicates and dicalcium silicates. Each of these has two or more metal oxides, CaO, per silicate molecule. An ionic bond forms between the metal oxide and the silicate.

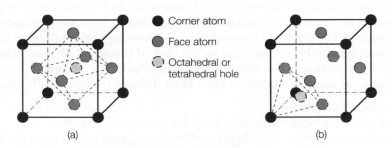

FIGURE 2.18   Lattice structure for simple ionic bonded ceramic materials: (a) octahedral hole and (b) tetrahedral hole.

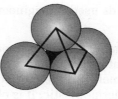

FIGURE 2.19   Silicate tetrahedron.

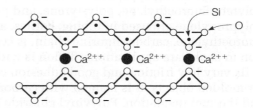

FIGURE 2.20   Simple inorganic solid structure (calcium silicate).

Ceramics have only about one-fiftieth the fracture toughness of metals. Due to the nature of the ionic and covalent bonds, ceramic compounds tend to fracture in a brittle manner rather than to have plastic deformation, as is the case for metals. Like metals, ceramic compounds form distinct grains as a result of multiple nuclei forming multiple crystals during the production of the compound. In addition, due to the production process, ceramic materials tend to have internal cracks and flaws. Stress concentrations occur at the cracks, flaws, and grain boundaries, all of which lower the strength and toughness of the material. As a result, ceramic materials lack tensile strength and must be reinforced when they are used for structural applications.

Glasses are a special type of inorganic solids; they do not develop a crystalline structure. Commercial glasses are based on silica. In glass, the silica tetrahedrons link at the corners, resulting in a random or amorphous structure. Glass can be made of pure silica; it has a high softening point and low thermal expansion. However, due to the high softening point, pure silica glass is hard to form. Metal oxides are added to reduce the cross-linking of the silicates, improving the ability to form the glass into the desired shape. Although glass has an amorphous structure, it is a very stable compound at atmospheric temperatures. It does not flow despite the often cited example of glass windows in European churches.

2.4   Organic Solids

All organic solids are composed of long molecules of covalent bonded carbon atoms. These molecules are chains of carbon and hydrogen combined with various radical components. The radical component can be a hydrogen atom, another hydrocarbon, or another element. These long molecules are bound together by secondary bonds; in many cases, the molecules are also cross-linked with covalent bonds. There are a wide range of organic solids used in engineering. These can be classified as follows (Ashby and Jones, 1986):

1. *Thermoplastics* are characterized by linear carbon chains that are not cross-linked; at low temperatures, secondary bonds adhere the chains. Upon heating, the secondary bonds melt and the thermoplastics become a viscous material. Asphalt is a natural thermoplastic. It is obtained primarily by refining petroleum. In addition, there are many manufactured thermoplastics that have broad engineering applications. These include polyethylene, polypropylene, polytetrafluoroethylene, polystyrene, and polyvinyl chloride. Polyethylene and polypropylene are used in tubing, bottles, and electrical insulation. Polytetrafluoroethylene, carbon–fluorine chain, is commonly known as Teflon®. In addition to cookware applications, Teflon is widely used for bearings and seals due to its very low friction and good adhesion characteristics. Polystyrene is used for molded objects. It is foamed with carbon dioxide to make packing materials and thermal insulation. Polyvinyl chloride is used for low-pressure waterlines.

2. *Thermosets* are characteristically made of a resin and a hardener that chemically react to harden. In the formation of the solid, the carbon chains are cross-linked to form stable compounds that do not soften upon heating. The three generic types of thermosets are epoxy, polyester, and phenol–formaldehyde. Epoxies are used as glues and as the matrix material in plastic composites. Polyester is a fibrous material used in the reinforcing phase of fiberglass. Phenol–formaldehydes are brittle plastics such as Bakelite® and Formica®.

3. *Elastomers or rubbers* are characterized as linear polymers with limited cross-linking. At atmospheric temperatures, the secondary bonds have melted. The cross-linking enables the material to return to its original shape when unloaded. Three forms of elastomers are polyisoprene (natural rubber), polybutadiene (synthetic rubber), and polychloroprene (Neoprene).

4. *Natural materials* are characterized as being grown in all plant matter. The primary material of interest is wood, which is composed of cellulose, lignin, and protein.

Other than the natural polymers, organic solids are produced from refining and processing crude oil. In general, these products are classified as plastics. The properties of these materials are highly variable. Mechanical properties depend on the length of the polymer chains, the extent of cross-linking, and the type of radical compound. All of these factors can be controlled and altered in the production process to alter the material properties.

2.4.1 ■ Polymer Development, Structure, and Cross-Linking

The physical structure of the polymer chain grossly affects the mechanical response of plastics. The word polymer literally means multiple "mer" units. The mer is a base molecule that can be linked together to form the polymers. Figure 2.21 shows the structure of a simple ethylene molecule and the development of a polymer. The square boxes are carbon atoms and the open circles are hydrogen. Forming the polymer

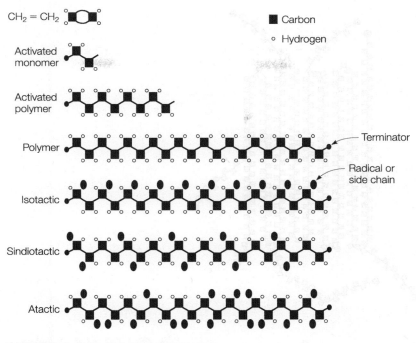

FIGURE 2.21   Polymer structures.

requires breaking the double bond, activating the monomer, and allowing it to link to others to form a long chain. The end of the chain either links to other chains or a terminator molecule, such as OH. Chains with useful mechanical properties require at least 500 monomers. The number of monomers in the chain defines the degree of polymerization; commercial polymers have a degree of polymerization of 10^3 to 10^5.

The complexity of the linear chain polymer is increased by replacing hydrogen atoms with side groups or radicals, as shown in Figure 2.21. The radicals or side groups can be aligned on one side of the chain (*isotactic*), symmetrically on alternate sides of the chain (*sindiotactic*), or in a random fashion (*atactic)*. The radicals can range from simple to complex molecules. For example, polyvinyl chloride has a Cl radical, polypropylene has a CH_3 radical, and polystyrene has C_6H_5. The ability of the polymer chains to stack together is determined by the arrangement of the side chains. The simple chains can fold together into an orderly arrangement, whereas the complex side groups prevent stacking, leading to the amorphous nature of these materials.

More complex linear polymers are formed when two of the hydrogen atoms are replaced by different radicals. Polymethylmethacrylate (Plexiglas®) has the radicals CH_3 and $COOCH_3$. As the complexity of the radicals and the substitution for hydrogen increase, it becomes more difficult to form regular patterns.

In thermoplastics, formed from linear polymers, the structure of the molecules is a blend of amorphous and crystal structure. When there are few side groups, an ordered structure is produced [Figure 2.22(a)]. As the number of side groups

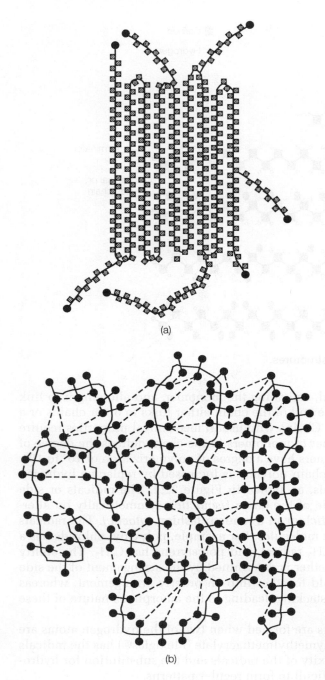

(a)

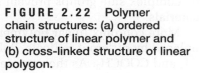

FIGURE 2.22   Polymer chain structures: (a) ordered structure of linear polymer and (b) cross-linked structure of linear polygon.

(b)

increases, the structure becomes increasingly random and cross-links develop. These structures are shown in Figure 2.22(b). Thermosets are formed from polyfunctional monomers. They are formed in a condensation reaction; in essence, the reaction bonds two chains together. Since the chains are formed from polyfunctional crystals, they have an amorphous structure with extensive cross-linking. Elastomers are formed with linear chains that have a limited number of cross-links.

2.4.2 ■ Melting and Glass Transition Temperature

The reaction of polymers to temperature depends on the degree to which the material has crystallized. Highly ordered polymers have a fairly well-defined transition between elastic and viscous behavior. As the percent of crystallization decreases, the melting point is not well defined. However, the point at which these polymers transition to a glass phase is well defined. At elevated temperatures, the motion of the molecules forces a separation between them, resulting in a volume that is greater than required for tightly packed, motionless molecules. This excess volume is termed the *free volume.* As the material cools, the motion of the molecules is reduced and the viscosity increases. At a sufficiently low temperature, the molecules are no longer free to rearrange; thus, their position is fixed and the free volume becomes zero. This is the glass transition temperature. Below this temperature, the secondary bonds bind the material into an amorphous solid; above this temperature, the material behaves in an elastic manner. Figure 2.23 illustrates the concept

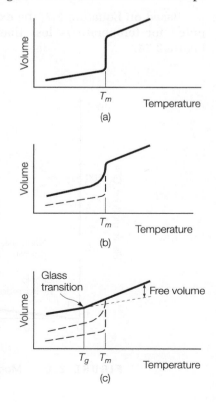

FIGURE 2.23   Melting point and glass transition temperatures: (a) perfect crystallization, (b) imperfect crystallization, and (c) glass formation.

of melting and glass transition temperatures, T_m and T_g. The glass transition temperature of Plexiglas is 100°C; at room temperature, it is a brittle solid. Above T_g, it becomes leathery and then rubbery. T_g for natural rubber is -70°C; it is flexible at all atmospheric temperatures. However, when frozen, say in liquid nitrogen, it becomes a brittle solid.

2.4.3 ■ Mechanical Properties

The mechanical behavior of polymers is directly related to the degree of orientation of the molecules and the amount of cross-linking by covalent bonds. The modulus of a polymer is the average of the stiffness of the bonds. The modulus can be estimated from volume fractions of the covalent bonding as (Ashby and Jones, 1986)

$$\varepsilon = f\frac{\sigma}{E_1} + (1 - f)\frac{\sigma}{E_2} \qquad (2.7)$$

where

ε = strain of the material
σ = stress of the material
f = fraction of covalent bonds
E_1 = stiffness of covalent bonds, about 1000 GPa for diamond
E_2 = stiffness of secondary bonds, about 1 GPa for paraffin wax

Based on Equation 2.7, the expected modulus of various polymers can be computed for temperatures less than the glass transition temperature, as shown in Figure 2.24.

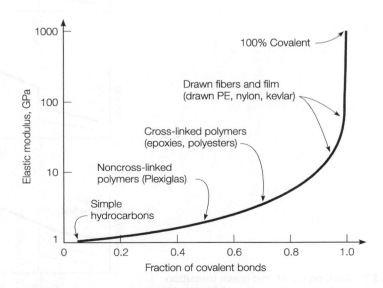

FIGURE 2.24   Modulus of polymers.

Summary

The behavior of materials important to engineering is directly related to their microscopic and macroscopic structure. Although our understanding of these materials is imperfect at this time, much of their behavior can be attributed to the bonding and arrangement of the materials at the atomic level. This chapter provides only a broad overview of the subject. For more information, consult references with more in-depth treatments of these subjects.

Questions and Problems

2.1   Define elastic and plastic behaviors at the micro and macro levels.

2.2   Describe the parts of an atom. Define proton, electron, atomic number, and atomic mass.

2.3   What are the valence electrons and why are they important?

2.4   Describe the order in which electrons fill the shells and subshells.

2.5   Briefly cite the main differences between ionic, covalent, and metallic bonds.

2.6   Why do atoms maintain specific separations?

2.7   Materials are generally classified into three categories based on the predominant types of bond. What are these three categories and what are the predominant types of bond in each category? For each category, provide two examples of common materials used by civil engineers.

2.8   What is the atomic packing factor? What information do you need to compute it?

2.9   Describe FCC, BCC, and HCP lattice structures.

2.10 Two hypothetical metals are created with different elements that have the same atomic mass (g/mole) and the same atomic radius. Metal A has a density of 9.50 g/cm^3 and metal B has a density of 8.73 g/cm^3. If one of these metals has a BCC lattice structure and the other has an FCC lattice structure, identify the structure that corresponds to each of one of them. Justify your answer.

2.11 Show for the face-center cubic crystal structure that the unit cell edge length a and the atomic radius r are related through $a = 2\sqrt{2}r$

2.12 Referring to the BCC lattice structure shown in Figure P2.12

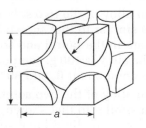

FIGURE P2.12

   a. Determine the number of equivalent whole atoms in the unit cell.
   b. Calculate the relation between a and r.
   c. Calculate the atomic packing factor of the BCC lattice structure.

2.13   Calculate the volume of the unit cell of iron in cubic meters, given that iron has a body-center cubic crystal structure and an atomic radius of 0.124 nm.

2.14   If aluminum has an FCC crystal structure and an atomic radius of 0.143 nm, calculate the volume of its unit cell in cubic meters.

2.15   Using the information available in Table 2.3, calculate the volume of the unit cell of copper in cubic meters.

2.16   Calculate the density of iron, given that it has a BCC crystal structure, an atomic radius of 0.124 nm, and an atomic mass of 55.9 g/mole.

2.17   Using the information available in Table 2.3, calculate the density of molybdenum given that it has an atomic mass of 95.94 g/mole.

2.18   Determine the density of a hypothetical BCC metal with an atomic mass of 63.5 g/mole and atomic radius of 0.128 nm.

2.19   Determine the density of a hypothetical FCC metal with an atomic mass of 42.9 g/mole and atomic radius of 0.132 nm.

2.20   Calculate the density of aluminum, given that it has an fcc crystal structure, an atomic radius of 0.143 nm, and an atomic mass of 26.98 g/mole.

2.21   Calculate the radius of the copper atom, given that copper has an fcc crystal structure, a density of 8.89 g/cm^3, and an atomic mass of 63.55 g/mole.

2.22   Calcium has an FCC crystal structure, density of 1.55 Mg/m^3, and atomic mass of 40.08 g/mole.
   a. Calculate the volume of the unit cell in cubic meters.
   b. Calculate the radius of the calcium atom.

2.23   A hypothetical metal A with a BCC lattice structure has a density of 8.87 g/cm^3. If a hypothetical metal B with an FCC lattice structure is created with a different element that has almost the same atomic mass (g/mole) and almost the same atomic radius, what is the density of metal B? Justify your answer.

2.24   What are the classes of defects in crystal structures?

2.25   Why do grains form in crystal structures?

2.26   Explain the slipping of atoms and the effect on material deformation.

2.27   Sketch a phase diagram for two soluble components. Show the melting temperature of each element. Label all axes, curves, and zones.

2.28   What is the eutectic composition and why is it important?

2.29   Considering an alloy of the two soluble components A and B described by a phase diagram similar to that shown in Figure 2.14, determine the masses of the alloy that are in the liquid and solid phases at a given temperature if the total mass of the alloy is 100 g, component B represents 65% of the alloy, 30% of the liquid is component B, and 80% of solid is component B.

2.30   Considering an alloy of the two soluble components A and B described by a phase diagram similar to that shown in Figure 2.14, determine the masses of the alloy that are in the liquid and solid phases at a given temperature if the total mass of the alloy is 100 g, component B represents 45% of the alloy, 17% of the liquid is component B, and 65% of solid is component B.

2.31 Considering an alloy of the two soluble components A and B described by a phase diagram Figure P2.31, determine the masses of the alloy that are in the liquid and solid phases at a given temperature if the total mass of the alloy is 100 g, component B represents 60% of the alloy, 25% of the liquid is component B, and 70% of solid is component B.

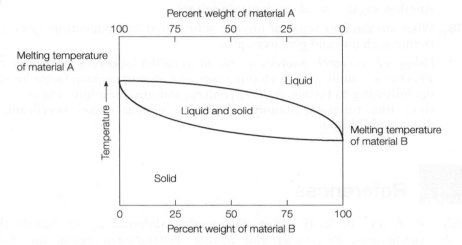

FIGURE P2.31

2.32 Considering an alloy of the two soluble components A and B described by a phase diagram Figure P2.31, determine the masses of the alloy that are in the liquid and solid phases at a given temperature if the total mass of the alloy is 100 g, component B represents 40% of the alloy, 20% of the liquid is component B, and 50% of solid is component B.

2.33 Figure P2.33 shows a portion of the $H_2O - NaCl$ phase diagram.

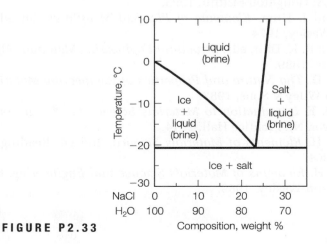

FIGURE P2.33

    a. Using the diagram, briefly explain how spreading salt on ice causes the ice to melt. Show numerical examples in your discussion.

    b. At a salt composition of 10%, what is the temperature at which ice will start melting?

    c. What is the eutectic temperature of the ice and salt combination?

2.34   What are the five classes of ceramic materials?

2.35   Name two common ceramic materials used in Civil Engineering structures. Are they organic solid or inorganic solids?

2.36   What are the four types of organic solids used in engineering applications? Define each one and give examples.

2.37   Using online search, prepare a table to show the following properties: crystal structures, atomic radius, atomic mass, and atomic packing factor for each of the following 15 metals: lead, zirconium, sodium, cadmium, iridium, magnesium, iron, tungsten, titanium, aluminum, copper, nickel, beryllium, vanadium, and lithium.

 References

Ashby M. F. and D. R. H. Jones. *Engineering Materials 2, an Introduction to Microstructures, Processing and Design.* International Series on Materials Science and Technology, Vol. 39. Oxford: Pergamon Press, 1986.

Budinski, K. G. and M. K. Budinski. *Engineering Materials, Properties and Selection.* 9th ed. Upper Saddle River, NJ: Pearson, 2010.

Callister, W. D., Jr. *Materials Science and Engineering—an Introduction.* 7th ed. New York: John Wiley and Sons, 2006.

Derucher, K. N., G. P. Korfiatis, and A. S. Ezeldin. *Materials for Civil & Highway Engineers.* 4th ed. Upper Saddle River, NJ: Prentice Hall, 1998.

Flinn, R. A., and P. K. Trojan. *Engineering Materials and Their Applications.* 4th ed. Boston, MA: Houghton Mifflin, 1995.

Guy, A. G. and J. Hren. *Elements of Physical Metallurgy.* 3d ed. Reading, MA: Addison-Wesley, 1974.

Jackson, N. and R. K. Dhir, eds. *Structural Engineering Materials.* 4th ed. New York: Hemisphere, 1989.

Jastrzebski, Z. D. *The Nature and Properties of Engineering Materials.* 3d ed. New York: John Wiley & Sons, 1987.

Shackelford, J. F. *Introduction to Materials Science for Engineers.* 8th ed. Upper Saddle River, NJ: Prentice Hall, 2014.

Van Vlack, L. H. *Elements of Materials Science.* 2nd ed. Reading, MA: Addison-Wesley, 1964.

Van Vlack, L. H. *Elements of Materials Science and Engineering.* 6th ed. Reading, MA: Addison-Wesley, 1989.

CHAPTER

3

STEEL

The use of iron dates back to about 1500 B.C., when primitive furnaces were used to heat the ore in a charcoal fire. Ferrous metals were produced on a relatively small scale until the blast furnace was developed in the 18th century. Iron products were widely used in the latter half of the 18th century and the early part of the 19th century. Steel production started in mid-1800s, when the Bessemer converter was invented. In the second half of the 19th century, steel technology advanced rapidly due to the development of the basic oxygen furnace and continuous casting methods. More recently, computer-controlled manufacturing has increased the efficiency and reduced the cost of steel production.

There are many organizations throughout the world that provide standards and test methods for steel. In this chapter the American Society for Testing and Materials (ASTM) and the American Institute for Steel Construction (AISC) standards and test methods are presented. However, when possible, dimensional quantities have been converted to SI.

The structural steel industry consists of four components (AISC, 2015):

1. producers of structural steel including hot-rolled structural shapes and hollow sections
2. service centers that function as warehouses and provide limited preprocessing of structural material
3. structural steel fabricators that prepare the steel for the building process
4. erectors that construct steel frames on the project site

Civil and construction engineers will primarily interact with the service centers to determine the availability of products required for a project, the fabricators for detail design drawings and preparation of the product, and the erectors on the job site.

Currently, steel and steel alloys are used widely in civil engineering applications. In addition, wrought iron is still used on a smaller scale for pipes, as well as for general blacksmith work. Cast iron is used for pipes, hardware, and machine parts not subjected to tensile or dynamic loading.

Steel products used in construction can be classified as follows:

1. *structural steel* (Figure 3.1 vertical columns) produced by continuous casting and hot rolling for large structural shapes, plates, and sheet steel

FIGURE 3.1   Steel trusses and columns for the structural support of a building.

2. *cold-formed steel* (Figure 3.1 trusses and decking) produced by cold-forming of sheet steel into desired shapes
3. *fastening products* used for structural connections, including bolts, nuts and washers
4. *reinforcing steel* (rebars) for use in concrete reinforcement (Figure 3.2)
5. miscellaneous products for use in such applications as forms and pans

Civil and construction engineers rarely have the opportunity to formulate steel with specific properties. Rather, they must select existing products from suppliers. Even the shapes for structural elements are generally restricted to those readily available from manufacturers. While specific shapes can be made to order, the cost to fabricate low-volume members is generally prohibitive. Therefore, the majority of civil engineering projects, with the exception of bridges, are designed using standard steel types and structural shapes. Bridges are a special case, in that the majority of bridge structures are fabricated from plate steel rather than hot-rolled sections or hollow structural sections (AISC 2015).

Even though civil and construction engineers are not responsible for formulating steel products, it is beneficial to understand how steel is manufactured and treated and how it responds to loads and environmental conditions. This chapter reviews steel production, the iron–carbon phase diagram, heat treatment, steel alloys, structural steel, steel fasteners, and reinforcing steel. This chapter also presents common tests used to characterize the mechanical properties of steel. The topics of welding, corrosion, and sustainability of steel are also introduced.

FIGURE 3.2   Steel rebars used to reinforce portland cement concrete wall.

3.1   Steel Production

The overall process of steel production is shown in Figure 3.3. This process consists of the following three phases:

1. reducing iron ore to pig iron
2. refining pig iron (and scrap steel from recycling) to steel
3. forming the steel into products

The materials used to produce pig iron are coal, limestone, and iron ore. The coal, after transformation to coke, supplies carbon used to reduce iron oxides in the ore. Limestone is used to help remove impurities. Prior to reduction, the concentration of iron in the ore is increased by crushing and soaking the ore. The iron is magnetically extracted from the waste, and the extracted material is formed into pellets and fired. The processed ore contains about 65% iron.

Reduction of the ore to pig iron is accomplished in a blast furnace. The ore is heated in the presence of carbon. Oxygen in the ore reacts with carbon to form gases. A flux is used to help remove impurities. The molten iron, with an excess of carbon in solution, collects at the bottom of the furnace. The impurities, slag, float on top of the molten pig iron.

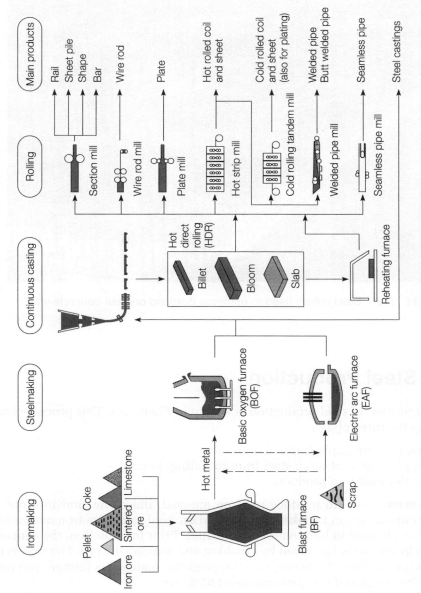

FIGURE 3.3 Conversion of raw material into different steel shapes.

The excess carbon, along with other impurities, must be removed to produce high-quality steel. Using the same refining process, scrap steel can be recycled. Two types of furnaces are used for refining pig iron to steel:

1. basic oxygen
2. electric arc

The basic oxygen furnaces remove excess carbon by reacting the carbon with oxygen to form gases. Lances circulate oxygen through the molten material. The process is continued until all impurities are removed and the desired carbon content is achieved. The basic oxygen furnace process is used to produce construction products that require more drawability, such as hollow structural sections, studs, decking, and plate (Calkins, 2008).

Electric furnaces use an electric arc between carbon electrodes to melt and refine the steel. These plants require a tremendous amount of energy and are used primarily to recycle scrap steel. Electric furnaces are frequently used in minimills, which produce a limited range of products. In this process, molten steel is transferred to the ladle. Alloying elements and additional agents can be added either in the furnace or the ladle. Structural members are primarily produced with the electric arc furnace process (Calkins, 2008).

During the steel production process, oxygen may become dissolved in the liquid metal. As the steel solidifies, the oxygen can combine with carbon to form carbon monoxide bubbles that are trapped in the steel and can act as initiation points for failure. Deoxidizing agents, such as aluminum, ferrosilicon, and manganese, can eliminate the formation of the carbon monoxide bubbles. Completely deoxidized steels are known as *killed steels*. Steels that are generally killed include:

- Those with a carbon content greater than 0.25%
- All forging grades of steels
- Structural steels with carbon content between 0.15 and 0.25%
- Some special steel in the lower carbon ranges

Regardless of the refining process, the molten steel, with the desired chemical composition, is then either cast into ingots (large blocks of steel) or cast continuously into a desired shape. Continuous casting with hot rolling is becoming the standard production method, since it is more energy efficient than casting ingots, as the ingots must be reheated prior to shaping the steel into the final product. Cold-formed steel is produced from sheets or coils of hot rolled steel which is formed into shape either through press-braking blanks sheared from sheets or coils, or more commonly, by roll- forming the steel through a series of dies. No heat is required to form the shapes (unlike hot-rolled steel) and thus the name cold-formed steel. Cold-formed steel members and other products are thinner, lighter, and easier to produce, and typically cost less than their hot-rolled counterparts (IBS Digest, 2005).

3.2 Iron–Carbon Phase Diagram

In refining steel from iron ore, the quantity of carbon used must be carefully controlled in order for the steel to have the desired properties. The reason for the strong relationship between steel properties and carbon content can be understood by examining the iron–carbon phase diagram.

Figure 3.4 shows a commonly accepted iron–carbon phase diagram. One of the unique features of this diagram is that the abscissa extends only to 6.7%, rather than 100%. This is a matter of convention. In an iron-rich material, each carbon atom bonds with three iron atoms to form iron carbide, Fe_3C, also called cementite. Iron carbide is 6.7% carbon by weight. Thus, on the phase diagram, a carbon weight of 6.7% corresponds to 100% iron carbide. A complete iron–carbon phase diagram should extend to 100% carbon. However, only the iron-rich portion, as shown in Figure 3.4, is of practical significance (Callister, 2006). In fact, structural steels have a maximum carbon content of less than 0.3%, so only a very small portion of the phase diagram is significant for civil engineers.

The left side of Figure 3.4 demonstrates that pure iron goes through two transformations as temperature increases. Pure iron below 912°C has a BCC crystalline structure called ferrite. At 912°C, the ferrite undergoes a polymorphic change to an FCC structure called austenite. At 1394°C, another polymorphic change occurs, returning the iron to a BCC structure. At 1539°C, the iron melts into a liquid. The high- and low-temperature ferrites are identified as δ and α ferrite, respectively.

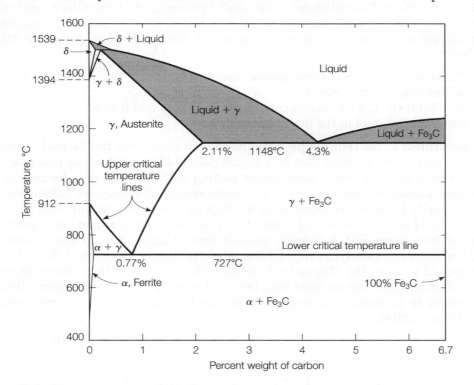

FIGURE 3.4 The iron-carbon carbide phase diagram.

Since δ ferrite occurs only at very high temperatures, it does not have practical significance for this book.

Carbon goes into solution with α ferrite at temperatures between 400°C and 912°C. However, the solubility limit is very low, with a maximum of 0.022% at 727°C. At temperatures below 727°C and to the right of the solubility limit line, α ferrite and iron carbide coexist as two phases. From 727°C to 1148°C, the solubility of carbon in the austenite increases from 0.77% to 2.11%. The solubility of carbon in austenite is greater than in a ferrite because of the crystalline structure of the austenite.

At 0.77% carbon and 727°C, a eutectoid reaction occurs; that is, a solid phase change occurs when either the temperature or carbon content changes. At 0.77% carbon, and above 727°C, the carbon is in solution as an interstitial element, within the FCC structure of the austenite. A temperature drop to below 727°C, which happens slowly enough to allow the atoms to reach an equilibrium condition, results in a two-phase material, a ferrite and iron carbide. The α ferrite will have 0.022% carbon in solution, and the iron carbide will have a carbon content of 6.7%. The ferrite and iron carbide will form as thin plates, a lamellae structure. This eutectoid material is called pearlite.

At carbon contents less than the eutectoid composition, 0.77% carbon, *hypoeutectoid* alloys are formed. Consider a carbon content of 0.25%. Above approximately 860°C, solid austenite exists with carbon in solution. The austenite consists of grains of uniform material that were formed when the steel was cooled from a liquid to a solid. Under equilibrium temperature drop from 860°C to 727°C, α ferrite is formed and accumulates at the grain boundaries of the austenite. This is a proeutectoid ferrite. At temperatures slightly above 727°C, the ferrite will have 0.022% carbon in solution and austenite will have 0.77% carbon. When the temperature drops below 727°C, the austenite will transform to pearlite. The resulting structure consists of grains of pearlite surrounded by a skeleton of α ferrite.

When the carbon content is greater than the eutectoid composition, 0.77% carbon, hypereutectoid alloys are formed. Iron carbide forms at the grain boundaries of the austenite at temperatures above 727°C. The resulting microstructure consists of grains of pearlite surrounded by a skeleton of iron carbide.

The lever rule for the analysis of phase diagrams can be used to determine the phases and constituents of steel.

Sample Problem 3.1

Calculate the amounts and compositions of phases and constituents of steel composed of iron and 0.25% carbon just above and below the eutectoid isotherm.

Solution

At a temperature just higher than 727°C, all the austenite will have a carbon content of 0.77% and will transform to pearlite. The ferrite will remain as primary ferrite. The proportions can be determined by using the lever rule:
Primary α: 0.022% C,

$$\text{Percent primary } \alpha = \left[\frac{0.77 - 0.25}{0.77 - 0.022} \right] \times 100 = 69.5\%$$

$$\text{Percent pearlite} = \left[\frac{0.25 - 0.022}{0.77 - 0.022} \right] \times 100 = 30.5\%$$

At a temperature just below 727°C, the phases are ferrite and iron carbide. The ferrite will have 0.022% carbon, so we have

$$\text{Percent ferrite, } \alpha\text{: (0.022\% C)} = \left[\frac{6.67 - 0.25}{6.67 - 0.022} \right] \times 100 = 96.6\%$$

$$\text{Percent pearlite} = \left[\frac{0.25 - 0.022}{6.67 - 0.022} \right] \times 100 = 3.4\%$$

Figure 3.5 shows an optical 50× photomicrograph of a hot-rolled mild steel plate with a carbon content of 0.18% by weight that was etched with 3% nitol. The light etching phase is proeutectoid ferrite and the dark constituent is pearlite. Note the banded structure resulting from the rolling processes. Figure 3.6 shows the same material as Figure 3.5, except that the magnification is 400×. At this magnification, the alternating layers of ferrite and cementite in the pearlite can be seen.

The significance of ferrite, pearlite, and iron carbide formation is that the properties of the steel are highly dependent on the relative proportions of ferrite and iron carbide. Ferrite has relatively low strength but is very ductile. Iron carbide has high strength but has virtually no ductility. Combining these materials in different proportions alters the mechanical properties of the steel. Increasing the carbon content increases strength and hardness but reduces ductility. However, the modulus of elasticity of steel does not change by altering the carbon content.

All of the preceding reactions are for temperature reduction rates that allow the material to reach equilibrium. Cooling at more rapid rates greatly alters the

FIGURE 3.5   Optical photomicrograph of hot rolled mild steel plate (magnification: 50×).

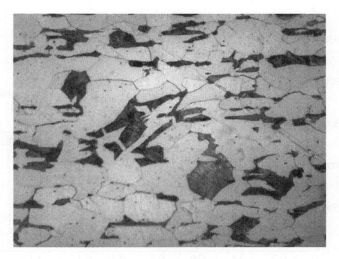

FIGURE 3.6   Optical photomicrograph of hot rolled mild steel plate (magnification: 400×).

microstructure. Moderate cooling rates produce bainite, a fine-structure pearlite without a proeutectoid phase. Rapid quenching produces martensite; the carbon is supersaturated in the iron, causing a body center tetragonal lattice structure. Time–temperature transformation diagrams are used to predict the structure and properties of steel subjected to heat treatment. Rather than going into the specifics, the different types of heat treatments are described.

3.3   Heat Treatment of Steel

Properties of steel can be altered by applying a variety of heat treatments. For example, steel can be hardened or softened by using heat treatment; the response of steel to heat treatment depends upon its alloy composition. Common heat treatments employed for steel include annealing, normalizing, hardening, and tempering. The basic process is to heat the steel to a specific temperature, hold the temperature for a specified period of time, then cool the material at a specified rate. The temperatures used for each of the treatment types are shown in Figure 3.7.

3.3.1 Annealing

The objectives of annealing are to refine the grain, soften the steel, remove internal stresses, remove gases, increase ductility and toughness, and change electrical and magnetic properties. Four types of annealing can be performed, depending on the desired results of the heat treatment:

Full annealing requires heating the steel to about 50°C above the austenitic temperature line and holding the temperature until all the steel transforms into either

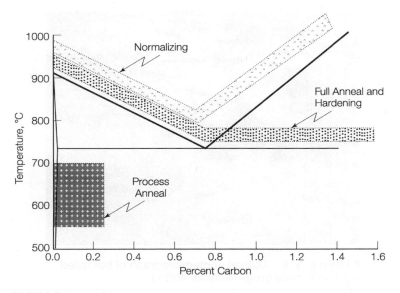

FIGURE 3.7 Heat treatment temperatures.

austenite or austenite–cementite, depending on the carbon content. The steel is then cooled at a rate of about 20°C per hour in a furnace to a temperature of about 680°C, followed by natural convection cooling to room temperature. Due to the slow cooling rate, the grain structure is a coarse pearlite with ferrite or cementite, depending on the carbon content. The slow cooling rate ensures uniform properties of the treated steel. The steel is soft and ductile.

Process annealing is used to treat work-hardened parts made with low carbon steel (i.e., less than 0.25% carbon). The material is heated to about 700°C and held long enough to allow recrystallization of the ferrite phase. By keeping the temperature below 727°C, there is not a phase shift between ferrite and austenite, as occurs during full annealing. Hence, the only change that occurs is refinement of the size, shape, and distribution of the grain structure.

Stress relief annealing is used to reduce residual stresses in cast, welded, and cold-worked parts and cold-formed parts. The material is heated to 600 to 650°C, held at temperature for about 1 hour, and then slowly cooled in still air.

Spheroidization is an annealing process used to improve the ability of high carbon (i.e., more than 0.6% carbon) steel to be machined or cold worked. It also improves abrasion resistance. The cementite is formed into globules (spheroids) dispersed throughout the ferrite matrix.

3.3.2 ■ Normalizing

Normalizing is similar to annealing, with a slight difference in the temperature and the rate of cooling. Steel is normalized by heating to about 60°C above the austenite line and then cooling under natural convection. The material is then air cooled. Normalizing produces a uniform, fine-grained microstructure. However, since the rate of cooling is faster than that used for full annealing, shapes with varying thicknesses results in the normalized parts having less uniformity than could be achieved with annealing. Since structural plate has a uniform thickness, normalizing is an effective process and results in high fracture toughness of the material.

3.3.3 ▮ Hardening

Steel is hardened by heating it to a temperature above the transformation range and holding it until austenite is formed. The steel is then quenched (cooled rapidly) by plunging it into, or spraying it with, water, brine, or oil. The rapid cooling changes the grain structure forming martensite rather than allowing the transformation to the ferrite BCC structure. Martensite has a very hard and brittle structure. Since the cooling occurs more rapidly at the surface of the material being hardened, the surface of the material is harder and more brittle than the interior of the element, creating nonhomogeneous characteristics. Due to the rapid cooling, hardening puts the steel in a state of strain. This strain sometimes causes steel pieces with sharp angles or grooves to crack immediately after hardening. Thus, hardening must be followed by tempering.

3.3.4 ▮ Tempering

The predominance of martensite in quench-hardened steel results in an undesirable brittleness. Tempering is performed to improve ductility and toughness. Martensite is a somewhat unstable structure. Heating causes carbon atoms to diffuse from martensite to produce a carbide precipitate and formation of ferrite and cementite. After quenching, the steel is cooled to about 40°C then reheated by immersion in either oil or nitrate salts. The steel is maintained at the elevated temperature for about 2 hours and then cooled in still air.

3.3.5 ▮ Example of Heat Treatment

In the quest to produce high-strength low-alloy steels economically, the industry has developed specifications for several new steel products, such as A913. This steel is available with yield stresses ranging from 345 to 510 MPa. The superior properties of A913 steel are obtained by a quench self-tempering process. Following the last hot rolling pass for shaping, for which the temperature is typically 850°C, an intense water-cooling spray is applied to the surface of the beam to quench (rapidly cool) the skin. Cooling is stopped before the core on the material is affected. The outer layers are then self-tempered as the internal heat of the beam flows to the surface. The self-tempering temperature is 600°C (Bouchard and Axmann, 2000).

3.4   Steel Alloys

Alloy metals can be used to alter the characteristics of steel. By some counts, there are as many as 250000 different alloys of steel produced. Of these, as many as 200 may be used for civil engineering applications. Rather than go into the specific characteristics of selected alloys, the general effect of different alloying agents will be presented. Alloy agents are added to improve one or more of the following properties:

1. hardenability
2. corrosion resistance
3. machinability
4. ductility
5. strength

Common alloy agents, their typical percentage range, and their effects are summarized in Table 3.1.

TABLE 3.1   Common Steel Alloying Agents (Budinski and Budinski, 2010) (© Pearson Education, Inc. Used by permission.)

	Typical Ranges in Alloy Steels (%)	Principal Effects
Aluminum	<2	Aids nitriding Restricts grain froth Removes oxygen in steel melting
Sulfur	<0.05	Adds machinability Reduces weldability and ductility
Chromium	0.3 to 4	Increases resistance to corrosion and oxidation Increases hardenability Increases high-temperature strength Can combine with carbon to form hard, wear-resistant microconstituents
Nickel	0.3 to 5	Promotes an austenitic structure Increases hardenability Increases toughness
Copper	0.2 to 0.5	Promotes tenacious oxide film to aid atmospheric corrosion resistance
Manganese	0.3 to 2	Increases hardenability Promotes an austenitic structure Combines with sulfur to reduce its adverse effects
Silicon	0.2 to 2.5	Removes oxygen in steel making Improves toughness Increases hardenability
Molybdenum	0.1 to 0.5	Promotes grain refinement Increases hardenability Improves high-temperature strength
Vanadium	0.1 to 0.3	Promotes grain refinement Increases hardenability Will combine with carbon to form wear-resistant microconstituents

By altering the carbon and alloy content and by using different heat treatments, steel can be produced with a wide variety of characteristics. These are classified as follows:

1. Low alloy

   - Low carbon
     Plain
     High strength–low alloy
   - Medium carbon
     Plain
     Heat treatable
   - High carbon
     Plain
     Tool

2. High Alloy

   - Tool
   - Stainless

Steels used for construction projects are predominantly low- and medium-carbon plain steels. Stainless steel has been used in some highly corrosive applications, such as dowel bars in concrete pavements and steel components in swimming pools and drainage lines. The Specialty Steel Industry of North America, SSINA, promotes the use of stainless steel for structural members where corrosion resistance is an important design consideration (SSINA, 1999).

The use and control of alloying agents is one of the most significant factors in the development of steels with better performance characteristics. The earliest specification for steel used in building and bridge construction, published in 1900, did not contain any chemical requirements. In 1991, ASTM published the specification which controls content of 10 alloying elements in addition to carbon (Hassett, 2003).

3.5 Structural Steel

Structural steel is used in hot-rolled structural shapes, plates, and bars. Structural steel is used for various types of structural members, such as columns, beams, bracings, frames, trusses, bridge girders, and other structural applications (see Figure 3.8).

3.5.1 Structural Steel Grades

Due to the widespread use of steel in many applications, there are a wide variety of systems for identifying or designating steel, based on grade, type and class. Virtually, every country with an industrial capacity has specifications for steel. In the United States, there are several associations that write specifications for steel, such as the Society of Automotive Engineers, SAE, the American Iron and Steel Institute, AISI, and the American Society for Testing and Materials, ASTM. The most widely used

FIGURE 3.8   Structural and cold-formed steel used to make columns, beams, and floors for the structural support of a building.

designation system was developed cooperatively by SAE and AISI based on chemical composition (www.key-to-steel.com). However, the materials and products used in building design and construction in the United States are almost exclusively designated by ASTM specifications (Carter, 2004). ASTM specification names consist of a letter, generally an A for ferrous materials, followed by an arbitrary, serially assigned number. For example, ASTM A7 was a specification for structural steel written in 1900 and ASTM A992 was published in 1999 (Carter, 2004). The designation or specification number does not contain any meaningful information other than to serve as a reference. Within ASTM specifications, the terms *grade, type,* and *class* are used in an inconsistent manner. In some ASTM steel specifications, the term grade identifies the yield strength, while in other specifications, the term grade can indicate requirements for both chemical compositions and mechanical properties. ASTM and SAE have developed the Unified Numbering System, UNS (ASTM E527), based on chemical composition. This system uses a letter to identify the broad class of alloys, and a five-digit number to define specific alloys within the class.

Several grades of structural steel are produced in the United States. Table 3.2 is a summary of selected information from various sources. The American Institute of Steel Construction, AISC, Manual for Steel Construction is an excellent reference on the types of steel used for structural applications. However, the best sources of information for structural steels are the various ASTM specifications. Of particular note is the fact that additional requirements are frequently included, dependent on the geometry of the product made with a particular steel.

TABLE 3.2 Designations, Properties, and Composition of ASTM Structural Steel

Steel Type	ASTM Designation		F_y^1 MPa	F_u^1 MPa	Minimum Elongation[2] (%)	C	Cu[5]	Mn	P	S	Ni	Cr	Si	Mo	V
Carbon	A36		250	400–550	23	0.26	0.2	0.8–1.2[6]	0.04	0.05					
	A53 Gr. B		240	415		0.25	0.4	0.95	0.05	0.045	0.4	0.4		0.15	0.08
	A500	Gr. B	290	400	23	0.3	0.18		0.045	0.045					
		Gr. B	320	400											
		Gr. C	320	430	21	0.27	0.18	1.4	0.045	0.045					
		Gr. C	345	430											
	A501		250	400	23	0.3	0.18		0.045	0.045					
	A529	Gr. 50	345	450–690	19	0.27	0.2	1.35	0.04	0.05					
		Gr. 55	380	485–690											
	A572	Gr. 42	290	415	24	0.21	-	1.35	0.04	0.05			0.15–0.4		
		Gr. 50	345	450	21	0.23	-	1.35	0.04	0.05			0.15–0.4		
		Gr. 55	380	485		0.25	-	1.35	0.04	0.05			0.15–0.5		
		Gr. 60	415	520	18	0.26	-	1.35	0.04	0.05			0.4		
High-strength Low-alloy		Gr. 65	450	550	17	0.23	-	1.65	0.04	0.05			0.4		
	A618	Gr. I&II	345	485	22	0.2	0.2	1.35	0.04	0.05			0.3		
		Gr. III	320	460	22	0.23	-	1.35	0.04	0.05					
	A913	50	345	450	21	0.12	0.45	1.6	0.04	0.03	0.25	0.25	0.4	0.07	0.06
		65	450	550	17	0.16	0.35	1.6	0.03	0.03	0.25	0.25	0.4	0.07	0.06
	A992[4]		345–450	450	18	0.23	0.6	0.5–1.5	0.04	0.05			0.4	0.15	0.11
Corrosion resistant, High-strength low-alloy	A242	50	345	485	21	0.15	0.2	1	0.15	0.05					
	A588		345	485	21	0.19	0.25–0.4	0.8–1.25	0.04	0.05	0.4	0.4–0.65		0.02–0.1	

[1] Minimum values unless range or other control noted

[2] Two-inch gauge length

[3] Maximum values unless range or other control noted

[4] A maximum yield to tensile strength ratio of 0.85 and carbon equivalent formula are included as mandatory in ASTM A992

[5] Several steel specifications can include a minimum copper content to provide weather resistance

[6] Range for plate given in table, bar range 0.6–0.9

Historically, dating back to 1900, only two types of structural steel were used in the United States: A7 for bridges and A9 for buildings. The specifications for these materials were very similar and in 1938, they were combined into a single specification, A7. The specification for A7 and A9 were limited to requirements for the tensile strength and yield point only; there were no chemical specifications. The chemical composition, particularly carbon content, became an issue during the 1950s, as welding gained favor for making structural connections. By 1964, AISC adopted five grades of steel for structural applications. The 2005 AISC Load and Resistance Factor Design Specification for Structural Steel Buildings (AISC 2005) identifies 28 different ASTM steel designations for structural applications.

3.5.2 ▇ Sectional Shapes

Figure 3.9 illustrates structural cross-sectional shapes commonly used in structural applications. These shapes are produced in different sizes and are designated with the letters W, HP, M, S, C, MC, and L. W shapes are doubly symmetric wide-flange shapes whose flanges are substantially parallel. HP shapes are also wide-flange shapes whose flanges and webs are of the same nominal thickness and whose depth and width are essentially the same. The S shapes are doubly symmetric shapes whose inside flange surfaces have approximately 16.67% slope. The M shapes are doubly symmetric shapes that cannot be classified as W, S, or HP shapes. C shapes are channels with inside flange surfaces having a slope of approximately 16.67%. MC shapes are channels that cannot be classified as C shapes. L shapes are angle

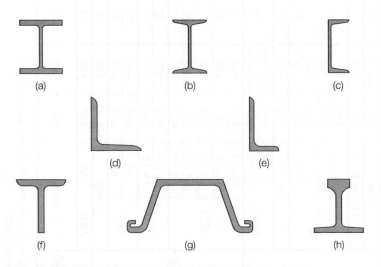

FIGURE 3.9 Shapes commonly used in structural applications: (a) wide-flange (W, HP, and M shapes), (b) I-beam (S shape), (c) channel (C and MC shapes), (d) equal-legs angle (L shape), (e) unequal-legs angle (L shape), (f) tee, (g) sheet piling, and (h) rail.

shapes with either equal or unequal legs. In addition to these shapes, other structural sections are available, such as tee, sheet piling, and rail, as shown in Figure 3.9.

The W, M, S, HP, C, and MC shapes are designated by a letter, followed by two numbers separated by an ×. The letter indicates the shape, while the two numbers indicate the nominal depth and the weight per linear unit length. For example, W 1100 × 548 means W shape with a nominal depth of 1100 mm and a weight of 548 kg/m. An angle shape is designated with the letter L, followed by three numbers that indicate the leg dimensions and thickness in millimeters, such as L 102 × 102 × 12.7. Dimensions of these structural shapes are controlled by ASTM A6/A6M.

W shapes are commonly used as beams and columns, HP shapes are used as bearing piles, and S shapes are used as beams or girders. Composite sections can also be formed by welding different shapes to use in various structural applications. Sheet piling sections are connected to each other and are used as retaining walls.

Tables 3.3 and 3.4 summarize the applicable ASTM specifications/designations for structural steels, and plates and bars, respectively (Carter, 2004). These tables are guides only; specific information should be sought from the applicable specifications for each material and application. In particular, the dimension and application of a member can affect some finer points about material selection, which are not covered in these tables. In general, the materials identified as "preferred" in the tables are generally available in the market place. Those identified as "other applicable materials" may or may not be readily available.

3.5.3 ■ Specialty Steels in Structural Applications

As the ability to refine steels improves, it is possible to produce special products with sufficient economy to permit their use in construction projects. The Federal Highway Administration, US Navy, and AISI have taken a leading role in the development and application of high-performance steels. These are defined as materials that possess the optimum combination of properties required to build cost-effective structures that will be safe and durable throughout their service life (Lane et al., 1998). One of the products developed through this effort is high-performance steels, HPS. Currently, two products are available: HPS 50W and HPS 70W. These are weathering steels that form a corrosion barrier on the surface of the steel when first exposed to the environment. This surface resists further corrosion, and hence reduces the need for maintenance. HPS 70W has stronger tensile properties than steel traditionally used for bridge construction, and hence bridges can be designed with a reduced quantity of material. These savings are somewhat offset by the cost of the material, but there is still a net reduction in construction costs. The tensile requirements of HPS 70W are yield strength 480 MPa, tensile strength of 580 to 750 MPa, and an elongation of 19% (50 mm gauge length). In addition, HPS must pass impact tests. HPS 70W is manufactured to tight and extensive alloy content requirements, as shown in Table 3.5 (ISG Plate, 2003).

Comparing the chemical requirements in Table 3.5 with those in Table 3.2 demonstrates that HPS 70W has more extensive chemical requirements, lower carbon content, and tighter controls on phosphorus and sulfur, which are detrimental alloy elements. The lower carbon content improves the weldability of the steel.

TABLE 3.3 Applicable ASTM Structural Shapes

Steel Type	ASTM Designation		W	M	S	HP	C	MC	L	Rect (HSS)	Round (HSS)	Pipe
Carbon	A36		▒	■	■	■	■	■	■			
	A53 Gr. B											■
	A500	Gr. B-42									■	
		Gr. B-46								■		
		Gr. C-46								■		
		Gr. C-50									■	
	A501									▒	▒	
	A529	Gr. 50	▒	▒	▒	▒	▒	▒	▒			
		Gr. 55	▒	▒	▒	▒	▒	▒	▒			
High-strength Low-alloy	A572	Gr. 42	▒	▒	▒	▒	▒	▒	▒			
		Gr. 50	▒	▒	▒	▒	▒	▒	▒			
		Gr. 55	▒	▒	▒	▒	▒	▒	▒			
		Gr. 60	▒	▒	▒	▒	▒	▒	▒			
		Gr. 65	▒	▒	▒	▒	▒	▒	▒			
	A618	Gr. I&II								▒	▒	
		Gr. III								▒	▒	
	A913	50	▒	▒	▒	▒	▒	▒	▒			
		65	▒	▒	▒	▒	▒	▒	▒			
	A992		■									
Corrosion Resistant, High-strength Low-alloy	A242	50	▒	▒	▒	▒	▒	▒	▒			
	A588		▒	▒	▒	▒	▒	▒	▒			
	A847									▒	▒	

■ = Preferred material specification

▒ = Other applicable material specification, the availability of which should be confirmed prior to specification

☐ = Material specification does not apply

HSS = hollow structural shape

TABLE 3.4   Applicable ASTM Specifications for Plates and Bars

Steel Types	ASTM Designations	F_y (MPa) or Grade	Thickness Ranges (mm)									
			19	19–32	32–38	38–51	51–64	64–102	102–127	127–152	152–203	>203
Carbon	A36	220										■
	A36	250	■	■	■	■	■	■	■	■	■	
	A529	Gr. 50	▨	b	b							
		Gr. 55	▨	b	b	b	b					
High-strength Low-alloy	A572	Gr. 42	▨	▨	▨	▨	▨	▨	▨	▨		
		Gr. 50	▨	▨	▨	▨	▨	▨				
		Gr. 55	▨	▨	▨	▨						
		Gr. 60	▨	▨	▨							
		Gr. 65	▨	▨								
Corrosion Resistant, High-strength Low-alloy	A242	290				▨	▨	▨				
		320		▨	▨							
		345	▨									
	A588	290								▨	▨	
		320							▨			
		345	▨	▨	▨	▨	▨	▨				
Quenched and tempered alloy	A514[a]	620						▨	▨	▨	▨	
		690	▨	▨	▨	▨	▨					
Quenched and tempered low-alloy	A852	485	▨	▨	▨	▨	▨	▨	▨			

■ = Preferred material specification

▨ = Other applicable material specification, the availability of which should be confirmed prior to specification

☐ = Material specification does not apply

a = Available as plates only

b = Applicable to bars only above 1 in. thickness

TABLE 3.5 Chemical Requirements of HPS 70W

Element	Composition (% by Weight)
Carbon	0.11 max
Manganese	1.10–1.35
Phosphorus	0.020 max
Sulfur*	0.006 max
Silicon	0.30–0.50
Copper	0.25–0.40
Nickel	0.25–0.40
Chromium	0.45–0.70
Molybdenum	0.02–0.08
Vanadium	0.04–0.08
Aluminum	0.01–0.04
Nitrogen	0.015 max

* All HPS 70 must be calcium treated for sulfide shape control

The industry has recognized the advantages of designing with the high-performance steels. The first HPS bridge went into service in 1997. As of 2011, more than 250 bridges with HPS components have been constructed and 150 more are in the design or construction stage (fhwa.dot.gov/hfl/innovations/hps.cfm).

The desire to improve the appearance and durability of steel structures has produced an interest in designing structural members with stainless steel. The durability of stainless steel has long been recognized, but the cost of the material was prohibitive. The ability of stainless steel to resist corrosion rests in the high chromium content. Whereas common structural steels have 0.3 to 0.4% chromium, stainless steel has in excess of 10% chromium, by definition. Five AISI grades of stainless steel are used for structural applications (SSINA, 1999):

- 304: the most readily available stainless steel, containing 18% chromium and 8% nickel. Excellent corrosion resistance and formability.
- 316: similar to 304, but with the addition of 3–4% molybdenum for greater corrosion resistance. Generally specified for highly corrosive environments such as industrial, chemical, and seacoast atmospheres.
- 409: a straight chrome alloy, 11 to 12% chromium. Primarily used for interior applications.
- 410–3: a dual phase alloy with micro alloy element control that permits welding in up to 32 mm.
- 2205: a duplex structure with about equal parts of austenite and ferrite. Excellent corrosion resistance and about twice the yield strength of conventional grades.

The chemical and tensile properties of these grades are summarized in Table 3.6.

TABLE 3.6   Properties of Stainless Steels Used for Structural Applications

AISI Type	F_y^1 (MPa)	F_u^1 (MPa)	Percent Elong[2]	Components (Typically Maximum Percent by Weight)								
				C	Cr	Mn	Mo	N	Ni	P	S	Si	Ti
304	215	505	70	0.08	18–20	2			8–10.5	0.045	0.03	1	
316	205	515	40	0.08	16–18	2				0.045	0.03	0.75	
409	240	450	25	0.08	11.13	1				0.045	0.045	1	0.75
410	1230	1525	45	0.15	12.5	1				0.04	0.03		
2205	515	760	35	0.02	22.4	0.7	3.3	0.16	5.8	0.25	0.001	0.4	

[1] Minimum values
[2] Percent elongation is the percentage of plastic strain at fracture (2" gauge length), minimum

3.6   Cold-Formed Steel

Cold-formed steel is used for structural framing of floors, walls, and roofs as well as interior partitions and exterior curtain wall applications. The thickness of cold-formed steel framing members ranges from 0.455 mm to 3.000 mm. Cold-formed steel was formerly known as "light gauge" steel; however, the reference nomenclature "gauge" became obsolete with the adoption of a Universal Designator System for all generic cold-formed steel framing members in 2000.

Cold-formed steel used for steel framing members is predominately manufactured from scrap steel using either electric arc or basic oxygen furnaces to cast slabs. The slabs are passed through a machine with a series of rollers that reduce the slab to sheets of the desired thicknesses, strengths, and other physical properties. The sheets are treated for corrosion resistance, usually hot-dipped galvanizing, and then rolled into coils that weigh approximately 900 kg.

The primary method of manufacturing steel framing members is roll forming. At the roll former, the coils are slit into the required width and fed through a series of dies, to form the stud, joist, angle, or other cold-formed member, as shown in Figure 3.10. Steel framing members may be manufactured with holes in the member webs to facilitate utility runs during construction. The fabrication of all cold-formed steel construction materials is governed by industry standards including the American Iron and Steel Institute's Specification for the Design of Cold-Formed Steel Framing Members (NASPEC) and ASTM.

3.6.1   Cold-Formed Steel Grades

Structural and non structural cold-formed steel members are manufactured from sheet steel in compliance with ASTM A1003/A1003M but limited to the material types and grades listed in Table 3.7. While multiple grades are acceptable for the

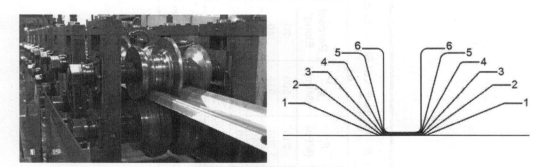

FIGURE 3.10   Roll-forming of cold-formed steel framing member showing the stages of rolling.

TABLE 3.7 Structural Grades of Steel Used in Cold-Formed Products

Type	Structural Grade[1]	Designation US	Designation Metric
H – high ductility	230	ST33H	ST230H
	340	ST50H	ST340H
L – low ductility	230	ST33L	ST230L
	340	ST50L	ST340L
NS – nonstructural	230	NS33	NS230

[1] Yield strength, MPa

different steel types, the North American Standard for Cold-Formed Steel Framing recognizes two yield strengths 228 and 345 MPa (AISI S201-07).

The large deformations caused by the cold-forming process results in local strain-hardening at the corners. The plastic deformation of the steel results in strain-hardening at the bends that increases the yield strength, tensile (ultimate) strength, and hardness but reduces ductility. Strain hardening can almost double the yield strength and increase the tensile strength by 40% (Karren and Winter, 1967).

3.6.2 ▨ Cold-Formed Steel Shapes

A wide variety of shapes can be produced by cold-forming and manufacturers have developed a wide range of products to meet specific applications. Figure 3.11 shows the common shapes of typical cold-formed steel framing members. Figure 3.12 shows common shapes for profiled sheets and trays used for roofing and wall cladding and for load bearing deck panels.

For common applications, such as structural studs, industry organizations, such as the Steel Framing Alliance (SFA) and the Steel Stud Manufacturers Association (SSMA) have developed standard shapes and nomenclature to promote uniformity of product availability across the industry. Figure 3.11 shows the generic shapes covered by the Universal Designator System. Note that the designator system is set up using inches. It is not reasonable to convert the designator system to a metric equivalent. The designator consists of four sequential codes. The first code is a three- or four-digit number indicating the member web depth in 1/100 in. The second is a single letter indicating the type of member, as follows:

S = Stud or joist framing member with stiffening lips
T = Track section
U = Cold-rolled channel
F = Furring channels
L = Angle or L-header

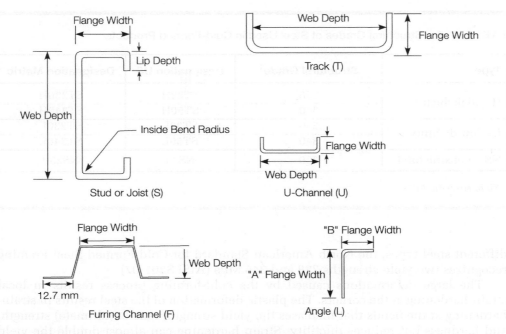

FIGURE 3.11 Generic cold-formed steel framing shapes.

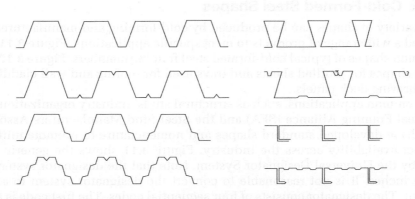

FIGURE 3.12 Common panel and deck shapes (TPU, 2009).

The third is a three-digit numeral indication flange width in 1/100 in. followed by a dash. The fourth is a two- or three-digit numeral indicating the base steel thickness in 1/1000 in. (mils).

As an example, the designator system for a 6″, C-shape with 1-5/8″ (1.62″) flanges and made with 0.054″ thick steel is 600S162-54.

3.6.3 ■ Special Design Considerations for Cold-Formed Steel

Structural design of cold-formed members is in many respects more challenging than the design of hot rolled, relatively thick, structural members. A primary difference is cold-formed members are more susceptible to buckling due to their limited thickness. The fact that the yield strength of the steel is increased in the cold-forming process creates a dilemma for the designer. Ignoring the increased strength is conservative, but results in larger members, hence more costly, than is needed if the increased yield strength is considered.

Corrosion creates a greater percent loss of cross section than is the case for thick members. All cold-formed steel members are coated to protect steel from corrosion during the storage and transportation phases of construction as well as for the life of the product. Because of its effectiveness, hot-dipped zinc galvanizing is most commonly used. Structural and nonstructural framing members are required to have a minimum metallic coating that complies with ASTM A1003/A1003M, as follows:

- structural members – G60 and
- nonstructural members G40 or equivalent minimum.

To prevent galvanic corrosion, special care is needed to isolate the cold-formed members from dissimilar metals, such as copper. The design, manufacture, and use of cold-formed steel framing is governed by standards that are developed and maintained by the American Iron and Steel Institute along with organizations such as ASTM, and referenced in the building codes. Additional information is available at www.steelframing.org.

3.7   Fastening Products

Fastening products include (Carter, 2004)

- Conventional bolts
- Twist-off-type tension control bolt assemblies
- Nuts
- Washers
- Compressible-washer-type direct tension indicators
- Anchor rods
- Threaded rods
- Forged steel structural hardware

Table 3.8 summarizes the applicable ASTM specifications for each type of fastener (Carter, 2004). High-strength bolts have a tensile strength in excess of 690 MPa. Common bolts have a tensile strength of 410 MPa. The preferred material for anchor rods, F1554 Grade 36, has a yield stress of 250 MPa and an ultimate strength in the range of 400 to 550 MPa. A36, with a yield stress of 250 MPa, is preferred for threaded rods. Nuts, washers, and direct tension indicators are made with materials that do not have a minimum required strength.

TABLE 3.8 Applicable ASTM Specifications for Structural Fasteners

Legend for shaded cells: ■ = Preferred material specification; ▨ = Other applicable material specification, availability should be confirmed before specifying; blank = Material specification does not apply.

ASTM Designation		F_y Yield Stress (MPa)	F_u Tensile Stress[a] (MPa)	Diameter Range (mm)	High Strength Bolts	Common Bolts	Nuts	Washers	Direct Tension Indicators	Threaded Rods	Hooked	Headed	Threaded & Nutted
A325		-	725	>25.4–38.1	■								
A325		-	830	12.7–38.1	■								
A490		-	1035	12.7–38.1	■								
F1852		-	725	28.6	■								
F1852		-	830	12.7–25.4	■								
A194 Gr. 2H		-	-	6.4–101.6			▨						
A563		-	-	6.4–101.6			■						
F436[b]		-	-	12.7–38.1				■					
F959		-	-	<254					■				
A36		250	400–550	>101.6–177.8						■	▨		
A193 Gr. B7		-	690	>63.5–101.6							▨		▨
A193 Gr. B7		-	795	≤63.5							▨		▨
A193 Gr. B7		-	860	6.4–101.6						▨	▨		▨
A307	Gr. A	-	415	6.4–101.6		■						▨	
A307	Gr. C	-	400–550	63.5–101.6						▨	▨		▨
A354	Gr. BD	-	965	6.4–63.5						▨	▨	▨	▨
A354	Gr. BD		1034	38.1–76.2						▨	▨	▨	▨
A449		-	620	38.6–38.1	c					▨	▨	▨	▨
A449			725	6.4–25.4	c					▨	▨	▨	▨
A449			830	<152.4	c					▨	▨	▨	▨
A572	Gr. 42	290	415	<101.6								▨	
A572	Gr. 50	345	450	<50.8								▨	
A572	Gr. 55	380	485	< 31.8								▨	
A572	Gr. 60	415	520	<31.8								▨	
A572	Gr. 65	450	550	>127–203								▨	
A588		290	435	>101.6–127								▨	
A588		320	460	<101.6								▨	
A588		345	485	15.9–76.2								▨	
A687		725	1034 max	6.4–101.6								▨	
F1554	Gr. 36	250	400–550	6.4–101.6						■	▨	▨	■
F1554	Gr. 55	380	520–655	6.4–76.2						▨	▨	▨	▨
F1554	Gr. 105	725	860–1034							▨	▨	▨	▨

■ = Preferred material specification

▨ = Other applicable material specification, availability should be confirmed before specifying

blank = Material specification does not apply

- Indicates that a value is not specified in the material specification
[a] Minimum values unless range or max. is indicated
[b] Special washer requirements apply for some steel to steel bolting, check design documents
[c] LRFD has limitations on use of ASTM A449 bolts

Structural connections are made by riveting, bolting, or welding. Rivet connections were used extensively in the past, but modern bolt technology has made riveting obsolete. Bolted connections may be snug tightened, pretensioned, or slip critical (Miller, 2001). Snug-tightened joints are accomplished by either a few impacts of an impact wrench or the full effort of an ironworker using an ordinary spud wrench to bring the members into firm contact. Pretensioned joints require tightening the bolt to a significant tensile stress with a corresponding compressive stress in the attached members. Four methods are used to ensure that the bolt is tightened to a sufficient stress level: turn-of-nut, calibrated wrench, twist-off-type tension-control bolts, and direct tension indicators. Bolts in slip-critical joints are also installed to pretensioned requirements, but these joints have "faying surfaces that have been prepared to provide a calculable resistance against slip." When the joint is placed under load, the stresses may be transmitted through the joint by the friction between the members. However, if slip occurs, the bolts will be placed in shear, in addition to the tension stresses from the installation—hence the need for high-strength bolts.

3.8   Reinforcing Steel

Since concrete has negligible tensile strength, structural concrete members subjected to tensile and flexural stresses must be reinforced. Either conventional or prestressed reinforcing can be used, depending on the design situation. In conventional reinforcing, the stresses fluctuate with loads on the structure. This does not place any special requirements on the steel. On the other hand, in prestressed reinforcement, the steel is under continuous tension. Any stress relaxation will reduce the effectiveness of the reinforcement. Hence, special steels are required for prestress applications.

3.8.1   Conventional Reinforcing

Reinforcing steel (rebar) is manufactured in three forms: *plain bars, deformed bars,* and *plain and deformed wire fabrics.* Plain bars are round, without surface deformations. Plain bars provide only limited bond with the concrete and therefore are not typically used in sections subjected to tension or bending. Deformed bars have protrusions (deformations) at the surface, as shown in Figure 3.13; thus, they ensure a good bond between the bar and the concrete. The deformed surface of the bar prevents slipping, allowing the concrete and steel to work as one unit. Wire fabrics are flat sheets in which wires pass each other at right angles, and one set of elements is parallel to the fabric axis. Plain wire fabrics develop the anchorage in concrete at the welded intersections, while deformed wire fabrics develop anchorage through deformations and at the welded intersections.

Deformed bars are used in concrete beams, slabs, columns, walls, footings, pavements, and other concrete structures, as well as in masonry construction. Welded wire fabrics are used in some concrete slabs and pavements, mostly to resist temperature

FIGURE 3.13   Steel rebars used to reinforce PCC columns.

and shrinkage stresses. Welded wire fabrics can be more economical to place and thus allow for closer spacing of bars than is practical with individual bars.

Reinforcing steel is produced in the standard sizes shown in Table 3.9. Bars are made of four types of steel: A615/A615M (billet), A996/A996M, and A706/A706M (low-alloy), as shown in Table 3.10. Billet steel is the most widely used. A706 steel is often used when the rebar must be welded to structural steel. Reinforcing steel is produced in four grades. Note the conversion from the US grades to metric trades is a "soft" conversion as the conversion of the yield stress requirements is approximations.

To permit ready identification of the different bar types in the field, marking symbols are rolled into the bars as they are being produced. As shown in Figure 3.14, there are four marking symbols:

1. Letter code for manufacturer
2. Numerical code for bar size, this code may be in either millimeters or "standard bar numbers," which indicates the number of eighths of an inch of the nominal diameter of the bar
3. Letter code for type of steel (bars marked with both S and W have steel that meets all the requirements of types S and W steel)
   a. S for billet steel—A615/A615M
   b. I Rail steel—A996/A996M
   c. A Axle steel—A996/A996M
   d. W Low alloy steel—A706/A706M

TABLE 3.9 Standard-Size Reinforcing Bars According to ASTM A615*

Bar Designation Number**	Nominal Mass (kg/m)	Nominal Dimensions***			Deformation Requirements (mm)****		
		Diameter (mm)	Cross-Sectional Area (mm²)	Perimeter (mm)	Maximum Average Spacing	Minimum Average Height	Maximum Gap*****
10 [3]	0.560	9.5	71	29.9	6.7	0.38	3.6
13 [4]	0.994	12.7	129	39.9	8.9	0.51	4.9
16 [5]	1.552	15.9	199	49.9	11.1	0.71	6.1
19 [6]	2.235	19.1	284	59.8	13.3	0.97	7.3
22 [7]	3.042	22.2	387	69.8	15.5	1.12	8.5
25 [8]	3.973	25.4	510	79.8	17.8	1.27	9.7
29 [9]	5.059	28.7	645	90.0	20.1	1.42	10.9
32 [10]	6.404	32.3	819	101.3	22.6	1.63	12.4
36 [11]	7.907	35.8	1006	112.5	25.1	1.80	13.7
43 [14]	11.38	43.0	1452	135.1	30.1	2.16	16.5
57 [18]	20.24	57.3	2581	180.1	40.1	2.59	21.9

* Copyright ASTM, Printed with Permission.

** Bar numbers approximate the number of millimeters of the nominal diameter of the bars.

*** The nominal dimensions of a deformed bar are equivalent to those of a plain round bar having the same weight per meter as the deformed bar.

**** Requirements for protrusions on the surface of the bar.

***** Chord 12.5% of Nominal Perimeter

TABLE 3.10 Types and Properties of Reinforcing Bars According to ASTM (Somayaji, 2001) (© Pearson Education, Inc. Used by permission.)

ASTM* Steel	Type	Grade	Tensile Strength Min., MPa	Yield Strength** Min., MPa	Size Availability (US No.)
		Metric			
A615	Billet steel bars (plain and deformed)	280	483	276	3–6
		420	620	414	3–18
		520	689	517	11–18
A616	Rail steel (plain and deformed)	420	620	474	3–11
A617	Axle steel (plain and deformed)	280	483	276	3–11
		420	620	414	3–11
A706	Low-alloy steel (deformed Bars)	420	552	414–538	3–18

* ASTM types are per the cited reference. These types have been replaced: (1) A615 is now A615/A615M, A616 and A617 are now A996/A996M, and A706 is now A706/A706M.

** When the steel does not have a well-defined yield point, yield strength is the stress corresponding to a strain of 0.005 m/m (0.5% extension) for grades 40, 50, and 60, and a strain of 0.0035 m/m (0.35% extension) for grade 75 of A615, A616, and A617 steels. For A706 steel, grade point is determined at a strain of 0.0035 m/m.

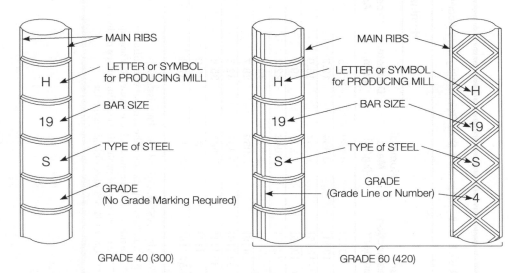

GRADE 40 (300)　　　　　GRADE 60 (420)

FIGURE 3.14 ASTM reinforcing bar identification codes.

4. Grade of steel designated by either grade lines or numerical code (a grade line is smaller and is located between the two main ribs which are on opposite sides of all bars made in the United States)
   a. Grade 40 or 300—no designation
   b. Grade 60 or 420—one grade line between the main ribs or the number 4
   c. Grade 75 or 520—two grade lines between the main ribs or the number 5

3.8.2 ■ Steel for Prestressed Concrete

Prestressed concrete requires special wires, strands, cables, and bars. Steel for prestressed concrete reinforcement must have high strength and low relaxation properties. High-carbon steels and high-strength alloy steels are used for this purpose. Properties of prestressed concrete reinforcement are presented in ASTM specification A416/A416M and AASHTO specification M203. These specifications define the requirements for a seven-wire uncoated steel strand. The specifications allow two types of steel: stress-relieved (normal-relaxation) and low relaxation. Relaxation refers to the percent of stress reduction that occurs when a constant amount of strain is applied over an extended time period. Both stress-relieved and low-relaxation steels can be specified as Grade 250 [1725] or Grade 270 [1860], with ultimate strengths of 1725 MPa and 1860 MPa, respectively. The specifications for this application are based on mechanical properties only; the chemistry of wires is not pertinent to this application. After stranding, low-relaxation strands are subjected to a continuous thermal–mechanical treatment to produce the required mechanical properties. Table 3.11 shows the required properties for seven-wire strand.

T A B L E   3 . 1 1   Required Properties for Seven-Wire Strand

	Stress-Relieved		Low-Relaxation	
Property	Grade 250	Grade 270	Grade 250	Grade 270
Breaking strength,* MPa	1725	1860	1725	1860
Yield strength (1% extension)	85% of breaking strength		90% of breaking strength	
Elongation (min. percent)	3.5		3.5	
Relaxation** (max. percent)				
Load = 70% min. breaking strength	—		2.5	
Load = 80% min. breaking strength	—		3.5	

* Breaking strength is the maximum stress required to break one or more wires.

** Relaxation is the reduction in stress that occurs when a constant strain is applied over an extended time period. The specification is for a load duration of 1000 hours at a test temperature of 20 ± 2°C.

3.9   Mechanical Testing of Steel

Many tests are available to evaluate the mechanical properties of steel. This section summarizes some laboratory tests commonly used to determine properties required in product specifications. Test specimens can take several shapes, such as bar, tube, wire, flat section, and notched bar, depending on the test purpose and the application.

Certain methods of fabrication, such as bending, forming, and welding, or operations involving heating, may affect the properties of the material being tested. Therefore, the product specifications cover the stage of manufacture at which mechanical testing is performed. The properties shown by testing before the material is fabricated may not necessarily be representative of the product after it has been completely fabricated. In addition, flaws in the specimen or improper machining or preparation of the test specimen will give erroneous results (ASTM A370).

3.9.1   Tension Test

The tension test (ASTM E8/E8M) on steel is performed to determine the yield strength, yield point, ultimate (tensile) strength, elongation, and reduction of area. Typically, the test is performed at temperatures between 10°C and 35°C.

The test specimen can be either full sized or machined into a shape, as prescribed in the product specifications for the material being tested. It is desirable to use a small cross-sectional area at the center portion of the specimen to ensure fracture within the gauge length. Several cross-sectional shapes are permitted, such as round and rectangular, as shown in Figure 3.15. Plate, sheet, round rod, wire, and tube specimens may be used. A 12.5 mm diameter round specimen is used in many cases. The gauge length over which the elongation is measured typically is four times the diameter for most round-rod specimens.

Various types of gripping devices may be used to hold the specimen, depending on its shape. In all cases, the axis of the test specimen should be placed at the center of the testing machine head to ensure axial tensile stresses within the gauge length without bending. An extensometer with a dial gauge (Figure 1.27) or an

FIGURE 3.15   Tension test specimens with round and rectangular cross sections.

LVDT (Figure 1.31) is used to measure the deformation of the entire gauge length. The test is performed by applying an axial load to the specimen at a specified rate. Figure 3.16 shows a tensile test being performed on a round steel specimen using an LVDT extensometer to measure the deformation.

As discussed in Chapter 1, mild steel has a unique stress–strain relationship (Figure 3.17). Here, a linear elastic response is displayed up to the proportion limit. As the stress is increased beyond the proportion limit, the steel will yield, at which time the strain will increase without an increase in stress (actually the stress will slightly decrease). As tension increases past the yield point, strain increases following a nonlinear relation up to the point of failure.

Note that the decrease in stress after the peak does not mean a decrease in strength. In fact, the actual stress continues to increase until failure. The reason for the apparent decrease is that a neck is formed in the steel specimen, causing an appreciable decrease in the cross-sectional area. The traditional, or engineering, way of calculating the stress and strain uses the original cross-sectional area and gauge length. If the stress and strain are calculated based on the instantaneous cross-sectional area and gauge length, a *true stress–strain curve* is obtained, which is different than the *engineering stress–strain curve* (Figure 3.17).

As shown in Figure 3.17, for small strain levels the true stress and strain are similar to the engineering stress and strain; the true stress is slightly greater than the engineering stress and the true strain is slightly less than the engineering strain. However, as the strain level increases, especially as the neck is formed, the true stress becomes much larger than the engineering stress, because of the reduced cross-sectional area at the neck. The necking also causes the true strain to be larger than the engineering

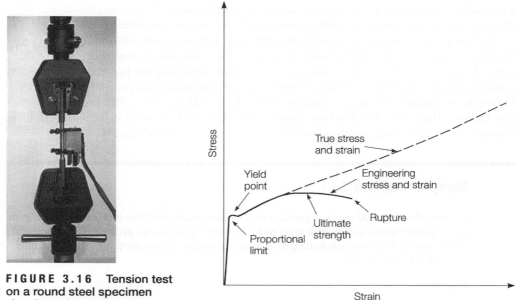

FIGURE 3.16 Tension test on a round steel specimen showing grips and an extensometer with an LVDT.

FIGURE 3.17 Typical stress–strain behavior of mild steel.

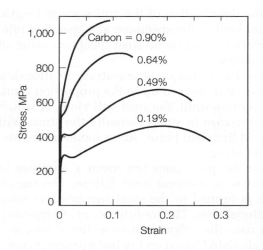

FIGURE 3.18 Tensile stress–strain diagrams of hot-rolled steel bars with different carbon contents.

strain, since the increase in length at the vicinity of the neck is much larger than the increase in length outside of the neck. The large increase in length at the neck increases the true strain to a large extent because the definition of true strain utilizes a ratio of the change in length in an infinitesimal gauge length. By decreasing the gauge length toward an infinitesimal size and increasing the length due to localization in the neck, the numerator of an expression is increased while the denominator stays small, resulting in a significant increase in the ratio of the two numbers. Note that when calculating the true strain, a small gauge length must be used at the neck, since the properties of the material (such as the cross section) at the neck represent the true material properties. For various practical applications, however, the engineering stresses and strains are used, rather than the true stresses and strains.

Different carbon-content steels have different stress–strain relationships. Increasing the carbon content in the steel increases the yield stress and reduces the ductility. Figure 3.18 shows the tension stress–strain diagram for hot-rolled steel bars containing carbons from 0.19% to 0.90%. Increasing the carbon content from 0.19% to 0.90% increases the yield stress from 280 to 620 MPa. Also, this increase in carbon content decreases the fracture strain from about 0.27 to 0.09 m/m. Note that the increase in carbon content does not change the modulus of elasticity.

Sample Problem 3.2

A steel alloy bar 100 mm long with a rectangular cross section of 10 mm × 40 mm is subjected to tension with a load of 80 kN and experiences an increase in length of 0.1 mm. If the increase in length is entirely elastic, calculate the modulus of elasticity of the steel alloy.

Solution

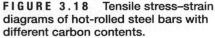

$$\sigma = \frac{80000}{0.01 \times 0.04} = 0.200 \times 10^9 \text{ Pa} = 0.200 \text{ GPa}$$

$$\varepsilon = \frac{0.1}{100} = 0.001 \text{ m/m}$$

$$E = \frac{\sigma}{\varepsilon} = \frac{0.2000}{0.001} = 200 \text{ GPa}$$

Sample Problem 3.3

A steel specimen is tested in tension. The specimen is 25 mm wide by 12.5 mm thick in the test region. By monitoring the load dial of the testing machine, it was found that the specimen yielded at a load of 160 kN and fractured at 214 kN.

a. Determine the tensile stress at yield and at fracture.
b. If the original gauge length was 100 mm, estimate the gauge length when the specimen is stressed to 1/2 the yield stress.

Solution

a. Yield stress (σ_y) = 160 × 10^3/(25 × 12.5) = 512 MPa

   Fracture stress (σ_f) = 214 × 10^3/(25 × 12.5) = 685 MPa

b. Assume E = 206 GPa

$$\varepsilon = (1/2) \times 512/(206 \times 10^3) = 1.243 \times 10^{-3}$$

$$\Delta L = L\varepsilon = 100 \times 1.243 \times 10^{-3} = 0.124 \text{ mm}$$

Final gauge length = 100 + 0.124 = 100.124 mm

Steel is generally assumed to be a homogeneous and isotropic material. However, in the production of structural members, the final shape may be obtained by cold rolling. This essentially causes the steel to undergo plastic deformations, with the degree of deformation varying throughout the member. As discussed in Chapter 1, plastic deformation causes an increase in yield strength and a reduction in ductility. Figure 3.19 demonstrates that the measured properties vary, depending on the orientation of the sample relative to the axis of rolling (Hassett, 2003). Thus, it is necessary to specify how the sample is collected when evaluating the mechanical properties of steel.

3.9.2 ▉ Torsion Test

The torsion test (ASTM E143) is used to determine the shear modulus of structural materials. The shear modulus is used in the design of members subjected to torsion, such as rotating shafts and helical compression springs. In this test, a cylindrical, or tubular, specimen is loaded either incrementally or continually by applying an external torque to cause a uniform twist within the gauge length

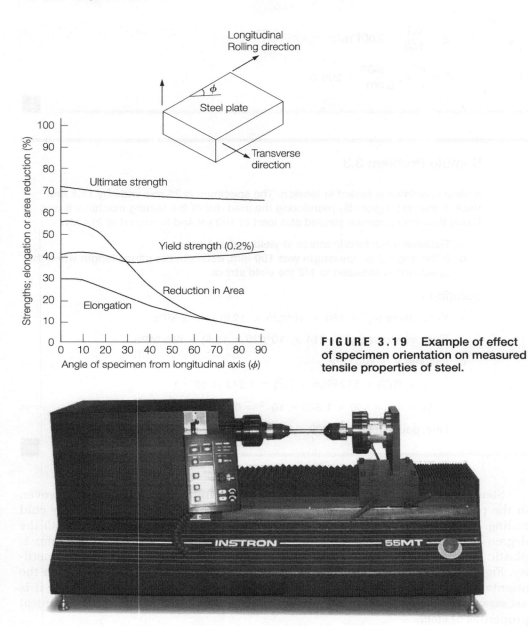

FIGURE 3.19 Example of effect of specimen orientation on measured tensile properties of steel.

FIGURE 3.20 Torsion test apparatus. (Courtesy of Instron®)

(Figure 3.20). The amount of applied torque and the corresponding angle of twist are measured throughout the test. Figure 3.21 shows the shear stress–strain curve. The shear modulus is the ratio of maximum shear stress to the corresponding shear strain below the proportional limit of the material, which is the slope of the straight line between R (a pretorque stress) and P (the proportional limit). For a

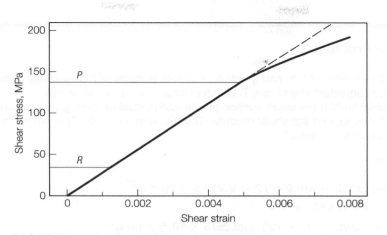

FIGURE 3.21   Typical shear stress–strain diagram of steel.
(Reprinted, with permission, from ASTM E143, Figure 1, copyright ASTM
International, 100 Barr Harbor Drive, West Conshohocken, PA 19428)

circular cross section, the maximum shear stress (τ_{max}), shear strain (γ), and the shear modulus (G) are determined by the equations

$$\tau_{max} = \frac{Tr}{J} \tag{3.1}$$

$$\gamma = \frac{\theta r}{L} \tag{3.2}$$

$$G = \frac{\tau_{max}}{\gamma} = \frac{TL}{JQ} \tag{3.3}$$

where

   T = torque
   r = radius
   J = polar moment of inertia of the specimen about its center, $\pi r^4 / 2$ for a solid
      circular cross section.
   θ = angle of twist in radians
   L = gauge length

   The test method is limited to materials and stresses at which creep is negligible compared with the strain produced immediately upon loading. The test specimen should be sound, without imperfections near the surface. Also, the specimen should be straight and of uniform diameter for a length equal to the gauge length plus two to four diameters. The gauge length should be at least four diameters. During the test, torque is read from a dial gauge or a readout device attached to the testing machine, while the angle of twist may be measured using a torsiometer fastened to the specimen at the two ends of the gauge length. A curve-fitting procedure can be used to estimate the straight-line portion of the shear stress–strain relationship of Figure 3.21 (ASTM E143).

Sample Problem 3.4

A rod with a length of 1 m and a radius of 20 mm is made of high-strength steel. The rod is subjected to a torque T, which produces a shear stress below the proportional limit. If the cross section at one end is rotated 45 degrees in relation to the other end, and the shear modulus G of the material is 90 GPa, what is the amount of applied torque?

Solution

$$J = \pi r^4/2 = \pi(0.02)^4/2 = 0.2513 \times 10^{-6} \text{ m}^4$$
$$\theta = 45(\pi/180) = \pi/4$$
$$T = GJ\theta/L = (90 \times 10^9) \times (0.2513 \times 10^{-6}) \times (\pi/4)/1$$
$$= 17.8 \times 10^3 \text{ N.m} = 17.8 \text{ kN.m}$$

3.9.3 ■ Charpy V Notch Impact Test

The Charpy V Notch impact test (ASTM E23) is used to measure the toughness of the material or the energy required to fracture a V-notched simply supported specimen. The test is used for structural steels in tension members.

The standard specimen is 55 × 10 × 10 mm with a V notch at the center of one side, as shown in Figure 3.22. Before testing, the specimen is brought to the specified temperature for a minimum of 5 minutes in a liquid bath or 30 minutes in a gas medium. The specimen is inserted into the Charpy V notch impact-testing machine (Figure 3.23) using centering tongs. The swinging arm of the machine has a striking tip that impacts the specimen on the side opposite the V notch. The striking head is released from the pretest position, striking and fracturing the specimen. By fracturing the test specimen, some of the kinetic energy of the striking head is absorbed, thereby reducing the ultimate height the strike head attains. By measuring the height the strike head attains after striking the specimen, the energy required to fracture the specimen is computed. This energy is measured in m · N as indicated on a gauge attached to the machine.

FIGURE 3.22   Charpy V notch specimens.

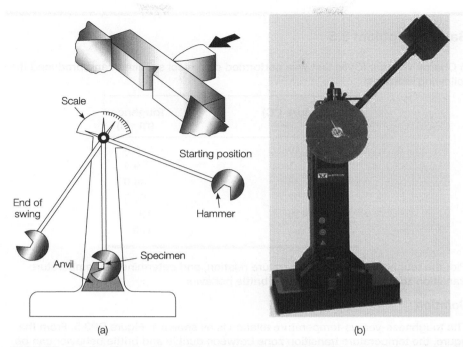

(a)   (b)

FIGURE 3.23   Charpy V notch impact testing machine (a) schematic and (b) apparatus. (Courtesy of Instron®)

The lateral expansion of the specimen is typically measured after the test using a dial gauge device. The lateral expansion is a measure of the plastic deformation during the test. The higher the toughness of the steel, the larger the lateral expansion.

Figure 1.14, in Chapter 1, shows the typical energy (toughness) required to fracture structural steel specimens at different temperatures. The figure shows that the required energy is high at high temperatures and low at low temperatures. This indicates that the material changes from ductile to brittle as the temperature decreases.

The fracture surface typically consists of a dull shear area (ductile) at the edges and a shiny cleavage area (brittle) at the center, as depicted in Figure 3.24. As the toughness of the steel decreases, due to lowering the temperature, for example, the shear area decreases while the cleavage area increases.

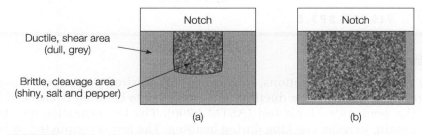

FIGURE 3.24   Fracture surface of Charpy V notch specimen: (a) at high temperature and (b) at low temperature.

Sample Problem 3.5

A Charpy V Notch (CVN) test was performed on a steel specimen and produced the following readings:

Temperature (°C)	Toughness (m · N)
− 40	6.8
1	9.5
38	38.0
77	89.5
116	107
154	108

Plot the toughness-versus-temperature relation, and determine the temperature transition zone between ductile and brittle behavior.

Solution

The toughness-versus-temperature relation is as shown in Figure SP3.5. From the figure, the temperature transition zone between ductile and brittle behavior can be seen to be 0 to 115°C.

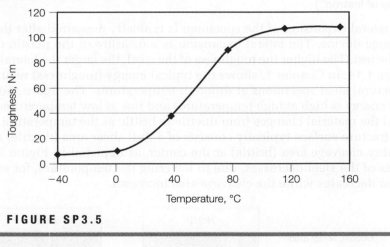

FIGURE SP3.5

3.9.4 ▓ Bend Test

In many engineering applications, steel is bent to a desired shape, especially in the case of reinforcing steel. The ductility to accommodate bending is checked by performing the semiguided bend test (ASTM E290). The test evaluates the ability of steel, or a weld, to resist cracking during bending. The test is conducted by bending the specimen through a specified angle and to a specified inside radius of curvature. When complete fracture does not occur, the criterion for failure is the number and size of cracks found on the tension surface of the specimen after bending.

The bend test is made by applying a transverse force to the specimen in the portion that is being bent, usually at midlength. Three arrangements can be used, as illustrated in Figure 3.25. In the first arrangement, the specimen is fixed at one end and bent around a reaction pin or mandrel by applying a force near the free end, as shown in Figure 3.25(a). In the second arrangement, the specimen is held at one end and a rotating device is used to bend the specimen around the pin or mandrel, as shown in Figure 3.25(b). In the third arrangement, a force is applied in the middle of a specimen simply supported at both ends, Figure 3.25(c).

3.9.5 ▨ Hardness Test

Hardness is a measure of a material's resistance to localized plastic deformation, such as a small dent or scratch on the surface of the material. A certain hardness is required for many machine parts and tools. Several tests are available to evaluate the hardness of materials. In these tests, an indenter (penetrator) is forced into the surface of the material with a specified load magnitude and rate of application. The depth, or the size, of the indentation is measured and related to a hardness index number. Hard materials result in small impressions, corresponding to high hardness numbers. Hardness measurements depend on test conditions and are, therefore, relative. Correlations and tables are available to convert the hardness measurements from one test to another and to approximate the tensile strength of the material (ASTM A370).

One of the methods commonly used to measure hardness of steel and other metals is the *Rockwell hardness test* (ASTM E18). In this test, the depth of penetration of a diamond cone, or a steel ball, into the specimen is determined under fixed conditions (Figure 3.26). A preliminary load of 10 kg is applied first, followed by

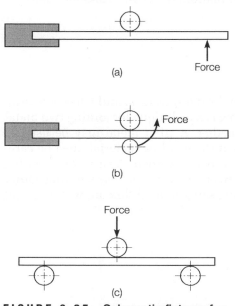

(a)

Force

(b)

Force

Force

(c)

FIGURE 3.25   Schematic fixtures for semi-guided bend test (ASTM E290). (Copyright ASTM. Reprinted with permission)

FIGURE 3.26
Rockwell hardness test machine.
(Courtesy of Phase II)

an additional load. The Rockwell number, which is proportional to the difference in penetration between the preliminary and total loads, is read from the machine by means of a dial, digital display, pointer, or other device. Two scales are frequently used, namely, B and C. Scale B uses a 1.588 mm steel ball indenter and a total load of 100 kg, while scale C uses a diamond spheroconical indenter with a 120° angle and a total load of 150 kg.

To test very thin steel or thin surface layers, the *Rockwell superficial hardness test* is used. The procedure is the same as the Rockwell hardness test except that smaller preliminary and total loads are used. The Rockwell hardness number is reported as a number, followed by the symbol HR, and another symbol representing the indenter and forces used. For example, 68 HRC indicates a Rockwell hardness number of 68 on Rockwell C scale.

Hardness tests are simple, inexpensive, nondestructive, and do not require special specimens. In addition, other mechanical properties, such as the tensile strength, can be estimated from the hardness numbers. Therefore, hardness tests are very common and are typically performed more frequently than other mechanical tests.

3.9.6 ■ Ultrasonic Testing

Ultrasonic testing is a nondestructive method for detecting flaws in materials. It is particularly useful for the evaluation of welds. During the test, a sound wave is directed toward the weld joint and reflected back from a discontinuity, as shown in Figure 3.27. A sensor captures the energy of the reflected wave and the results are displayed on an oscilloscope. This method is highly sensitive in detecting planar defects, such as incomplete weld fusion, delamination, or cracks (Hassett, 2003).

3.10 Welding

Many civil engineering structures such as steel bridges, frames, and trusses require welding during construction and repair. Welding is a technique for joining two metal pieces by applying heat to fuse the pieces together. A filler metal may be used to facilitate the process. The chemical properties of the welding material must be carefully selected to be compatible with the materials being welded. A variety of welding methods are available, but the common types are arc welding and gas welding. Other types of welding include flux-cored arc welding, self-shielded flux arc welding, and electroslag welding (Frank and Smith, 1990).

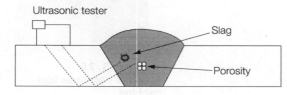

FIGURE 3.27   Ultrasonic test of welds.

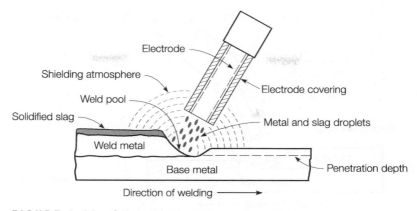

FIGURE 3.28   Schematic drawing of arc welding.

Arc welding uses an arc between the electrode and the grounded base metal to bring both the base metal and the electrode to their melting points. The resulting deposited weld metal is a cast structure with a composition dependent on the base metal, electrode, and flux chemistry. *Shielded metal arc welding (stick welding)* is the most common form of arc welding. It is limited to short welds in bridge construction. A consumable electrode, which is covered with flux, is used. The flux produces a shielding atmosphere at the arc to prevent oxidation of the molten metal. The flux is also used to trap impurities in the molten weld pool. The solidified flux forms a slag that covers the solidified weld, as shown in Figure 3.28. *Submerged arc welding* is a semiautomatic or automatic arc welding process. In this process, a bare wire electrode is automatically fed by the welding machine while a granular flux is fed into the joint ahead of the electrode. The arc takes place in the molten flux, which completely shields the weld pool from the atmosphere. The molten flux concentrates the arc heat, resulting in deep penetration into the base metal.

Gas welding (mig welding) is another type of welding in which no flux is used. An external shielding gas is used, which shields the molten weld pool and provides the desired arc characteristics (Figure 3.29). This welding process is normally used for small welds due to the lack of slag formation.

Care must be taken during welding to consider the distortion that is the result of the nonuniform heating of the welding process. When the molten weld metal cools, it shrinks, causing deformation of the material and introducing residual stresses into the structure. Frequently, these residual stresses cause cracks outside the weld area. The distortion produced by welding can be controlled by proper sequencing of the welds and predeforming the components prior to welding. Finally, care must be taken by the welder and the inspector to protect their eyes and skin from the intensive ultraviolet radiation produced during welding (Frank and Smith, 1990).

The relative ease with which steel can be welded is related to the hardness of the steel. In general, harder steels are more difficult to weld. Winterton developed the concept of using a carbon equivalent formula for estimating the carbon equivalent of steels and an associated zone chart for determining the need to preheat steel to

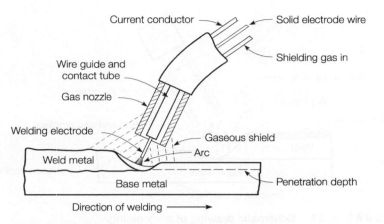

FIGURE 3.29 Schematic drawing of gas welding.

control the development of hydrogen in the welded steel. There are several different carbon equivalent formulas. The one used for structural steels is

$$CE = C + \frac{(Mn + Si)}{6} + \frac{(Cr + Mo + V)}{5} + \frac{(Ni + Cu)}{15} \tag{3.4}$$

Figure 3.30 shows the zones associated with the different combinations of carbon and carbon equivalent. The zones are used as a guide to determining the method used to determine preheat requirements (Hassett, 2003):

- Zone I—Cracking is unlikely but may occur with high hydrogen or high restraint. Use hydrogen control method to determine preheat.
- Zone II—The hardness control method and selected hardness shall be used to determine minimum energy input for single-pass fillet welds without preheat.
- Zone III—The hydrogen control method shall be used.

The hardness and hydrogen control methods are means of determining the level of energy used to preheat the weld area before the weld is performed. Preheating of the metal and the use of low hydrogen electrodes are the best means of avoiding hydrogen embrittlement. This occurs when steel is melted during welding, which

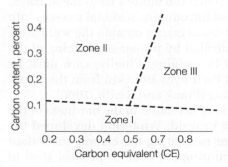

FIGURE 3.30 Welding zone classification of steel.

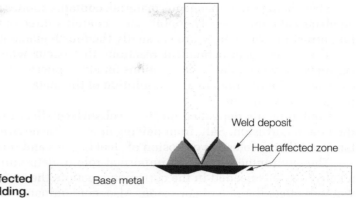

FIGURE 3.31 Heat affected zone produced during welding.

may allow hydrogen to dissolve in the molten metal and diffuse into the base metal adjacent to the weld.

Whenever metal is welded, the base material adjacent to the weld is heated to a temperature that may be sufficient to affect its metallurgy. The material affected in this manner is termed the heat-affected zone, HAZ (Figure 3.31). The material in the HAZ is a high-risk area for failure, especially if proper preheating and cooling procedures are not followed.

3.11   Steel Corrosion

Corrosion is defined as the destruction of a material by electrochemical reaction to the environment. For simplicity, corrosion of steel can be defined as the destruction that can be detected by rust formation. Corrosion of steel structures can cause serious problems and embarrassing and/or dangerous failures. For example, corrosion of steel bridges, if left unchecked, may result in lowering weight limits, costly steel replacement, or collapse of the structure. Other examples include corrosion of steel pipes, trusses, frames, and other structures. It is estimated that the cost of corrosion of the infrastructure in the United States alone is $22.6 billion each year (corrosion costs web site, 2009). The infrastructure includes (1) highway bridges, (2) gas and liquid transmission pipelines, (3) waterways and ports, (4) hazardous materials storage, (5) airports, and (6) railroads.

Corrosion is an electrochemical process; that is, it is a chemical reaction in which there is transfer of electrons from one chemical species to another. In the case of steel, the transfer is between iron and oxygen, a process called *oxidation reduction.* Corrosion requires the following four elements (without any of them corrosion will not occur):

1. an *anode*—the electrode where corrosion occurs
2. a *cathode*—the other electrode needed to form a corrosion cell
3. a *conductor*—a metallic pathway for electrons to flow
4. an *electrolyte*—a liquid that can support the flow of electrons

Steel, being a heterogeneous material, contains anodes and cathodes. Steel is also an electrical conductor. Therefore, steel contains three of the four elements needed for corrosion, while moisture is usually the fourth element (electrolyte).

The actual electrochemical reactions that occur when steel corrodes are very complex. However, the basic reactions for atmospherically exposed steel in a chemically neutral environment are dissolution of the metal at the anode and reduction of oxygen at the cathode.

Contaminants deposited on the steel surface affect the corrosion reactions and the rate of corrosion. Salt, from deicing or a marine environment, is a common contaminant that accelerates corrosion of steel bridges and reinforcing steel in concrete.

The environment plays an important role in determining corrosion rates. Since an electrolyte is needed in the corrosion reaction, the amount of time the steel stays wet will affect the rate of corrosion. Also, contaminants in the air, such as oxides or sulfur, accelerate corrosion. Thus, areas with acid rain, coal-burning power plants, and other chemical plants may accelerate corrosion.

3.11.1 ■ Methods for Corrosion Resistance

Since steel contains three of the four elements needed for corrosion, protective coatings can be used to isolate the steel from moisture, the fourth element. There are three mechanisms by which coatings provide corrosion protection (Hare, 1987):

1. *Barrier coatings* work solely by isolating the steel from the moisture. These coatings have low water and oxygen permeability.
2. *Inhibitive primer coatings* contain passivating pigments. They are low-solubility pigments that migrate to the steel surface when moisture passes through the film to passivate the steel surface.
3. *Sacrificial primers (cathodic protection)* contain pigments such as elemental zinc. Since zinc is higher than iron in the galvanic series, when corrosion conditions exist the zinc gives up electrons to the steel, becomes the anode, and corrodes to protect the steel. There should be close contact between the steel and the sacrificial primer in order to have an effective corrosion protection.

Cathodic protection can take forms other than coating. For example, steel structures such as water heaters, underground tanks and pipes, and marine equipment can be electrically connected to another metal that is more reactive in the particular environment, such as magnesium or zinc. Such reactive metal (sacrificial anode) experiences oxidation and gives up electrons to the steel, protecting the steel from corrosion. Figure 3.32 illustrates an underground steel tank that is electrically connected to a magnesium sacrificial anode (Fontana and Green, 1986).

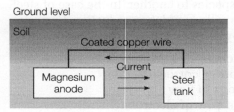

FIGURE 3.32 Cathodic protection of an underground pipeline using a magnesium sacrificial anode.

3.12 Steel Sustainability

3.12.1 ■ LEED Considerations

The relationship between LEED materials and resource categories and steel includes (Eckmann et al., 2004):

- Building Reuse is a design decision made at the planning stage of a project. An argument can be made that steel frame buildings can be designed with long clear-spans that would facilitate reconfiguration of the space.
- Construction Waste and Recycling is generally favorable to steel. Due to the high-value content of reclaimed steel and its ability to be recycled LEED points can generally be claimed for steel. However, the LEED points are computed for the entire waste stream for a project, not for specific products.
- Resource Reuse is generally not favorable to steel as the cost of recycled steel use of salvaged structural components is not cost effective.
- Recycled Content is very favorable to steel. The Steel Recycling Institute (SRI, 2012) has developed a bulletin that provides the information needed to claim LEED credit for recycled steel.
- Local/Regional Materials for steel consider two issues, the point of fabrication and the location of the raw materials used for the steel production. If the fabrication occurs within 800 km of the project (a weighted average based on mode of transportation, 800 km is for highway transportation, if the material is transported by rail, waterway or over a sea the mileage is greater) the local fabrication credit can be claimed. Since all steel used for construction has recycled content, it is difficult if not impossible to trace the source of the original reclaimed steel to earn the LEED credit for this category.

LEED design innovation credits may apply if the designer can demonstrate the structure has unique sustainable features. Examples include (Eckmann, 2004):

- Use of exposed steel as the finish rather than using conventional products such as tiles and dry wall.
- Minimizing the weight of steel required as compared to a conventional design.
- Design the structure for recyclability by using a high steel content credit may be earned for construction and demolition waste credits.
- Deconstruction and reuse by designing bolted rather than welded connectors allowing a building to be dismantled and reassembled for a new project.
- One designer used hollow structural steel members that were integrated with the heating system to provide radiant heat to the building.

3.12.2 ■ Other Sustainability Considerations

Steel production, even with reclaimed steel, is energy intensive as reduction of iron ore to pig iron and the subsequent processing in either the basic oxygen furnace or electric arc furnace requires melting the product to a molten state. However, through

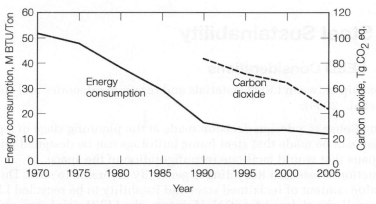

FIGURE 3.33   Energy consumption and carbon dioxide emissions by greenhouse gas for US steel production.

restructuring of the industry, technological and energy efficiency improvements and greater use of reclaimed steel emissions and energy use per ton of steel has decreased significantly as shown on Figure 3.33 (Warrell et al., 2010).

Summary

The history of civil engineering is closely tied to that of steel, and this will continue into the foreseeable future. With the development of modern production facilities, the availability of a wide variety of economical steel products is virtually assured. High strength, ductility, the ability to carry tensile as well as compressive loads, and the ability to join members either with welding or with mechanical fastening are the primary positive attributes of steel as a structural material. The properties of steel can be tailored to meet the needs of specific applications through alloying and heat treatments. The primary shortcoming of steel is its tendency to corrode. When using steel in structures, the engineer should consider the means for protecting the steel from corrosion over the life of the structure.

Questions and Problems

3.1   What is the chemical composition of steel? What is the effect of carbon on the mechanical properties of steel?

3.2   Why does the iron–carbon phase diagram go only to 6.7% carbon?

3.3 Draw a simple iron–carbon phase diagram showing the liquid, liquid–solid, and solid phases.

3.4 What is the typical maximum percent of carbon in steel used for structures?

3.5 Calculate the amounts and compositions of phases and constituents of steel composed of iron and 0.10% carbon just above and below the eutectoid isotherm.

3.6 Briefly discuss four heat treatment methods to enhance the properties of steel. What are the advantages of each treatment?

3.7 Define alloy steels. Explain why alloys are added to steel.

3.8 Name three alloying agents and their principal effects.

3.9 Specifically state the shape and size of the structural steel section: W 920 × 271.

3.10 What are the typical uses of structural steel?

3.11 What is the range of thicknesses of cold-formed steel structural members?

3.12 Why is coil steel used for cold-formed steel fabrication annealed prior to shaping?

3.13 If steel with a 228 MPa yield strength is used for a cold-formed shape, what would be the approximate yield strength of the deformed corner after cold-forming?

3.14 Why is reinforcing steel used in concrete? Discuss the typical properties of reinforcing steel.

3.15 What is high-performance steel? State two HPS products that are currently being used in structural applications and show their properties.

3.16 Name three mechanical tests used to measure properties of steel.

3.17 The following laboratory tests are performed on steel specimens:
a. Tension test
b. Charpy V notch test
c. Bend test

What are the significance and use of these tests?

3.18 Sketch the stress–strain behavior of steel, and identify different levels of strength. What is a typical value for yield strength of mild steel? What is the effect of increasing the carbon content in steel on each of the each of the following items?
a. Yield strength
b. Modulus of elasticity
c. Ductility

3.19 Three steel bars have a diameter of 25 mm and carbon contents of 0.2, 0.5, and 0.8%, respectively. The specimens were subjected to tension until rupture. The load versus deformation results were as shown in Table P3.19.

If the gauge length is 50 mm, determine the following:
a. The tensile stresses and strains for each specimen at each load increment.
b. Plot stresses versus strains for all specimens on one graph.

TABLE P3.19

Specimen No.	1	2	3
Carbon Content (%)	0.2	0.5	0.8
Deformation (mm)	Load (kN)		
0.00	0	0	0
0.07	133	133	133
0.10	137	191	191
0.15	142	196	285
0.50	147	201	324
1.00	140	199	383
2.50	155	236	447
5.00	196	295	491 (Rupture)
7.50	226	336	
10.00	241	341	
12.50	218	304 (Rupture)	
13.75	196 (Rupture)		

    c. The proportional limit for each specimen.
    d. The 0.2% offset yield strength for each specimen.
    e. The modulus of elasticity for each specimen.
    f. The strain at rupture for each specimen.
    g. Comment on the effect of increasing the carbon content on the following:
       i. Yield strength
      ii. Modulus of elasticity
      iii. Ductility

3.20 Draw a typical stress–strain relationship for steel subjected to tension. On the graph, show the modulus of elasticity, the yield strength, the ultimate strength, and the rupture stress.

3.21 Getting measurements from Figure 3.18, determine approximate values of the following items for steel with different carbon contents. Use one system of units only as specified by the instructor.
    a. Yield stress
    b. Ultimate strength
    c. Strain at failure
    d. Effect on modulus of elasticity
    e. Approximate toughness. Curves may be approximated with a series of straight lines.
    f. Comment on the effect of increasing carbon content on items a through e above.

3.22 A steel specimen is tested in tension. The specimen is 25 mm wide by 5 mm thick in the test region. By monitoring the load dial of the testing machine, it was found that the specimen yielded at a load of 55 kN and fractured at 78 kN.
   a. Determine the tensile stresses at yield and at fracture.
   b. Estimate how much elongation would occur at 60% of the yield stress in a 50-mm gauge length.

3.23 A steel bar with a 30 mm diameter is tested in tension. By monitoring the load reading of the testing machine, it was found that the specimen yielded at a load of 185 kN and fractured at 215 kN.
   a. Determine the tensile stresses at yield and at fracture.
   b. Estimate how much increase in length would occur at 70% of the yield stress in a 50 mm gauge length.

3.24 A mild steel specimen originally 300 mm long is pulled in tension with a stress of 500 MPa. If the deformation is entirely elastic, what will be the resultant elongation? The modulus of elasticity of steel is 200 GPa.

3.25 A tensile stress is applied along the long axis of a cylindrical steel rod with a diameter of 10 mm. If the deformation is entirely elastic, determine the magnitude of the load required to produce a 5×10^{-3} mm change in diameter. Assume a modulus of elasticity of 200 GPa and a Poisson's ratio of 0.27.

3.26 A round steel alloy bar with a diameter of 19 mm and a gauge length of 76 mm was subjected to tension, with the results shown in Table P3.26. Using a computer spreadsheet program, plot the stress–strain relationship. From the graph, determine the Young's modulus of the steel alloy and the deformation corresponding to a 37 kN load.

TABLE P3.26

Load, kN	Deformation, mm
9	0.0286
18	0.0572
27	0.0859
36	0.1145
45	0.1431
54	0.1718

3.27 A 19-mm reinforcing steel bar and a gauge length of 75 mm was subjected to tension, with the results shown in Table P3.27. Using a computer spreadsheet program, plot the stress–strain relationship. From the graph, determine the Young's modulus of the steel and the deformation corresponding to a 150-kN load.

TABLE P3.27

Load, kN	Deformation, mm
0	0
54	0.084
163	0.168
284	0.336
330	1.428
366	3.360

3.28 Testing a round steel alloy bar with a diameter of 15 mm and a gauge length of 250 mm produced the stress–strain relationship shown in Figure P3.28.

Determine
a. the elastic modulus
b. the proportional limit
c. the yield strength at a strain offset of 0.002
d. the tensile strength
e. the magnitude of the load required to produce an increase in length of 0.38 mm
f. the final deformation, if the specimen is unloaded after being strained by the amount specified in (e)
g. In designing a typical structure made of this material, would you expect the stress applied in (e) reasonable? Why?

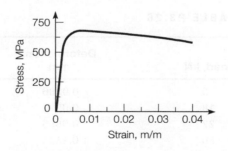

FIGURE P3.28

3.29 During the tension test on a steel rod within the elastic region, the following data were measured:

Applied load   = 102 kN
Original diameter   = 25 mm
Current diameter   = 24.99325 mm
Original gauge length = 100 mm
Current gauge length   = 100.1 mm

Calculate the Young's modulus and Poisson's ratio.

3.30   A round steel bar with a diameter of 12.5 mm and a gauge length of 50 mm was subjected to tension to rupture following ASTM E-8 test procedure. The load and deformation data were as shown in Table P3.30.

Using a spreadsheet program obtain the following:
a. A plot of the stress–strain relationship. Label the axes and show units.
b. A plot of the linear portion of the stress–strain relationship. Determine modulus of elasticity using the best fit approach.
c. Proportional limit.
d. Yield stress.
e. Ultimate strength.
f. When the applied load was 18 kN, the diameter was measured as 12.7 mm Determine Poisson's ratio.
g. After the rod was broken, the two parts were put together and the diameter at the neck was measured as 10.6 mm What is the true stress value at fracture? Is the true stress at fracture larger or smaller than the engineering stress at fracture? Why?
h. Do you expect the true strain at fracture to be larger or smaller than the engineering strain at fracture? Why?

TABLE P3.30

Load (kN)	Displacement (mm)	Load (kN)	Displacement (mm)
0	0	38.1	2.1
12.2	0.02	39.1	2.4
18.1	0.04	39.9	2.8
31.7	0.06	40.7	3.1
31.8	0.43	41.1	3.4
32.6	1.07	41.6	3.7
33.5	1.17	42.0	4.6
35.2	1.49	35.0	7.6
36.8	1.81		

3.31   A high-yield-strength alloy steel bar with a rectangular cross section that has a width of 40 mm, a thickness of 6 mm, and a gauge length of 200 mm was tested in tension to rupture, according to ASTM E8/E8M method. The load and deformation data were as shown in Table P3.31.

Using a spreadsheet program, obtain the following:
a. A plot of the stress–strain relationship. Label the axes and show units.
b. A plot of the linear portion of the stress–strain relationship. Determine modulus of elasticity using the best-fit approach.
c. Proportional limit.
d. Yield stress.
e. Ultimate strength.

f. If the specimen is loaded to 155 kN only and then unloaded, what is the permanent deformation?

g. In designing a typical structure made of this material, would you expect the stress applied in (f) safe? Why?

TABLE P3.31

Load (kN)	Displacement (mm)	Load (kN)	Displacement (mm)
0	0	159.4	8.560
18.5	0.102	161.2	9.856
84.3	0.379	162.8	11.217
147.4	0.623	164.2	12.588
147.6	1.744	165.0	13.789
149.2	4.327	165.8	15.201
150.8	4.742	166.6	18.767
153.9	6.030	153.6	30.746
157.0	7.339		

3.32 Estimate the cross-sectional area of a 350S125-27 cold-formed shape.

a. If the member is tested in tension, what would be the maximum force the sample could carry before reaching the yield strength if the steel has a yield strength of 225 MPa?

b. Would you expect a 2.5 m stud to carry the same load in compression? (explain)

3.33 A 32-mm rebar with a gauge length of 200 mm was subjected to tension to fracture according to ASTM E-8 method. The load and deformation data were as shown in Table P3.33.

Using a spreadsheet program obtain the following:

a. A plot of the stress–strain relationship. Label the axes and show units.

b. A plot of the linear portion of the stress–strain relationship. Determine modulus of elasticity using the best-fit approach.

c. Proportional limit.

d. Yield stress.

e. Ultimate strength.

f. If the rebar is loaded to 390 kN only and then unloaded, what is the permanent change in length?

TABLE P3.33

Load (kN)	Displacement (mm)	Load (kN)	Displacement (mm)
0	0	472.9	8.4
62.2	0.1	487.1	9.7
188.9	0.2	496.4	11.1
329.8	0.4	505.7	12.4
383.4	1.7	512.8	13.7
426.0	4.0	522.6	15.3
447.3	5.9	532.4	18.5
462.5	7.2	525.9	22.4

3.34 A steel pipe having a length of 1 m, an outside diameter of 0.2 m, and a wall thickness of 10 mm, is subjected to an axial compression of 200 kN. Assuming a modulus of elasticity of 200 GPa and a Poisson's ratio of 0.3, find
   a. the shortening of the pipe,
   b. the increase in the outside diameter, and
   c. the increase in the wall thickness.

3.35 A drill rod with a diameter of 10 mm is made of high-strength steel alloy with a shear modulus of 80 GPa. The rod is to be subjected to a torque T. What is the minimum required length L of the rod so that the cross section at one end can be rotated 90° with respect to the other end without exceeding an allowable shear stress of 300 MPa?

3.36 What is the shear modulus of the material whose shear stress–strain relationship is shown in Figure 3.21?

3.37 An engineering technician performed a tension test on a mild steel specimen to fracture. The original cross-sectional area of the specimen is 160 mm^2 and the gauge length is 100 mm. The information obtained from this experiment consists of applied tensile load (P) and increase in length (ΔL). The results are tabulated in Table P3.37. Using a spreadsheet program, complete the table by calculating the engineering stress (σ) and the engineering strain (ε). Determine the toughness of the material (u_t) by calculating the area under the stress–strain curve, namely,

$$u_t = \int_0^{\varepsilon_f} \sigma \, d\varepsilon$$

TABLE P3.37

Observation No.	ΔL (mm)	P (kg)	σ (MPa)	ε (mm/mm)	u_i (MPa)
0	0	0			N/A
1	0.013	25			
2	0.025	50			
3	0.038	75			
4	0.051	100			
5	0.064	125			
6	0.076	133			
7	0.089	133			
8	0.102	135			
9	0.114	146			
10	0.127	148			
11	0.254	150			
12	0.381	151			
13	0.508	153			
14	0.635	157			
15	0.762	161			
16	1.016	164			
17	1.143	170			
18	1.270	172			
19	2.540	179			
20	3.810	211			
21	5.080	228			
22	6.350	237			
23	7.62	243			
24	8.89	246			
25	10.16	248			
26	11.43	245			
27	12.70	242			
28	13.97	243			
29	15.24	223			
30	16.51	208			
31	17.78				

$$u_t =$$

where ε_f is the strain at fracture. The preceding integral can be approximated numerically using a trapezoidal integration technique:

$$u_t = \sum_{i=1}^{n} u_i = \sum_{i=1}^{n} \frac{1}{2}(\sigma_i + \sigma_{i-1})(\varepsilon_i - \varepsilon_{i-1})$$

3.38 A Charpy V Notch (CVN) test was performed on a steel specimen and produced the readings shown in Table P3.38. Plot the toughness-versus-temperature relation and determine the temperature transition zone between ductile and brittle behavior.

TABLE P3.38

Temperature, C	Toughness, Joule (N.m)
−40	7
0	15
50	54
100	109
150	120
200	122

3.39 A Charpy V Notch test was conducted for a steel specimen. The average values of the test results at four different test temperatures were found to be
15 J at −45°C
21 J at −18°C
60 J at 5°C
75 J at 40°C

A bridge will be located in a region where specifications require a minimum of 35 J fracture toughness at 0°C for welded fracture-critical members. If the bridge contains a welded flange in a fracture-critical member, does the steel have adequate Charpy V notch fracture toughness to be used for this bridge? Show your supporting calculations.

3.40 How can flaws in steel and welds be detected? Discuss the concept of a nondestructive test used for this purpose.

3.41 Determine the welding zone classification of A36/A36M and A922/A922M steel.

3.42   Briefly define steel corrosion. What are the four elements necessary for corrosion to occur?

3.43   Discuss the main methods used to protect steel from corrosion.

3.44   Using Internet resources, watch a few video clips on the Minnesota I35 bridge collapse on August 1, 2007, read material online about the collapse, and try to understand the reasons behind the failure of the bridge. Investigate if the failure was brittle or ductile overall.

3.45   The World Trade Center collapse on September 11, 2001, was due in large part to steel failure. Using the Internet, investigate what happened to the steel in this structure and relate it to fundamental steel properties.

3.46   The U.S. iron and steel industry has been implementing more energy efficient technologies and practices in order to reduce production cost and improve competitiveness. Using online resources, discuss some methods that have been used in recent years.

3.13   References

American Institute of Steel Construction (AISC). Load and Resistance Factor Design Specification for Structural Steel Buildings, ANSI/AISC 360-05, American Institute of Steel Construction, Inc., 2005.

American Institute of Steel Construction (AISC). *Structural Steel: An Industry Overview,* www.aisc.org, Chicago, IL, 2015.

Bouchard, S., and G. Axmann. ASTM A913 Grades 50 and 65: Steels for Seismic Applications. *STESSA Conference Proceedings,* 2000.

Budinski, K. G. and M. K. Budinski. *Engineering Materials, Properties and Selection.* 9th ed. Upper Saddle River, NJ: Prentice Hall, 2010.

Calkins, M. Materials for Sustainable Sites, John Wiley & Sons, 2008.

Callister, W. D., Jr. *Materials Science and Engineering—an Introduction.* 7th ed. New York: John Wiley and Sons, 2006.

Carter, C. J., Are You Properly Specifying Materials? *Modern Steel Construction,* January edition, 2004.

Cordon, W. A. *Properties, Evaluation, and Control of Engineering Materials.* New York: McGraw-Hill, 1979.

Corrosion Costs. http://www.corrosioncost.com/infrastructure/index.htm, 2009.

Eckmann, D., T. Harrison, and R. Ekman. Structural Steel Contributions Toward Obtaining a LEED Rating, *Modern Steel Construction,* May edition, 2003, Revised 2004.

Fontana, M. G. and N. D. Greene. *Corrosion Engineering.* 3rd ed. New York: McGraw-Hill, 1986.

Frank, K. H. and L. M. Smith. *Highway Materials Engineering: Steel, Welding and Coatings.* Publication No. FHWA-HI-90-006. Washington, DC: Federal Highway Administration, 1990.

Hare, C. H. *Protective Coatings for Bridge Steel.* National Cooperative Highway Research Program, Synthesis of Highway Practice No. 136. Washington, DC: Transportation Research Board, 1987.

Hassett, P. M. *Steel Construction in the New Millennium.* Hassett Engineering, Inc., Castro Valley, CA, 2003.

IBS Digest, https://www.cidb.gov.my/cidbv4/images/publication/digest/IBS%20Digest% 202%202005.pdf, April-June 2005.

ISG Plate. *High Performance Steel for Bridges.* Coatsville, PA, 2003.

Karren, K. W., and G. Winter, Effects of Cold-Straining on Structural Sheet Steels, Journal of the Structural Division, ASCE Proceedings, vol. 93, 1967.

Key-to-Steel. Designation of Carbon and Low-Alloy Steels, http://www.key-to-steel. com.

Lane, S., E. Mulley, B. Wright, M. Simon, and J. D. Cooper. Focus on High-Performance, *Better Roads,* May, 1998.

Miller, D. Mixing Welds and Bolts, Part 1. *Welding Innovations.* Vol. XVIII, No. 2, 2001.

Somayaji, S. *Civil Engineering Materials.* Upper Saddle River, NJ: Prentice Hall, 2001.

Specialty Steel Industry of North America. *Stainless Steel for Structural Applications, Designer Handbook.* Washington, DC, 1999.

Steel Stud Manufactures Association (SSMA). Product Technical Information, http:// www.ssma.com/technical-library-pages-9.php, Chicago, Il. 2012.

SRI, Steel Takes LEED with Recycled Content, Steel Recycling Institute, http://www. recycle-steel.org/~/media/Files/SRI/Media%20Center/LEED_Oct2012.pdf, 2012.

Timisoara Polytechnical University (TPU), Course 11, Introduction to Cold Formed Steel Design, http://cemsig.ct.upt.ro/dinu/metal3/Curs_11.pdf, Romania accesses 2009.

Warrell, E., P. Blinde, M. Neelis, E. Blomen, and E. Masanet. Energy Efficiency Improvement and Cost Saving Opportunities for the US Iron and Steel Industry, Berkeley National Laboratory, University of California, Report LBNL-4779E, 2010.

CHAPTER

4

ALUMINUM

Aluminum is the most plentiful metal on Earth, representing 8% of its crust. Although plentiful, aluminum exists primarily as oxides. The process of extracting aluminum from oxide is very energy intensive. In fact, approximately 2% to 3% of the electricity used in the United States is consumed in aluminum production. This high-energy requirement makes recycling of aluminum products economical. Of the 24 million tons of aluminum produced annually, approximately 75% is from ore reduction and 25% is from recycled materials.

The properties of pure aluminum are not suitable for structural applications. Some industrial applications require pure aluminum, but otherwise, alloying elements are almost always added. These alloying elements, along with cold working and heat treatments, impart characteristics to the aluminum that make this product suitable for a wide range of applications. Figures 4.1 and 4.2 show aluminum used in two construction applications. Here, the term *aluminum* is used to refer to both the pure element and alloys.

In terms of the amount of metal produced, aluminum is second only to steel. About 25% of aluminum produced is used for containers and packaging; 20% for architectural applications, such as doors, windows, and siding; and 10% for electrical conductors. The balance is used for industrial goods, consumer products, aircraft, and highway vehicles.

Aluminum accounts for 80% of the structural weight of aircraft, and its use in the automobile and light truck industry has increased 300% since 1971 (Reynolds Metals Company, 1996). However, use of aluminum for infrastructure applications has been limited. Of the approximately 600000 bridges in the United States, only nine have primary structural members made of aluminum. Two reasons for the limited use of aluminum are the relatively high initial cost when compared with steel and the lack of performance information on aluminum structures.

FIGURE 4.1   Aluminum frame used for structural support of a building.

FIGURE 4.2   Building facade made of aluminum.

Aluminum has many favorable characteristics and a wide variety of applications. The advantages of aluminum are that it (Budinski and Budinski, 2010)

- has one-third the density of steel
- has good thermal and electrical conductivity
- has high strength-to-weight ratio
- can be given a hard surface by anodizing and hard coating
- has alloys that are weldable
- will not rust
- has high reflectivity
- can be die cast
- is easily machined
- has good formability
- is nonmagnetic
- is nontoxic

Aluminum's high strength-to-weight ratio and its ability to resist corrosion are the primary factors that make aluminum an attractive structural engineering material. Although aluminum alloys can be formulated with strengths similar to steel products, the modulus of elasticity of aluminum is only about one-third that of steel. Thus, the dimensions of structural elements must be increased to compensate for the lower modulus of elasticity of aluminum.

Sample Problem 4.1

In many beam design problems, deflection is a limiting criteria. Assuming a rectangular simply supported beam loaded as shown in Figure SP4.1, and the height of the beam is fixed by other design considerations, determine the difference in width required for an aluminum beam compared to a steel beam.

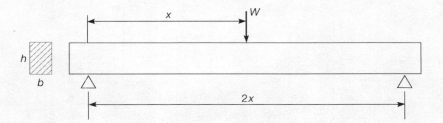

FIGURE SP4.1

Solution:

The equation for the deflection in a beam is

$$\delta = \frac{W(2x)^3}{48EI}$$

where

W = applied force
x = distance from support to load
E = modulus of elasticity
I = moment of inertia

Using the subscripts a and s for aluminum and steel, respectively, and using equal deflection for the aluminum and steel beam:

$$\delta = \frac{W(2x)^3}{48E_a I_a} = \frac{W(2x)^3}{48E_s I_s}$$

which reduces to

$$E_a I_a = E_s I_s$$

For a rectangular section, the moment of inertia is

$$I = \frac{1}{12}bh^3$$

By substitution

$$E_a \frac{1}{12}b_a h^3 = E_s \frac{1}{12}b_s h^3$$

Since the problem statement specified the height of the beams are equal, this equation reduces to

$$b_a = \frac{E_s}{E_a}b_s$$

The modulus of elasticity of steel and aluminum are 200 GPa and 69 GPa, respectively, so the aluminum beam must be 2.9 times wider than the steel beam.

4.1   Aluminum Production

Aluminum production uses processes that were developed in the 1880s. Bayer developed the sodium aluminate leaching process to produce pure alumina (Al_2O_3). Hall and Héroult, working independently, developed an electrolytic process for reducing the alumina to pure aluminum. The essence of the aluminum production process is shown in Figure 4.3.

The production of aluminum starts with the mining of the aluminum ore, bauxite. Commercial grade bauxite contains between 45% and 60% alumina. The bauxite is crushed, washed to remove clay and silica materials, and is kiln dried to remove most of the water. The crushed bauxite is mixed with soda ash and lime and passed through a digester, pressure reducer, and settling tank to produce a concentrated solution of sodium aluminate. This step removes silica, iron oxide, and other impurities from the

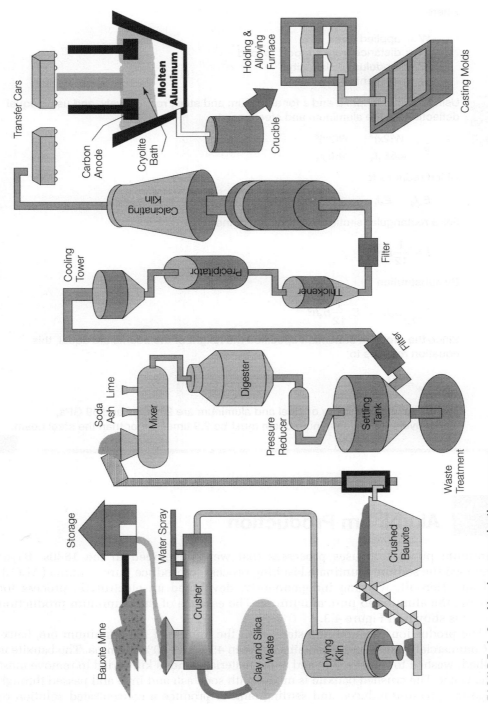

FIGURE 4.3 Aluminum production process.

sodium aluminate solution. The solution is seeded with hydrated alumina crystals in precipitator towers. The seeds attract other alumina crystals and form groups that are heavy enough to settle out of solution. The alumina hydrate crystals are washed to remove remaining traces of impurities and are calcined in kilns to remove all water. The resulting alumina is ready to be reduced with the Hall–Héroult process. The alumina is melted in a cryolite bath (a molten salt of sodium–aluminum–fluoride). An electric current is passed between anodes and cathodes of carbon to separate the aluminum and oxygen molecules. The molten aluminum is collected at the cathode at the bottom of the bath. The molten aluminum, with better than 99% purity, is siphoned off to a crucible. It is then processed in a holding furnace. Hot gases are passed through the molten material to further remove any remaining impurities. Alloying elements are then added.

The molten aluminum is either shipped to a foundry for casting into finished products or is cast into ingots. The ingots are formed by a direct-chill process that produces huge sheets for rolling mills, round loglike billets for extrusion presses, or square billets for production of wire, rod, and bar stock.

Final products are made by either casting, which is the oldest process, or deforming solid aluminum stock. Three forms of casting are used: die casting, permanent mold casting, and sand casting. The basic deformation processes are forging, impact extrusion, stamping, drawing, and drawing plus ironing. Many structural shapes are made with the extrusion process. Either cast or deformed products can be machined to produce the final shape and surface texture, and they can be heat treated to alter the mechanical behavior of the aluminum. Casting and forming methods are summarized in Table 4.1.

When recycling aluminum, the scrap stock is melted in a furnace. The molten aluminum is purified and alloys are added. This process takes only about 5% of the electricity that is needed to produce aluminum from bauxite.

In addition to these conventional processes, very high strength aluminum parts can be produced using powder metallurgy methods. A powdered aluminum alloy is compacted in a mold. The material is heated to a temperature that fuses the particles into a unified solid.

4.2 Aluminum Metallurgy

Aluminum has a face center cubic (FCC) lattice structure. It is very malleable, with a typical elongation over a 50-mm gauge length of over 40%. It has limited tensile strength, on the order of 28 MPa. The modulus of elasticity of aluminum is about 69 GPa. Commercially pure aluminum (i.e., more than 99% aluminum content) is limited to nonstructural applications, such as electrical conductors, chemical equipment, and sheet metal work.

Although the strength of pure aluminum is relatively low, aluminum alloys can be as much as 15 times stronger than pure aluminum, through the addition of small amounts of alloying element, strain hardening by cold working, and heat treatment. For example, typical yield and tensile strengths of 6061-T6 aluminum are 276 MPa and 310 MPa, respectively, The common alloying elements are copper, manganese, silicon,

T A B L E 4 . 1 Casting and Forming Methods for Aluminum Products (Extracted from *Reynolds Infrastructure,* 1996)

Casting Methods

Sand Casting	Sand with a binder is packed around a pattern. The pattern is removed and molten aluminum is poured in, reproducing the shape. Produces a rough texture which can be machined or otherwise surfaced if desired. Economical for low volume production and for making very large parts. Also applicable when an internal void must be formed in the product.
Permanent Mold Casting	Molten aluminum is poured into a reusable metal mold. Economical for large volume production.
Die Casting	Molten aluminum is forced into a permanent mold under high pressure. Suitable for mass production of precisely formed castings.

Forming Methods

Extrusion	Aluminum heated to 425 to 540°C is forced through a die. Complex cross sections are possible, including incompletely or completely enclosed voids. A variety of architectural and structural members are formed by extrusion, including tubes, pipes, I-beams, and decorative components, such as window and door frames.
Rolling	Rollers compress and elongate heated aluminum ingots, producing plates (more than 6 mm thick), sheets (0.15 to 6 mm) thick, and foil (less than 0.15 mm).
Roll Forming	Shaping of sheet aluminum by passing stock between a series of special rollers, usually in stages. Used for mass production of architectural products, such as moldings, gutters, downspouts, roofing, siding, and frames for windows and screens.
Brake Forming	Forming of sheet products with a brake press. Uses simpler tooling than roll forming but production rates are lower and the size of the product is limited.
Cutting Operations	Production of outline shapes by blanking and cutting. In blanking, a punch with the desired shape is pressed through a matching die. Used for mass production of flat shapes. Holes through a sheet are produced by piercing and perforating. Stacks of sheets can be trimmed or cut to an outline shape by a router or sheared in a guillotine-action shear.
Embossing	Shaping an aluminum sheet by pressing between mated rollers or dies, producing a raised pattern on one side and its negative indent on the other side.
Drawing	Shaping an aluminum sheet by drawing it through the gap between two mated dies in a press.
Superplastic Forming	An aluminum sheet is heated and forced over or into a mold by air pressure. Complex and deep contour shapes can be produced, but the process is slow.

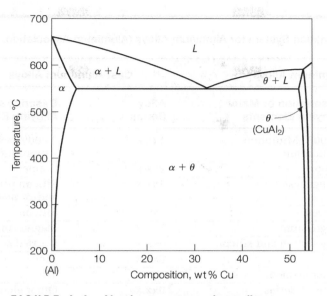

FIGURE 4.4   Aluminum–copper phase diagram.

magnesium, and zinc. Cold working increases strength by causing a disruption of the slip planes in the material that resulted from the production process.

Figure 4.4 shows the two-phase diagram for aluminum and copper. This diagram is typical of the phase diagrams of other two-phase aluminum alloys. The alloying elements have low solubility in aluminum, and the solubility reduces as temperature drops.

As described previously, the properties of metals with this characteristic are very sensitive to heat treatments, which affect the grain size of the material and the distribution of the alloying element throughout the matrix of the lattice structures. Heat treatments typically used on aluminum alloys include annealing, hardening, aging, and stabilizing.

4.2.1   Alloy Designation System

Aluminum classification starts by separating the product according to its production method, either casting or wrought methods. Aluminum alloys designed for casting are formulated to flow into the mold. Wrought aluminum alloys are used for products fabricated by deforming the aluminum into its final shape. The Aluminum Association has developed an aluminum alloy classification system shown in Table 4.2.

The designation system for wrought alloys consists of a four-digit code. The first digit indicates the alloy series. The second digit, if different from 0, indicates a modification in the basic alloy. The third and fourth digits identify the specific alloy in the series; these digits are arbitrarily assigned, except for the 1xxx series, in which the final two digits indicate the minimum aluminum content. For the 1xxx series, the aluminum content is 99% plus the last two digits of the code, expressed as a decimal fraction. For example, a 1060 contains a minimum aluminum content of 99.60%.

T A B L E 4 . 2 Designation System for Aluminum Alloys (Aluminum Association, 1993)

Wrought Aluminum Alloys		Cast Aluminum Alloys	
Alloy Series	Description or Major Alloying Elements	Alloy Series	Description or Major Alloying Elements
1xxx	99.00% Minimum Aluminum	1xx.x	99.00% Minimum Aluminum
2xxx	Copper	2xx.x	Copper
3xxx	Manganese	3xx.x	Silicon plus copper and/or magnesium
4xxx	Silicon	4xx.x	Silicon
5xxx	Magnesium	5xx.x	Magnesium
6xxx	Magnesium and silicon	6xx.x	Unused series
7xxx	Zinc	7xx.x	Zinc
8xxx	Other element	8xx.x	Tin
9xxx	Unused Series	9xx.x	Other element

Cast alloys are assigned a three-digit number followed by one digit after the decimal point, as shown in Table 4.2. The first digit represents the alloy series. Note that series 3, 6, 8, and 9 have different meanings for cast versus wrought alloys. The second and third digits are arbitrarily assigned to identify specific alloys. The digit after the decimal indicates whether the alloy composition is for the final casting (xxx.0) or for ingot (xxx.1 and xxx.2).

4.2.2 ■ Temper Treatments

The mechanical properties of aluminum are greatly altered by both heat treatment and strain hardening. Therefore, specification of an aluminum material must include the manner in which the product was tempered. The processes described in Table 4.3 define the types of tempering aluminum products undergo.

Aluminum alloys used for structural applications are classified as being either heat treatable or not. Non–heat-treatable or "common" alloys contain elements that remain substantially in solid solution or that form insoluble constituents. Thus, heat treatment does not influence their mechanical properties. The properties of these alloys are dependent on the amount of cold working introduced after annealing. Heat-treatable or "strong" alloys contain elements, groups of elements, or constituents that have a considerable solid solubility at elevated temperatures and limited solubility at lower temperatures. The strength of these alloys is increased primarily by heat treatment.

TABLE 4.3 Temper Designations for Aluminum Alloys

Symbol	Meaning	Comment
F	as fabricated	No special control over thermal conditions or strain hardening is employed.
O	annealed	Wrought products—annealed to the lowest strength temper Cast products—annealed to improve ductility and dimensional stability. The "O" may be followed by a digit other than zero, indicating a variation with special characteristics.
H	strain hardened	Wrought products only. Strength is increased by strain hardening, with or without supplemental thermal treatments. The "H" is always followed by two or more numerical digits. The first digit indicates a specific combination of basic operations. The second digit indicates the degree of strain hardening. (Codes for the second digit are 2—quarter hard, 4—half hard, 8—full hard, 9—extra hard.) When used, the third digit indicates a variation of the two digit temper. The basic operations identified by the first digit are as follows:
		H1—strain hardening only. Applies to products that are strain hardened to obtain the desired strength, without supplementary thermal treatment.
		H2—strain hardened and partial annealed. Applies to products that are strain hardened more than the desired final amount, and then reduced in strength to the desired level by partial annealing.
		H3—strain hardened and stabilized. Applies to products that are strain hardened and whose mechanical properties are stabilized either by a low temperature thermal treatment or as a result of heat introduced during fabrication. Stabilization usually improves ductility.
W	solution heat treated	An unstable temper applicable only to alloys that spontaneously age at room temperature after solution heat treatment. This designation is specific only when the period of natural aging is indicated—for example, $W\frac{1}{2}$ hr.
T	thermally treated to produce stable tempers other than F, O, or H	Applies to thermally treated products, with or without supplementary strain hardening to produce stable tempers. The "T" is always followed by one or two digits: T1—cooled from an elevated temperature shaping process and naturally aged to a substantially stable condition. Products not cold worked after cooling from an elevated temperature shaping process, or in which the effect of cold working flattening or straightening may not be recognized in mechanical property limits.
		T2—cooled from an elevated temperature shaping process, cold worked, and naturally aged to a substantially stable condition. Products cold worked to improve strength after cooling from an elevated temperature shaping process, or in which the effect of cold work in flattening or straightening is recognized in mechanical property limits.

(Continued)

TABLE 4.3 (Continued)

Symbol	Meaning	Comment
		T3—solution heat treated, cold worked, and naturally aged to a substantially stable condition. Products cold worked to improve strength after solution heat treatment, or in which the effect of cold work in flattening or straightening is recognized in mechanical property limits.
		T4—solution heat treated and naturally aged to a substantially stable condition. Products not cold worked after solution heat treatment, or in which the effect of cold work in flattening or straightening is recognized in mechanical property limits.
		T5—cooled from an elevated temperature shaping process, then artificially aged. Products not cold worked after cooling from an elevated temperature shaping process, or in which the effect of cold work in flattening or straightening may not be recognized in mechanical property limits.
		T6—solution heat treated and then artificially aged. Products which are not cold worked after solution heat treatment, or in which the effect of cold work in flattening or straightening may not be recognized in mechanical property limits.
		T7—solution heat treated and overaged/stabilized. Wrought products artificially aged after solution heat treatment to carry them beyond a point of maximum strength to provide control of some significant characteristic. Cast products artificially aged after solution heat treatment to provide dimensional and strength stability.
		T8—solution heat treated, cold worked, and then artificially aged. Products cold worked to improve strength, or in which the effect of cold work in flattening or straightening is recognized in mechanical property limits.
		T9—solution heat treated, artificially aged, and then cold worked. Products cold worked to improve strength.
		T10—cooled from an elevated temperature shaping process, cold worked, and then artificially aged. Products worked to improve strength, or in which the effect of cold work in flattening and straightening is recognized in mechanical property limits.
		Additional digits can be appended to the preceding temper designations to indicate significant variations.

4.3   Aluminum Testing and Properties

Typical properties are provided in Tables 4.4 and 4.5 for non–heat-treatable and heat-treatable wrought aluminum alloys, respectively. Typical properties for cast aluminum alloys that may be used for structural applications are given in Table 4.6. These values are only an indication of the properties of cast aluminum alloys. Material properties of cast members can vary throughout the body of the casting due to differential cooling rates.

Tests performed on aluminum are similar to those described for steel. These typically include stress–strain tensile tests to determine elastic modulus, yield strength, ultimate strength, and percent elongation. In contrast to steel, aluminum alloys do not display an upper and lower yield point. Instead, the stress–strain curve is linear up to the proportional limit and then is a smooth curve up to the ultimate strength. Yield strength is defined based on the 0.20% strain offset method, as shown in Figure 4.5. As indicated earlier, the modulus of elasticity of aluminum alloys is on the order of 69 GPa and is not very sensitive to types of alloys or temper treatments.

Sample Problem 4.2

An aluminum alloy rod with 10 mm diameter was subjected to a 5-kN tensile load. After the load was applied, the diameter was measured and found to be 9.997 mm. If the yield strength is 139 MPa, calculate the Poisson's ratio of the material.

Solution

$$\sigma = \frac{5000}{\pi d^2 / 4} = 63.7 \times 10^6 \text{ Pa} = 63.7 \text{ MPa}$$

It is clear that the applied stress is well below the yield stress and, as a result, the deformation is elastic. Hence, assume that

$$E = 69 \text{ GPa}$$

$$\varepsilon_{axial} = \frac{\sigma}{E} = \frac{63.7 \times 10^6}{69 \times 10^9} = 0.000923 \text{ m/m}$$

$$\Delta d = 9.997 - 10.000 = 0.003 \text{ m}$$

$$\varepsilon_{lateral} = -0.003 / 10.000 = -0.0003 \text{ m/m}$$

$$\nu = \frac{-\varepsilon_{lateral}}{\varepsilon_{axial}} = \frac{0.0003}{0.000923} = 0.33$$

Aluminum's coefficient of thermal expansion is 0.000023/°C, about twice as large as that of steel and concrete. Thus, joints between aluminum and steel or concrete must be designed to accommodate the differential movement.

TABLE 4.4 Properties of Select Non-Heat-Treatable Wrought Aluminum Alloys

Alloy		Tension				Hardness[2]	Shear Ultimate	Fatigue[3] Endurance Limit	Nominal Chemical Composition
		Ultimate	Yield	Elongation[1] (thickness)					
		MPa	MPa	1.6 mm	12.5 mm		MPa	MPa	
1060	O	69	28	43	45	19	48	21	99.6 Al
	H-12	83	76	16	25	23	55	28	
	H-14	97	90	12	50	26	62	34	
	H-16	110	103	5	17	30	69	45	
	H-18	131	124	6	15	35	76	45	
1100	O	90	34	35	45	23	62	34	99. Al
	H-12	110	103	12	25	28	69	41	
	H-14	124	117	9	50	32	76	48	
	H-16	145	138	6	17	38	83	62	
	H-18	165	152	5	15	44	90	62	
3003	O	110	41	30	40	28	76	48	1.2 Mn
	H-12	131	124	10	20	35	83	55	
	H-14	152	145	8	16	40	97	62	
	H-16	179	172	5	14	47	103	69	
	H-18	200	186	4	10	55	110	69	

Alloy	Temper			[1]	[1]		[2]	[3]	
5005	O	124	41	25		28	76		0.8 Mg
	H-12	138	131	10			97		
	H-14	159	152	6			97		
	H-16	179	172	5			103		
	H-18	200	193	4			110		
	H-32	138	117	11		36	97		
	H-34	159	138	8		41	97		
	H-36	179	165	6		46	103		
	H-38	200	186	5		51	110		
5086	O	262	117	22	30	60	159	145	Mg
	H-32	290	207	12	16	72	172	152	0.45 Mn
	H-34	324	255	10	14	82	186	159	
	H-111	276	186		17	65	159	145	
	H-112	269	131	14		64	159	145	
	H-116	290	207		16	72	173	152	
5456	O	310	159	20	24	70	186	152	5.1 Mg
	H-111	324	228		18	75	186	165	0.7 Mn
	H-112	310	156		22	70	186		0.12 Cr
	H-116	352	255	16	16	90	207	159	

[1] Percent elongation over 50 mm.

[2] Brinell number, 500-kg load

[3] 500,000,000 cycles of complete stress reversal using R.R. Moore type of machine and specimen.

TABLE 4.5 Properties of Select Heat-Treatable Wrought Aluminum Alloys

Alloy		Tension				Hardness[2]	Shear Ultimate	Fatigue[3] Endurance Limit	Nominal Chemical Composition
		Ultimate	Yield	Elongation[1] (thickness)					
		MPa	MPa	1.6 mm	12.5 mm		MPa	MPa	
2014	O	186	97		18	45	124	90	4.5 Cu, 0.8 Mn
	T4/T451	427	290		50	105	262	138	0.8 Si, 0.4 M
	T6/T651	483	414		13	135	290	124	
6053	O	110	55		35	26	76	55	1.2 Mg,
	T6	255	221		13	80	159	90	0.25 CR
6061	O	124	55	25	30	30	83	62	1.0 Mg, 0.6 SI
	T4/T451	241	145	22	25	65	165	97	0.25 Cu,
	T6/T651	310	276	12	17	95	207	97	0.25 Cr
6063	O	90	48			25	69	55	0.7 Mg
	T1	152	90	20	33	42	97	62	0.4 Si
	T4	172	90	22			110		
	T5	186	145	12	22	60	117	69	
	T6	241	214	12	18	73	152	69	
	T83	255	241	9		82	152		
	T831	207	186	10		70	124		
	T832	290	269	12		95	186		
7178	O	228	103	15	16	60	152		6.8 Zn, 2.0 Cu
	T6/T651	607	538	10	11	160	359	152	2.7 Mg, 0.3 Mn
	T76/	572	503		11				
	T765								

[1] Percent elongation over 50 mm.

[2] Brinell number, 500-kg load

[3] 500,000,000 cycles of complete stress reversal using R.R. Moore type of machine and specimen.

TABLE 4.6   Typical Properties of Select Cast Aluminum Alloys

Cast Alloy Designation	Tension			Hardness[2]	Shear Ultimate (MPa)	Fatigue[3] Endurance Limit (MPa)
	Ultimate (MPa)	Yield (MPa)	Elongation[1]			
356.0-T6[4]	276	186	5	90	221	90
356.0-T7[4]	228	165	5	70	172	76
A356.0-T61[4]	283	207	10	80		
A357.0-T6[4]	345	276	10	85	296	110
A444.0-T4[4]	159	69	21	45		
356.0-T6[5]	228	165	3.5	70	179	59
356.0-T7[5]	234	207	2.0	75	165	62
Almag 35 535.0[5]	276	145	13	70	193	69

[1] Percent elongation over 50 mm.
[2] Brinell number, 500-kg load
[3] 500,000,000 cycles of complete stress reversal using R.R. Moore type of machine and specimen.
[4] Permanent mold
[5] Sand casting

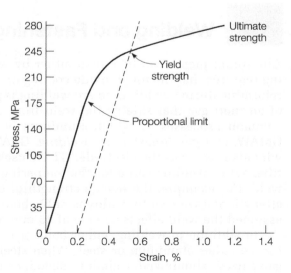

FIGURE 4.5   Aluminum stress–strain diagram.

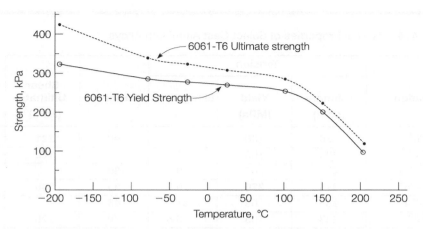

FIGURE 4.6   **Tensile strength of aluminum at different temperatures.**

Strengths of aluminum are considerably affected by temperature, as shown in Figure 4.6. At temperatures above 150°C, tensile strengths are reduced considerably. The temperature at which the reduction begins and the extent of the reduction depends on the alloy. At temperatures below room temperature, aluminum becomes stronger and tougher as the temperature decreases.

4.4   Welding and Fastening

Aluminum pieces can be joined either by welding or by using fasteners. Welding requires that the tough oxide coating on aluminum be broken and kept from reforming during welding, so arc welding is generally performed in the presence of an inert gas that shields the weld from oxygen in the atmosphere. The two common processes by which aluminum is welded are *gas metal arc welding,* GMAW, and *gas tungsten arc welding,* GTAW. In the GMAW process, the filler wire also serves as the electrode. GTAW uses a tungsten electrode and a separate filler wire. Welding can alter the tempering of the aluminum in the area of the weld. For example, the tensile strength of 6061-T6 is 290 MPa, but the tensile strength of a weld in this alloy is only about 165 MPa. For design purposes, it is assumed the weld affects an area of 25 mm on each side of the weld.

In addition to welding, either bolts or rivets can join aluminum pieces. Bolts can be either aluminum or steel. When steel bolts are used, they must be either galvanized, aluminized, cadmium plated, or made of stainless steel to prevent the development of galvanic corrosion. Rivet fasteners are made of aluminum and are cold driven. Both bolt and rivet joints are designed based on the shear strength of the fastener and the bearing strength of the material being fastened.

4.5   Corrosion

Aluminum develops a thin oxidation layer immediately upon exposure to the atmosphere. This tough oxide film protects the surface from further oxidation. The alloying elements alter the corrosion resistance of the aluminum. The alloys used for airplanes are usually given extra protection by painting or "cladding" with a thin coat of a corrosion-resistant alloy. Painting is generally not needed for medium-strength alloys used for structural applications.

Galvanic corrosion occurs when aluminum is in contact with any of several metals in the presence of an electrical conductor, such as water. The best protection for this problem is to break the path of the galvanic cell by painting, using an insulator, or keeping the dissimilar metals dry.

4.6   Aluminum Sustainability

4.6.1   LEED Considerations

The Aluminum Association (2015) LEED fact sheet only identifies the recyclability of aluminum as a potential way to earn credits for LEED. The recycled content of flat rolled products for the building and construction market was approximately 85%. Relative to the steel industry there is much less information about using aluminum to earn LEED credits.

4.6.2   Other Sustainability Considerations

The aluminum industry has made substantial sustainability improvements in the production of their products. Relative to the production of aluminum from bauxite ore, recycled aluminum only requires 5% of the energy, every ton of recycled aluminum saves four tons of bauxite, and the air and water pollution are reduced by 97% (Aluminum Association, 2015). The Aluminum Association (2014) reports industry efficiencies are improving with a 26% reduction in the amount of energy needed to produce a ton of aluminum since 1995 and an 85 percent in perfluorocarbons (a greenhouse gas) emissions since 1990.

Summary

Although aluminum has many desirable attributes, its use as a structural material in civil engineering has been limited, primarily by economic considerations and a lack of performance information. Aluminum alloys and heat treatments provide products with a wide range of characteristics. The advantages of aluminum relative to steel include lightweight, high strength-to-weight ratio, and corrosion resistance.

Questions and Problems

4.1 Name the two primary factors that make aluminum an attractive structural engineering material.

4.2 Complete Table P4.2. For a member where a deflection criterion controls, which material would require a larger cross section? What about for a member where tension controls?

TABLE P4.2

	Mild Steel	7178 T76 Aluminum
Yield Strength		
Ultimate Strength		
Modulus of Elasticity		

4.3 An aluminum alloy bar with a radius of 7 mm was subjected to tension until fracture and produced results shown in Table P4.3.
   a. Using a spreadsheet program, plot the stress–strain relationship.
   b. Calculate the modulus of elasticity of the aluminum alloy.
   c. Determine the proportional limit.
   d. What is the maximum load if the stress in the bar is not to exceed the proportional limit?
   e. Determine the 0.2% offset yield strength.
   f. Determine the tensile strength.
   g. Determine the percent of elongation at failure.

TABLE P4.3

Stress, MPa	Strain, 10^{-3} m/m	Stress, MPa	Strain, 10^{-3} m/m
0	0.0	399.9	5.2
55.2	0.6	427.5	5.8
117.2	1.5	441.3	6.2
186.2	2.4	448.2	6.5
241.3	3.2	462.0	7.3
296.5	4.0	468.9	8.1
344.8	4.6	482.7	9.7

4.4 Decode the characteristics of a 6063 T831 aluminum.

4.5 A round aluminum alloy bar with a 15 mm. diameter and 50 mm gauge length was subjected to tension to fracture. The load and deformation data were as shown in Table P4.5.

TABLE P4.5

Load (kN)	ΔL (mm)	Load (kN)	ΔL (mm)
0	0	70	0.345
9	0.036	74	0.427
18	0.071	78	0.559
27	0.107	81	0.787
36	0.139	83	1.067
45	0.178	84	1.341
54	0.211	85	Fracture
63	0.262		

Using a spreadsheet program, obtain the following:
a. A plot of the stress–strain relationship. Label the axes and show units.
b. A plot of the linear portion of the stress–strain relationship. Determine modulus of elasticity using the best fit approach.
c. Proportional limit.
d. Yield stress at an offset strain of 0.002 m/m.
e. Tangent modulus at a stress of 415 MPa.
f. Secant modulus at a stress of 415 MPa.

4.6 An aluminum alloy bar with a rectangular cross section that has a width of 12.5 mm, thickness of 6.25 mm, and a gauge length of 50 mm was tested in tension to fracture according to ASTM E-8 method. The load and deformation data were as shown in Table P4.6.

Using a spreadsheet program, obtain the following:
a. A plot of the stress–strain relationship. Label the axes and show units.
b. A plot of the linear portion of the stress–strain relationship. Determine the modulus of elasticity using the best fit approach.
c. Proportional limit.
d. Yield stress at an offset strain of 0.002 m/m.
e. Tangent modulus at a stress of 450 MPa.
f. Secant modulus at a stress of 450 MPa.

TABLE P4.6

Load (kN)	ΔL (mm)	Load (kN)	ΔL (mm)
0	0	33.5	1.486
3.3	0.025	35.3	2.189
14.0	0.115	37.8	3.390
25.0	0.220	39.8	4.829
29.0	0.406	40.8	5.961
30.6	0.705	41.6	7.386
31.7	0.981	41.2	8.047
32.7	1.245		

4.7   A round aluminum alloy bar with a 6 mm diameter and a 25 mm gauge length was tested in tension to fracture according to ASTM E-8 method. The load and deformation data were as shown in Table P4.7.

TABLE P4.7

Load (kN)	Displacement (mm)	Load (kN)	Displacement (mm)
0	0	13	0.743
1	0.013	14	1.095
5	0.057	15	1.695
10	0.109	15.9	2.415
11	0.203	16.4	2.980
12	0.353	17	3.693
12	0.490	16.5	4.024
13	0.623		

Using a spreadsheet program, obtain the following:
a. A plot of the stress–strain relationship. Label the axes and show units.
b. A plot of the linear portion of the stress–strain relationship. Determine modulus of elasticity using the best fit approach.
c. Proportional limit.
d. Yield stress at an offset strain of 0.002 m/m.
e. Initial tangent modulus.
f. If the specimen is loaded to 14.23 kN only and then unloaded, what is the permanent change in gauge length?
g. When the applied load was 5.5 kN, the diameter was measured as 6 mm Determine Poisson's ratio.

4.8   An aluminum alloy rod has a circular cross section with a diameter of 8 mm. This rod is subjected to a tensile load of 4 kN. Assume that the material is within the elastic region and $E = 69$ GPa.
a. What will be the lateral strain if Poisson's ratio is 0.33?
b. What will be the diameter after load application?

4.9   An aluminum alloy cylinder with a diameter of 76 mm and a height of 150 mm. is subjected to a compressive load of 220 kN. Assume that the material is within the elastic region and a modulus of elasticity of 75 GPa.
a. What will be the lateral strain if Poisson's ratio is 0.33?
b. What will be the diameter after load application?
c. What will be the height after load application?

4.10   A 3003-H14 aluminum alloy rod with 15 mm diameter is subjected to a 9-kN tensile load. Calculate the resulting diameter of the rod. If the rod is subjected to a compressive load of 9 kN, what will be the diameter of the rod? Assume that the modulus of elasticity is 70 GPa, Poisson's ratio is 0.33, and the yield strength is 145 MPa.

4.11   The stress–strain relation of an aluminum alloy bar having a length of 2 m and a diameter of 10 mm is expressed by the equation

$$\varepsilon = \frac{\sigma}{70,000}\left[1 + \frac{3}{7}\left(\frac{\sigma}{270}\right)^9\right]$$

where σ is in MPa. If the rod is axially loaded by a tensile force of 20 kN and then unloaded, what is the permanent deformation of the bar?

4.12   An aluminum specimen originally 300 mm long is pulled in tension with a stress of 275 MPa. If the deformation is entirely elastic, what will be the resultant elongation? The aluminum modulus of elasticity is 75 GPa.

4.13   A tension test was performed on an aluminum alloy specimen to fracture. The original diameter of the specimen is 15 mm and the gauge length is 50 mm. The information obtained from this experiment consists of applied tensile load (P) and increase in length (ΔL). The results are tabulated in Table P4.13. Using a spreadsheet program, complete the table by calculating engineering stress (σ) and engineering strain (ε). Determine the toughness of the material (u_t) by calculating the area under the stress–strain curve, namely,

$$u_t = \int_0^{\varepsilon_f} \sigma\,d\varepsilon$$

where ε_f is the strain at fracture. The preceding integral can be approximated numerically using a trapezoidal integration technique:

$$u_t = \sum_{i=1}^{n} u_i = \sum_{i=1}^{n} \frac{1}{2}(\sigma_i + \sigma_{i-1})(\varepsilon_i - \varepsilon_{i-1})$$

TABLE P4.13

Observation No.	P (kN)	ΔL (mm)	σ (MPa)	ε (m/m)	u_i (MPa)
0	0	0			N/A
1	5	0.038			
2	10	0.076			
3	16	0.114			
4	21	0.150			
5	27	0.191			
6	32	0.226			
7	38	0.279			
8	42	0.371			
9	44	0.457			
10	46	0.597			
11	48	0.843			
12	50	1.140			
13	51	1.435			
14	51	1.725			
					$u_t =$

4.14   In Problem 4.13, plot the stress–strain relationship and determine the following:
a. the elastic modulus
b. the proportional limit
c. the yield strength at a strain offset of 0.002
d. the tensile strength
e. the magnitude of the load required to produce an increase in length of 0.4 mm
f. the final deformation, if the specimen is unloaded after being strained by the amount specified in (e)
g. in designing a typical structure made of this material, would you expect the stress applied in (e) to be reasonable? Why?

4.15   Referring to Figure 4.5, determine approximate values of the following:
a. modulus of elasticity
b. proportional limit
c. yield strength at a strain offset of 0.002
d. tangential modulus at a stress of 230 MPa
e. secant modulus at a stress of 230 MPa

4.16   Referring to Figure 4.6, determine approximate values of the following:
a. percent change in yield strength of aluminum when temperature increases from −150°C to 150°C
b. percent change in ultimate strength of aluminum when temperature increases from −150°C to 150°C

4.17   A tensile stress is applied along the long axis of a cylindrical aluminum rod with a diameter of 10 mm. If the deformation is entirely elastic, determine the magnitude of the load required to produce a 2.5×10^{-3} mm change in diameter. Assume a modulus of elasticity of 70 GPa and a Poisson's ratio of 0.33.

4.18   Assume a uniformly loaded simply supported beam with a rectangular cross section. The width of the beam is fixed by other design considerations. Determine the difference in depth required for an aluminum beam compared to a steel beam in order to obtain the same maximum deflection.

4.19   Discuss galvanic corrosion of aluminum. How can aluminum be protected from galvanic corrosion?

 ## **References**

Aluminum Association. *Aluminum Design Manual*. 10th ed. Washington, DC: The Aluminum Association, 2015.

Aluminum Association. *Structural Design with Aluminum*. Washington, DC: The Aluminum Association, 1987.

Aluminum Association. *Aluminum Standards and Data, 2013*. Washington, DC: The Aluminum Association, 2013.

Aluminum Association, LEED Fact Sheet, web document, The Aluminum Association, http://www.aluminum.org/sites/default/files/LEED_Fact_Sheet_8_13_08.pdf, 2015.

Aluminum Association, *Aluminum Industry Life-Cycle Assessment Report Briefing*, The Aluminum Association, 2014.

ASM Handbooks Online, http://products.asminternational.org/hbk/index.jsp.

Budinski, K. G. and M. K. Budinski. *Engineering Materials, Properties and Selection*. 9th ed. Upper Saddle River, NJ: Prentice Hall, 2009.

Davis, J.(ed.), Aluminum and Aluminum Alloys, ASM International, 1993, ISBN 978-0-87170-496-2.

Reynolds Metals Company. *Reynolds Infrastructure*. Richmond, VA: Reynolds Metals Company, 1996.

CHAPTER

5

AGGREGATES

There are three main uses of aggregates in civil engineering: as an underlying material for foundations and pavements, as riprap for erosion protection, and as ingredients in portland cement and asphalt concretes. The National Stone, Sand and Gravel Association, NSSGA, reports more than 2 billion metric tons of aggregate were produced in the United States in 2013 (NSSGA, 2015). If you use this amount and the population of the U.S., then over 6 thousand kilograms were used per person. By dictionary definition, aggregates are a combination of distinct parts gathered into a mass or a whole. Generally, in civil engineering, the term *aggregate* means a mass of crushed stone, gravel, sand, etc, predominantly composed of individual particles, but in some cases including clays and silts. The largest particle size in aggregates may have a diameter as large as 150 mm, and the smallest particle can be as fine as 5 to 10 μm.

Due to the differences in the size of the aggregate particles, four terms are defined to facilitate the discussion of the general size characteristics of aggregates:

Coarse aggregates: aggregate particles that are retained on a 4.75 mm sieve. A 4.75 mm sieve has openings equal to 4.75 mm between the sieve wires.
Fine aggregates: aggregate particles that pass a 4.75 mm sieve.
Maximum aggregate size: The smallest sieve through which 100% of the aggregates pass.
Nominal maximum aggregate size: The largest sieve that retains any of the aggregate, but generally not more than 10%.

These are the traditional definitions for the maximum and nominal maximum aggregate sizes. However, they are not entirely satisfactory. To be definitive, terms such as "generally" should not be used. The other problem with these definitions is the possibility that the maximum aggregate size may be two or more standard sieve sizes larger than the nominal maximum aggregate size. To overcome these problems, these terms were redefined when the Superpave mix design method was introduced in the 1990s (McGennis et al., 1995). The Superpave definitions are as follows:

Maximum aggregate size: One sieve size larger than the nominal maximum aggregate size.

Nominal maximum aggregate size: One sieve larger than the first sieve to retain more than 10% of the aggregates.

The traditional and Superpave definitions usually yield the same results, but on occasion, the differences in the definitions will result in different size classifications of an aggregate stockpile. In these cases, the engineer must take care that the proper specifications and definitions are used per the job requirements.

The balance of this chapter presents information about aggregates as used in construction. Information is not presented about the characteristics and properties of soils, as this is the purview of textbooks on geotechnical engineering.

5.1 Aggregate Sources

Natural sources for aggregates include gravel pits, river run deposits, and rock quarries. Generally, *gravel* comes from pits and river deposits, whereas *crushed stones* are the result of processing rocks from quarries. Usually, gravel deposits must also be crushed to obtain the needed size distribution, shape, and texture (Figure 5.1).

Manufactured aggregates can use slag waste from iron and steel mills and expanded shale and clays to produce lightweight aggregates. Heavyweight concrete, used for radiation shields, can use steel slag and bearings for the aggregate. Styrofoam

FIGURE 5.1   Aggregate stockpiling. (Courtesy of FHWA)

beads can be used as an aggregate in lightweight concrete used for insulation. Natural lightweight aggregates include pumice, scoria, volcanic cinders, tuff, and diatomite.

5.2   Geological Classification

All natural aggregates result from the breakdown of large rock masses. Geologists classify rocks into three basic types: *igneous, sedimentary,* and *metamorphic.* Volcanic action produces igneous rocks by hardening or crystallizing molten material, magma. The magma cools either at the Earth's surface, when it is exposed to air or water, or within the crust of the earth. Cooling at the surface produces *extrusive* igneous rocks, while cooling underground produces *intrusive* igneous rocks. In general, the extrusive rocks cool much more rapidly than the intrusive rocks. Therefore, we would expect extrusive igneous rocks to have a fine grain size and potentially to include air voids and other inclusions. Intrusive igneous rocks have larger grain sizes and fewer flaws. Igneous rocks are classified based on grain size and composition. Coarse grains are larger than 2 mm and fine grains are less than 0.2 mm. Classification based on composition is a function of the silica content, specific gravity, color, and the presence of free quartz.

Sedimentary rocks coalesce from deposits of disintegrated existing rocks or inorganic remains of marine animals. Wind, water, glaciers, or direct chemical precipitation transport and deposit layers of material that become sedimentary rocks, resulting in a stratified structure. Natural cementing binds the particles together. Classification is based on the predominant mineral present: calcareous (limestone, chalk, etc.), siliceous (chert, sandstone, etc.), and argillaceous (shale, etc.).

Metamorphic rocks form from igneous or sedimentary rocks that are drawn back into the Earth's crust and exposed to heat and pressure, reforming the grain structure. Metamorphic rocks generally have a crystalline structure, with grain sizes ranging from fine to coarse.

All three classes of rock are used successfully in civil engineering applications. The suitability of aggregates from a given source must be evaluated by a combination of tests to check physical, chemical, and mechanical properties, and must be supplemented by mineralogical examination. The best possible prediction of aggregate suitability for a given application is that based on historical performance in a similar design.

5.3   Evaluation of Aggregate Sources

Civil engineers select aggregates for their ability to meet specific project requirements, rather than their geologic history. The physical and chemical properties of the rocks determine the acceptability of an aggregate source for a construction project. These characteristics vary within a quarry or gravel pit, making it necessary to sample and test the materials continually as the aggregates are being produced.

Due to the quantity of aggregates required for a typical civil engineering application, the cost and availability of the aggregates are important when selecting an aggregate source. Frequently, one of the primary challenges facing the materials engineer on a project is how to use the locally available material in the most cost-effective manner.

Potential aggregate sources are usually evaluated for quality of the larger pieces, the nature and amount of fine material, and the gradation of the aggregate. The extent and quality of rock in the quarry is usually investigated by drilling cores and performing trial blasts (or shots) to evaluate how the rock breaks and by crushing some materials in the laboratory to evaluate grading, particle shape, soundness, durability, and amount of fine material. Cores are examined petrographically for general quality, suitability for various uses, and amount of deleterious materials. Potential sand and gravel pits are evaluated by collecting samples and performing sieve analysis tests. The amount of large gravel and cobble sizes determines the need for crushing, while the amount of fine material determines the need for washing. Petrographic examinations evaluate the nature of aggregate particles and the amount of deleterious material (Meininger and Nichols, 1990).

Price and availability are universal criteria that apply to all uses of aggregates. However, the required aggregate characteristics depend on how they will be used in the structure.

5.4 Aggregate Uses

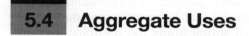

Aggregates are used primarily as an underlying material for foundations and pavements as rip-rap for erosion control, and as ingredients in portland cement and asphalt concretes. Aggregate underlying materials, or base courses, can add stability to a structure, provide a drainage layer, and protect the structure from frost damage (Figure 5.2). Stability is a function of the interparticle friction between the aggregates and the amount of clay and silt "binder" material in the voids between the aggregate particles. However, increasing the clay and silt content will block the drainage paths between the aggregate particles, thereby inhibiting the ability of the material to act as a drainage layer.

In portland cement concrete, 60% to 75% of the volume and 79% to 85% of the weight are made up of aggregates. The aggregates act as a filler to reduce the amount of cement paste needed in the mix. In addition, aggregates have greater volume stability than the cement paste. Therefore, maximizing the amount of aggregate, to a certain extent, improves the quality and economy of the mix.

In asphalt concrete, aggregates constitute 75% to 85% of the volume and 92% to 96% of the mass. The asphalt cement acts as a binder to hold the aggregates together, but does not have enough strength to lock the aggregate particles into position. As a result, the strength and stability of asphalt concrete depends mostly on interparticle friction between the aggregates and, to a limited extent, on the binder. Due to the importance of aggregate properties on the performance of asphalt concrete, the Superpave mix design method (McGennis et al., 1995), introduced in the 1990s established two sets of aggregate requirements: source and consensus. Source properties

FIGURE 5.2   Constructing aggregate base before placing the hot-mix asphalt or portland cement concrete layer of a paved road.

and specifications are established by each owner agency to meet local conditions. The consensus properties were established at the national level with the intention that all agencies would implement the test methods and criteria into their specifications. The consensus properties for Superpave are:

Coarse aggregate
    angularity
    flat and elongated

Fine aggregate
    sand equivalency
    fine aggregate angularity

 5.5   Aggregate Properties

Aggregates' properties are defined by the characteristics of both the individual particles and the characteristics of the combined material. These properties can be further described by their physical, chemical, and mechanical characteristics, as shown

TABLE 5.1   Basic Aggregate Properties (Meininger and Nichols, 1990)

| Property | Relative Importance for End Use* | | |
	Portland Cement Concrete	Asphalt Concrete	Base
PHYSICAL			
Particle shape (angularity)	M	V	V
Particle shape (flakiness, elongation)	M	M	M
Particle size—maximum	M	M	M
Particle size—distribution	M	M	M
Particle surface texture	M	V	V
Pore structure, porosity	V	M	U
Specific gravity, absorption	V	M	M
Soundness—weatherability	V	M	M
Unit weight, voids—loose, compacted	V	M	M
Volumetric stability—thermal	M	U	U
Volumetric stability—wet/dry	M	U	M
Volumetric stability—freeze/thaw	V	M	M
Integrity during heating	U	M	U
Deleterious constituents	V	M	M
CHEMICAL			
Solubility	M	U	U
Surface charge	U	V	U
Asphalt affinity	U	V	M
Reactivity to chemicals	V	U	U
Volume stability—chemical	V	M	M
Coatings	M	M	U
MECHANICAL			
Compressive strength	M	U	U
Toughness (impact resistance)	M	M	U
Abrasion resistance	M	M	M
Character of products of abrasion	M	M	U
Mass stability (stiffness, resilience)	U	V	V
Polishability	M	M	U

* V = Very important; M = Moderately important; U = Unimportant or importance unknown

in Table 5.1 (Meininger and Nichols, 1990). Several individual particle characteristics are important in determining if an aggregate source is suitable for a particular application. Other characteristics are measured for designing portland cement and asphalt concrete mixes (Goetz and Wood, 1960).

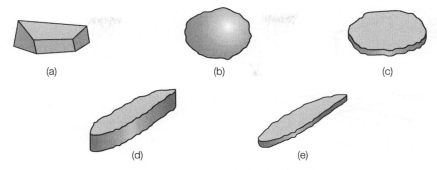

FIGURE 5.3   Particle shapes: (a) angular, (b) rounded, (c) flaky, (d) elongated, and (e) flaky and elongated.

5.5.1 ▨ Particle Shape and Surface Texture

The shape and surface texture of the individual aggregate particles determine how the material will pack into a dense configuration and also determines the mobility of the stones within a mix. There are two considerations in the shape of the material: *angularity* and *flakiness.* Crushing rocks produces angular particles with sharp corners and rough texture. Due to weathering, the corners of the aggregates break down, creating *subangular* particles and smooth texture. When the aggregates tumble while being transported in water, the corners can become completely *rounded.* Generally, angular and rough-textured aggregates produce bulk materials with higher stability than rounded, smooth-textured aggregates. However, the angular aggregates will be more difficult to work into place than rounded aggregates, since their shapes make it difficult for them to slide across each other. Due to the size differences between coarse and fine aggregates, different test methods are used for their evaluation.

Particle Shape of Coarse Aggregates   Figures 5.3 and 5.4 show the different shapes of coarse aggregates: angular, rounded, flaky, elongated, and flaky and elongated. Flakiness, also referred to as flat and elongated, describes the relationship between

FIGURE 5.4   Angular and rounded aggregates.

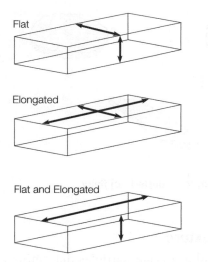

Flat

Elongated

Flat and Elongated

FIGURE 5.5 Concept of flakiness test.

the dimensions of the aggregate, ASTM D4791. Figure 5.5 shows the concept of the flakiness test. This is an evaluation of the coarse portion of the aggregates, but only aggregates retained on the 9.5 mm sieve are evaluated. Under the traditional definition, a flat particle is defined as one where the ratio of the "middle" dimension to the smallest dimension of the particle exceeds the 3 to 1. An elongated particle is defined as one where the ratio of the longest dimension to the middle dimension of the particle exceeds the 3 to 1. Under the Superpave criteria, particles are classified as "flat and elongated" if the ratio of the largest dimension to the smallest dimension exceeds 5 to 1. Figure 5.6 is the apparatus used for the Superpave flat and elongated test.

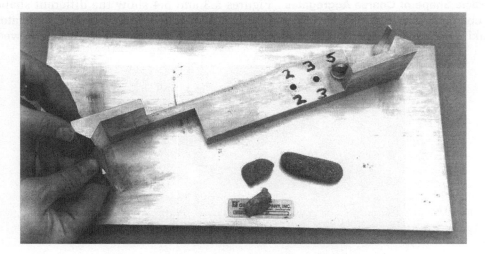

FIGURE 5.6 Superpave flat and elongated apparatus.

Texture of Coarse Aggregates The roughness of the aggregate surface plays an important role in the way the aggregate compacts and bonds with the binder material. Aggregates with a *rough* texture are more difficult to compact into a dense configuration than *smooth* aggregates. Rough texture generally improves bonding and increases interparticle friction. In general, natural gravel and sand have a smooth texture, whereas crushed aggregates have a rough texture.

Since the stability of portland cement concrete is mostly developed by the cementing action of the portland cement and by the aggregate interlock, it is desirable to use rounded and smooth aggregate particles to improve the workability of fresh concrete during mixing. However, the stability of asphalt concrete and base courses is mostly developed by the aggregate interlock. Therefore, angular and rough particles are desirable for asphalt concrete and base courses in order to increase the stability of the materials in the field and to reduce rutting. Flaky and elongated aggregates are undesirable for asphalt concrete, since they are difficult to compact during construction and are easy to break.

To meet the needs of angular aggregates with high texture, many specifications for coarse aggregates used in asphalt concrete require a minimum percentage of aggregates with crushed faces as a surrogate *angularity* and *texture requirement.* A crushed particle exhibits one or more mechanically induced fractured faces and typically has a rough surface texture. To evaluate the angularity and surface texture of coarse aggregate, the percentages of particles with one and with two or more crushed faces are counted in a representative sample, ASTM D5821.

Particle Shape and Texture of Fine Aggregates The angularity and texture of fine aggregates have a very strong influence on the stability of asphalt concrete mixes. The Superpave mix design method recognizes this by requiring a fine aggregate angularity test, ASTM C1252 method, Test Method for Uncompacted Void Content of Fine Aggregate. A sample of fine aggregate with a defined gradation is poured into a small cylinder by flowing it through a standard funnel, as shown in Figure 5.7. By determining the weight of the fine aggregate in the filled cylinder of known volume, the void content can be calculated as the difference between the cylinder volume and the fine aggregate volume collected in the cylinder. The volume of the fine aggregate is calculated by dividing the weight of the fine aggregate by its bulk density. The higher the amount of void content, the more angular and the rougher will be the surface texture of the fine aggregate. ASTM C1252 allows three different methods for the gradation of the sample, but Method A, which requires a specific gradation, is required for Superpave. It is necessary to use a specified gradation; as different gradations would alter the amount of voids between the aggregates and hence consistent criteria could not be developed.

5.5.2 ▨ Soundness and Durability

The ability of aggregate to withstand *weathering* is defined as *soundness* or *durability.* Aggregates used in various civil engineering applications must be sound and durable, particularly if the structure is subjected to severe climatic conditions. Water freezing in the voids of aggregates generates stresses that can fracture the stones.

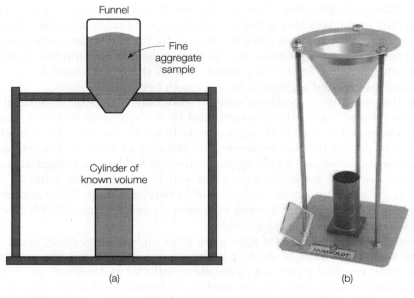

FIGURE 5.7   Apparatus used to measure angularity and surface texture of fine aggregate. (Courtesy of Humboldt Mfg. Co.)

The soundness test (ASTM C88) simulates weathering by soaking the aggregates in either a sodium sulfate or a magnesium sulfate solution. These sulfates cause crystals to grow in the aggregates, simulating the effect of freezing. The test starts with an oven-dry sample separated into different sized fractions. The sample is subjected to cycles of soaking in the sulfate for 16 hours, followed by drying. Typically, the samples are subjected to five cycles. Afterwards, the aggregates are washed and dried, each size is weighed, and the weighted average percentage loss for the entire sample is computed. This result is compared with allowable limits to determine whether the aggregate is acceptable. This is an empirical screening procedure for new aggregate sources when no service records are available.

The soundness by freeze thaw (AASHTO T103) and potential expansion from hydrated reactions (ASTM D4792) are alternative screening tests for evaluating soundness. The durability of aggregates in portland cement concrete can be tested by rapid freezing and thawing (ASTM C666), critical dilation by freezing (ASTM C671), and by frost resistance of coarse aggregates in air-entrained concrete by critical dilation (ASTM C682).

5.5.3 ■ Toughness, Hardness, and Abrasion Resistance

The ability of aggregates to resist the damaging *effect of loads* is related to the hardness of the aggregate particles and is described as the *toughness* or *abrasion resistance*. The aggregate must resist crushing, degradation, and disintegration when stockpiled, mixed as either portland cement or asphalt concrete, placed and compacted, and exposed to loads.

FIGURE 5.8   Los Angeles abrasion machine. (Courtesy of Humboldt Mfg. Co.)

The Los Angeles abrasion test (ASTM C131, C535) evaluates the aggregates' toughness and abrasion resistance. In this test, aggregates blended to a fixed size distribution are placed in a large steel drum with standard sized steel balls that act as an abrasive charge (see Figure 5.8). The drum is rotated, typically for 500 revolutions. The material is recovered from the machine and passed through a sieve that retains all of the original material. The percentage weight loss is the LA abrasion number. This is an empirical test; that is, the test results do not have a scientific basis and are meaningful only when local experience defines the acceptance criteria.

5.5.4 ■ Absorption

Although aggregates are inert, they can capture water and asphalt binder in surface voids. The amount of water the aggregates absorb is important in the design of portland cement concrete, since moisture captured in the aggregate voids is not available to react with the cement or to improve the workability of the plastic concrete. There is

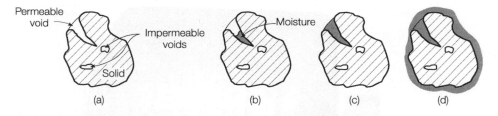

FIGURE 5.9   Voids and moisture absorption of aggregates: (a) bone dry, (b) air dry, (c) saturated surface-dry (SSD), and (d) moist.

no specific level of aggregate absorption that is desirable for aggregates used in portland cement concrete, but aggregate absorption must be evaluated to determine the appropriate amount of water to mix into the concrete.

Absorption is also important for asphalt concrete, since absorbed asphalt is not available to act as a binder. Thus, highly absorptive aggregates require greater amounts of asphalt binder, making the mix less economical. On the other hand, some asphalt absorption is desired to promote bonding between the asphalt and the aggregate. Therefore, low-absorption aggregates are desirable for asphalt concrete.

Figure 5.9 demonstrates the four moisture condition states for an aggregate particle. *Bone dry* means that the aggregate contains no moisture; this requires drying the aggregate in an oven to a constant mass. In an *air dry* condition, the aggregate may have some moisture but the saturation state is not quantified. In a *saturated surface–dry* (SSD) condition, the aggregate's voids are filled with moisture but the main surface area of the aggregate particles is dry. *Absorption* is defined as the moisture content in the SSD condition. Moist aggregates have moisture content in excess of the SSD condition. Free moisture is the moisture content in the aggregate in excess of the moisture content in the SSD condition.

The percent moisture content (MC) in the aggregate can be calculated as

$$\text{MC} = \frac{W_{\text{moist}} - W_{\text{dry}}}{W_{\text{dry}}} 100 \qquad (5.1)$$

where

W_{moist} = weight of moist aggregate
W_{dry} = weight of dry aggregate

Sample Problem 5.1

A sample of sand has the following properties:

Moist mass = 625.2 g
Dry mass = 589.9 g
Absorption = 1.6%

Determine (a) total moisture content and (b) free moisture content.

Solution

a. Mass of water $= 625.2 - 589.9 = 35.3$ g

Total moisture content $= \dfrac{35.3}{589.9} \times 100 = 6.0\%$

b. Free moisture $= 6.0 - 1.6 = 4.4\%$

Sample Problem 5.2

50 kg of gravel is mixed with 30 kg of sand. The gravel has a moisture content of 3.9% and absorption of 4.7%, whereas the sand has a moisture content of 3.5% and absorption of 4.9%. What is the amount of water required to increase the moisture contents of both gravel and sand to reach absorption? Why is it important to determine the amount of water required to increase the moisture content of aggregate to reach absorption?

Solution

Since moisture content and absorption are related to the aggregate dry weight, the first step of the solution is determine the dry weights of gravel and sand.

Dry weight of gravel $= 50/1.039 = 48.123$ kg

Dry weight of sand $= 30/1.035 = 28.986$ kg

Water required for gravel to reach absorption $= 48.123 (0.047 - 0.039) = 0.385$ kg

Water required for sand to reach absorption $= 28.986 (0.049 - 0.035) = 0.406$ kg

Water required for both gravel and sand to reach absorption $= 0.385 + 0.406 = 0.791$ kg

When mixing concrete, it is important to maintain the correct moisture content for strength and workability. Failure to properly compute absorbed and/or free water can affect all the desirable properties of concrete.

5.5.5 ■ Specific Gravity

The weight–volume characteristics of aggregates are not an important indicator of aggregate quality, but they are important for concrete mix design. *Density,* the mass per unit volume, could be used for these calculations. However, *specific gravity* (Sp. Gr.), the mass of a material divided by the mass of an equal volume of distilled water, is more commonly used. Four types of specific gravity are defined based on how voids in the aggregate particles are considered. Three of these types—*bulk-dry, bulk-saturated surface–dry*, and *apparent* specific gravity—are widely accepted and used in portland cement and asphalt concrete mix design. These are defined as

$$\text{Bulk Dry Sp. Gr.} = \frac{\text{Dry Weight}}{(\text{Total Particle Volume})\gamma_w} = \frac{W_s}{(V_s + V_i + V_p)\gamma_w} \tag{5.2}$$

$$\text{Bulk SSD Sp. Gr.} = \frac{\text{SSD Weight}}{(\text{Total Particle Volume})\gamma_w} = \frac{W_s + W_p}{(V_s + V_i + V_p)\gamma_w} \tag{5.3}$$

$$\text{Apparent Sp. Gr.} = \frac{\text{Dry Weight}}{(\text{Volume Not Accessible to Water})\gamma_w} = \frac{W_s}{(V_s + V_i)\gamma_w} \tag{5.4}$$

where

W_s = weight of solids
V_s = volume of solids
V_i = volume of water impermeable voids
V_p = volume of water permeable voids
W_p = weight of water in the permeable voids when the aggregate is in the SSD condition
γ_w = unit weight of water

Figure 5.10 shows that, when aggregates are mixed with asphalt binder, only a portion of the water-permeable voids are filled with asphalt. Hence, a fourth type of specific gravity—the *effective specific gravity*—is defined as

$$\text{Effective Sp.Gr.} = \frac{\text{Dry weight}}{(\text{Volume not accessible to asphalt})\gamma_w} = \frac{W_s}{(V_s + V_c)\gamma_w} \tag{5.5}$$

where V_c is volume of permeable voids not filled with asphalt cement.

At present, there is no standard method for directly determining the effective specific gravity of aggregates. The U.S. Corps of Engineers has defined a method for determining the effective specific gravity of aggregates that absorb more than 2.5% water.

The specific gravity and absorption of coarse aggregates are determined in accordance with ASTM C127. In this procedure, a representative sample of the aggregate is soaked for 24 hours and weighed suspended in water. The sample is then dried

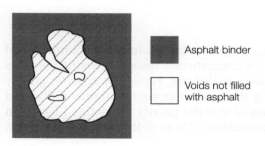

Asphalt binder

Voids not filled with asphalt

FIGURE 5.10 Aggregate particle submerged in asphalt cement; not all voids are filled with asphalt.

to the SSD condition and weighed. Finally, the sample is dried to a constant weight and weighed. The specific gravity and absorption are determined by

$$\text{Bulk Dry Sp. Gr.} = \frac{A}{B - C} \tag{5.6}$$

$$\text{Bulk SSD Sp. Gr.} = \frac{B}{B - C} \tag{5.7}$$

$$\text{Apparent Sp. Gr.} = \frac{A}{A - C} \tag{5.8}$$

$$\text{Absorption (\%)} = \frac{B - A}{A}(100) \tag{5.9}$$

where

A = dry weight
B = SSD weight
C = submerged weight

ASTM C128 defines the procedure for determining the specific gravity and absorption of fine aggregates. A representative sample is soaked in water for 24 hours and dried back to the SSD condition. A 500-g sample of the SSD material is placed in a *pycnometer,* a constant volume flask; water is added to the constant volume mark on the pycnometer and the weight is determined again. The sample is then dried and the weight is determined. The specific gravity and absorption are determined by

$$\text{Bulk Dry Sp. Gr.} = \frac{A}{B + S - C} \tag{5.10}$$

$$\text{Bulk SSD Sp. Gr.} = \frac{S}{B + S - C} \tag{5.11}$$

$$\text{Apparent Sp. Gr.} = \frac{A}{B + A - C} \tag{5.12}$$

$$\text{Absorption (\%)} = \frac{S - A}{A}(100) \tag{5.13}$$

where

A = dry weight
B = weight of the pycnometer filled with water
C = weight of the pycnometer filled with aggregate and water
S = saturated surface—dry weight of the sample

5.5.6 ▨ Bulk Unit Weight and Voids in Aggregate

The bulk unit weight of aggregate is needed for the proportioning of portland cement concrete mixtures. According to ASTM C29 procedure, a rigid container of known volume is filled with aggregate, which is compacted either by rodding, jigging, or shoveling. The bulk unit weight of aggregate (γ_b) is determined as

$$\gamma_b = \frac{W_s}{V} \qquad\qquad (5.14)$$

where W_s is the weight of aggregate (stone) and V is the volume of the container.

If the bulk dry specific gravity of the aggregate (G_{sb}) (ASTM C127 or C128) is known, the percentage of voids between aggregate particles can be determined as follows:

$$\% V_s = \frac{V_s}{V} \times 100 = \frac{W/\gamma_s}{W/\gamma_b} \times 100 = \frac{\gamma_b}{\gamma_s} \times 100 = \frac{\gamma_b}{G_{sb} \cdot \gamma_w} \times 100$$

$$\% \text{Voids} = 100 - \% V_s \qquad\qquad (5.15)$$

where

V_s = volume of aggregate
γ_s = unit weight of aggregate
γ_b = bulk unit weight of aggregate
γ_w = unit weight of water

Sample Problem 5.3

Coarse aggregate is placed in a rigid bucket and rodded with a tamping rod to determine its unit weight. The following data are obtained:

Volume of bucket = 0.00944 m³ = 9.44 L
Weight of empty bucket = 8.39 kg
Weight of bucket filled with dry rodded coarse aggregate = 25.36 kg

    a. Calculate the dry-rodded unit weight
    b. If the bulk dry specific gravity of the aggregate is 2.630, calculate the percent voids in the aggregate.

Solution

    a. Dry-rodded unit weight = (25.36 − 8.39)/0.00944 = 1798 kg/m³

    b. Percent volume of particles = $\dfrac{1798}{2.630 \times 1000} \times 100 = 68.4\%$

        Percent voids = 100 − 68.4 = 31.6%

5.5.7 ■ Strength and Modulus

The strength of portland cement concrete and asphalt concrete cannot exceed that of the aggregates. It is difficult and rare to test the strength of aggregate particles. However, tests on the parent rock sample or a bulk aggregate sample provide an indirect estimate of these values. Aggregate strength is generally important in high-strength concrete and in the surface course on heavily traveled pavements. The tensile strength of aggregates ranges from 0.7 to 16 MPa, while the compressive strength ranges from 35 to 350 MPa (Meininger and Nichols, 1990; The Aggregate Handbook, 2012). Field service records are a good indication of the adequacy of the aggregate strength.

The modulus of elasticity of aggregates is not usually measured. However, new mechanistic-based methods of pavement design require an estimate of the modulus of aggregate bases. The response of bulk aggregates to stresses is nonlinear and depends on the confining pressure on the material. Since the modulus is used for pavement design, dynamic loads are used in a test to simulate the magnitude and duration of stresses in a pavement base caused by a moving truck. During the test, as the stresses are applied to the sample, the deformation response has two components, a recoverable or resilient deformation and a permanent deformation. Only the resilient portion of the strain is used with the applied stress level to compute the modulus of the aggregate. Hence, the results are defined as the resilient modulus M_R.

In the resilient modulus test (AASHTO T292), a prepared cylindrical sample is placed in a triaxial cell, as shown in Figure 5.11. A specimen with large aggregates is typically 0.15 m in diameter by 0.30 m high, while soil samples are 71 mm in diameter by 142 mm high. The specimen is subjected to a specified confining pressure and a repeated axial load. Accurate transducers, such as LVDTs, measure the axial deformation. The test requires a determination of the modulus over a range of axial loads and confining pressures. The resilient modulus equals the repeated axial stress divided by the resilient strain for each combination of load level and confining pressure. The resilient modulus test requires the measurement of very small loads and deformations and is, therefore, difficult to perform. Currently, the test is mostly limited to research projects.

5.5.8 ■ Gradation

Gradation describes the particle size distribution of the aggregate. The particle size distribution is an important attribute of the aggregates. Large aggregates are economically advantageous in portland cement and asphalt concrete, as they have less surface area and, therefore, require less binder. However, large aggregate mixes, whether asphalt or portland cement concrete, are harsher and more difficult to work into place. Hence, construction considerations, such as equipment capability, dimensions of construction members, clearance between reinforcing steel, and layer thickness, limit the maximum aggregate size.

Sieve Analysis   Gradation is evaluated by passing the aggregates through a series of sieves, as shown in Figure 5.12 (ASTM C136, E11). The sieve retains particles larger than the opening, while smaller ones pass through. Metric sieve descriptions are based on the size of the square openings measured between the wires in millimeters. Sieves smaller than 0.6 mm can be described in either millimeters or micrometers.

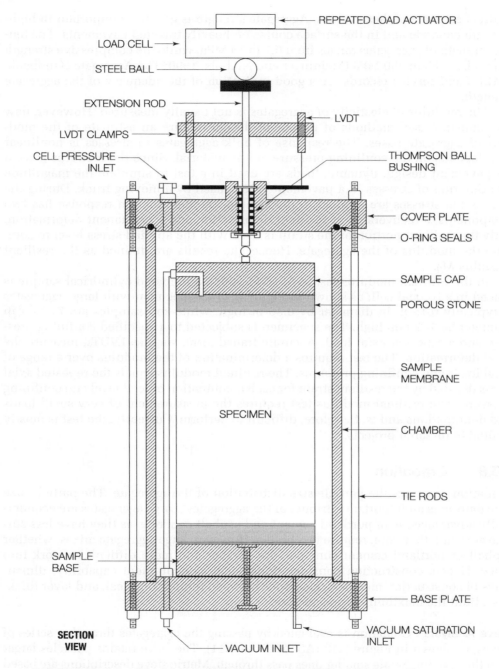

REPEATED LOAD ACTUATOR

LOAD CELL

STEEL BALL

EXTENSION ROD

LVDT

LVDT CLAMPS

CELL PRESSURE INLET

THOMPSON BALL BUSHING

COVER PLATE

O-RING SEALS

SAMPLE CAP

POROUS STONE

SAMPLE MEMBRANE

SPECIMEN

CHAMBER

TIE RODS

SAMPLE BASE

BASE PLATE

VACUUM SATURATION INLET

SECTION VIEW

VACUUM INLET

FIGURE 5.11   Triaxial chamber with external LVDTs and load cell.

FIGURE 5.12   Sieve shaker for large samples of aggregates.

Gradation results are described by the cumulative percentage of aggregates that either pass through or are retained by a specific sieve size. Percentages are reported to the nearest whole number, except that if the percentage passing the 0.075-mm sieve is less than 10%, it is reported to the nearest 0.1%. Gradation analysis results are generally plotted on a semilog chart, as shown in Figures 5.13 and A.24.

Aggregates are usually classified by size as coarse aggregates, fine aggregates, and mineral fillers (fines). ASTM defines coarse aggregate as particles retained on the 4.75-mm sieve, fine aggregate as those passing the 4.75-mm sieve, and mineral filler as material mostly passing the 0.075-mm sieve.

Maximum Density Gradation   The density of an aggregate mix is a function of the size distribution of the aggregates. In 1907, Fuller established the relationship for determining the distribution of aggregates that provides the maximum density or minimum amount of voids as

$$P_i = 100\left(\frac{d_i}{D}\right)^n \tag{5.16}$$

where

P_i = percent passing a sieve of size d_i
d_i = the sieve size in question
D = maximum size of the aggregate

The value of the exponent n recommended by Fuller is 0.5. In the 1960s, the Federal Highway Administration recommended a value of 0.45 for n and introduced the "0.45 power" gradation chart, Figures 5.14 and A.25, designed to produce a straight line for maximum density gradations (Federal Highway Administration 1988). Table 5.2 presents a sample calculation of the particle size distribution required for maximum density. Note that the gradation in Table 5.2 is plotted on both gradation charts in Figures 5.13 and 5.14.

TABLE 5.2 Sample Calculations of Aggregate Distribution Required to Achieve Maximum Density

Sieve Size, mm	$P_i = 100(d_i/D)^{0.45}$
25	100
19	88
12.5	73
9.5	64
4.75	47
2.36	34
0.60	19
0.30	14
0.075	7.3

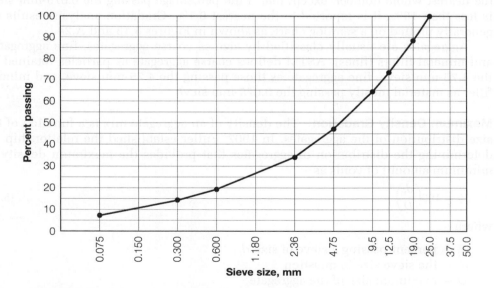

FIGURE 5.13 Semi-log aggregate gradation chart showing a gradation example. (See Table 5.2.)

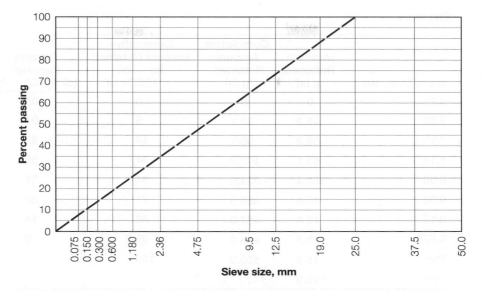

F I G U R E   5 . 1 4   Federal Highway Administration 0.45 power gradation chart showing the maximum density gradation for a maximum size of 25 mm. (See Table 5.2.)

Frequently, a *dense* gradation, but not necessarily the maximum possible density, is desired in many construction applications because of its high stability. Using a high-density gradation also means the aggregates occupy most of the volume of the material, limiting the binder content and thus reducing the cost. For example, aggregates for asphalt concrete must be dense but must also have sufficient voids in the mineral aggregate to provide room for the binder, plus room for voids in the mixture.

Sample Problem 5.4

A sieve analysis test was performed on a sample of fine aggregate and produced the following results:

Sieve, mm	4.75	2.36	2.00	1.18	0.60	0.30	0.15	0.075	pan
Amount retained, g	0	33.2	56.9	83.1	151.4	40.4	72.0	58.3	15.6

Calculate the percent passing each sieve, and draw a 0.45 power gradation chart with the use of a spreadsheet program. The specific steps or functions selected in the following description can vary somewhat depending on the version of the spreadsheet program being used.

Solution

Sieve Size, mm	Amount Retained, g (a)	Cumulative Amount Retained, g (b)	Cumulative Percent Retained (c) = (b) × 100/*Total*	Percent Passing* (d) = 100 − (c)
4.75	0	0	0	100
2.36	33.2	33.2	6	94
2.00	56.9	90.1	18	82
1.18	83.1	173.2	34	66
0.60	151.4	324.6	64	36
0.30	40.4	365.0	71	29
0.15	72.0	437.0	86	14
0.075	58.3	495.3	96.9	3.1
Pan	15.6	510.9	100	
Total	510.9			

*Percent passing is computed to a whole percent, except for the 0.075 mm material, which is computed to 0.1%.

The first step in drawing the graph is to compute the sieve size to the 0.45 power, using the metric sieve sizes.

1	2	3	4	5	6	7
Sieve Size (mm)	Sieve to the 0.45 Power	Percent Passing	Columns Used for Plotting the Power 45 Graph in Excel			
4.75	2.02	100	2.02	2.02	0	100
2.36	1.47	94	1.47	1.47	0	100
2	1.37	82	1.37	1.37	0	100
1.18	1.08	66	1.08	1.08	0	100
0.6	0.79	36	0.79	0.79	0	100
0.3	0.58	29	0.58	0.58	0	100
0.15	0.43	14	0.43	0.43	0	100
0.075	0.31	3.1	0.31	0.31	0	100

Then the *x–y* scatter graph function is used to plot the percent passing on the *y*-axis versus the sieve size to the 0.45 power.

Since the sieve size raised to the 0.45 power is not a meaningful number, the values on the axis are deleted by right clicking on the x-axis and selecting clear. However, the x-axis does need to be labeled. A good way to do this is to add a series for each of the sieve sizes. Right click on the graph and select **source data**. Starting with the row for the largest sieve, add a series to the graph. Use the sieve size as the label (4.75 for the first new series), then use columns 4 and 5 for the row as the x-values and columns 6 and 7 as the y-values. Repeat for each row, then select the **OK** button. For each of these sieve size series, the format for the series. Right click on a sieve

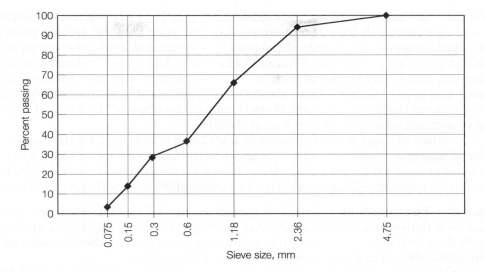

FIGURE SP5.4

size data point and select **format data series.** Under the patterns tab select a line color, black is always a good choice, and select **None** for the marker. Select the data labels tab and select the **series name.** Click on **OK** and repeat for each of the sieve sizes. There will now be vertical lines for each of the sieves and labels at the top and bottom of each line. Slowly click twice on the label at the top of the line; the label will be displayed as a textbox; delete it. Right click on the label at the bottom on the sieve line and select **format data label.** Select **below** for the label position and **-90 degrees** for the orientation. Click **OK** and repeat for each of the sieve size labels. The spreadsheet program may automatically set the y-axis to 120. Since it is impossible for the percent passing to be greater than 100%, the y-axis should be reset to 100%. Right click on the y-axis, select Format Axis. Under the Scale tab, enter 100 in the maximum value box. The resulting graph is as shown in Figure SP5.4.

Other Types of Gradation In addition to maximum density (i.e., *well-graded*), aggregates can have other characteristic distributions, as shown in Figure 5.15. A *one-sized* distribution has the majority of aggregates passing one sieve and being retained on the next smaller sieve. Hence, the majority of the aggregates have

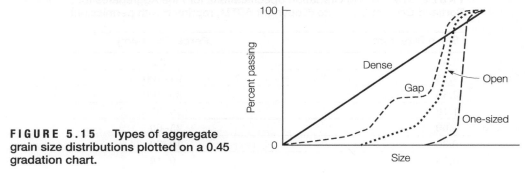

FIGURE 5.15 Types of aggregate grain size distributions plotted on a 0.45 gradation chart.

essentially the same diameter; their gradation curve is nearly vertical. One-sized graded aggregates will have good permeability, but poor stability, and are used in such applications as chip seals of pavements. *Gap-graded* aggregates are missing one or more sizes of material. Their gradation curve has a near horizontal section indicating that nearly the same portions of the aggregates pass two different sieve sizes. *Open-graded* aggregates are missing small aggregate sizes that would block the voids between the larger aggregate. Since there are a lot of voids, the material will be highly permeable but may not have good stability.

As shown in Table 5.3, the amount of fines has a major effect on the characteristics of aggregate base materials. Aggregates with the percentage of fines equal to the amount required for maximum density have excellent stability and density but may have a problem with permeability, frost susceptibility, handling, and cohesion.

Gradation Specifications Gradation specifications define maximum and minimum cumulative percentages of material passing each sieve. Aggregates are commonly described as being either coarse or fine, depending on whether the material is predominantly retained on or passes through a 4.75-mm sieve.

Portland cement concrete requires separate specifications for coarse and fine aggregates. The ASTM C33 specifications for fine aggregates for concrete are given in Table 5.4. Table 5.5 shows the ASTM C33 gradation specifications for coarse concrete aggregates.

TABLE 5.3 Effect of Amount of Fines on the Relative Properties of Aggregate Base Material

Characteristic	No Fines (Open or Clean)	Well-Graded (Dense)	Large Amount of Fines (Dirty or Rich)
Stability	Medium	Excellent	Poor
Density	Low	High	Low
Permeability	Permeable	Low	Impervious
Frost Susceptibility	No	Maybe	Yes
Handling	Difficult	Medium	Easy
Cohesion	Poor	Medium	Large

TABLE 5.4 ASTM Gradation Specifications for Fine Aggregates for Portland Cement Concrete (Copyright ASTM, reprinted with permission)

Sieve Size, mm	Percent Passing
9.5	100
4.75	95–100
2.36	80–100
1.18	50–85
0.60	25–60
0.30	10–30
0.15	0–10

TABLE 5.5 Coarse Aggregate Grading Requirements for Concrete (Reprinted, with permission, from ASTM C33, Table 2, copyright ASTM International, 100 Barr Harbor Drive, West Conshohocken, PA 19428).

Size No.	Nominal Size	Amounts Finer Than Each Laboratory Sieve (Square Openings), Weight Percent													
		100 mm	90 mm	75 mm	63 mm	50 mm	37.5 mm	25.0 mm	19.0 mm	12.5 mm	9.5 mm	4.75 mm	2.36 mm	1.18 mm	0.3 mm
1	90 to 37.5 mm	100	90 to 100	...	25 to 60	...	0 to 15	...	0 to 5	...	...	...	...	...	...
2	63 to 37.5 mm	...	...	100	90 to 100	35 to 70	0 to 15	...	0 to 5	...	...	...	...	...	...
3	50 to 25.0 mm	...	...	...	100	90 to 100	35 to 70	0 to 15	...	0 to 5	...	...	...	...	...
357	50 to 4.75 mm	...	...	...	100	95 to 100	...	35 to 70	...	10 to 30	...	0 to 5	...	...	...
4	37.5 to 19 mm	...	...	...	...	100	90 to 100	20 to 55	0 to 15	...	0 to 5	...	...	...	...
467	37.5 to 4.75 mm	...	...	...	...	100	95 to 100	...	35 to 70	...	10 to 30	0 to 5	0 to 5	...	...
5	25.0 to 12.5 mm	...	...	...	...	...	100	90 to 100	20 to 55	0 to 10	0 to 5	...	...	...	...
56	25.0 to 9.5 mm	...	...	...	...	...	100	90 to 100	40 to 85	10 to 40	0 to 15	0 to 5	...	...	...
57	25.0 to 4.75 mm	...	...	...	...	...	100	95 to 100	...	25 to 60	...	0 to 10	0 to 5	...	...
6	19.0 to 9.5 mm	...	...	...	...	...	...	100	90 to 100	20 to 55	0 to 15	0 to 5	...	...	...
67	19.0 to 4.75 mm	...	...	...	...	...	...	100	90 to 100	...	20 to 55	0 to 10	0 to 5	...	...
7	12.5 to 4.75 mm	...	...	...	...	...	...	...	100	90 to 100	40 to 70	0 to 15	0 to 5	...	...
8	9.5 to 2.36 mm	...	...	...	...	...	...	...	...	100	85 to 100	10 to 30	0 to 10	0 to 5	...
89	9.5 to 1.18 mm	...	...	...	...	...	...	...	...	100	90 to 100	20 to 55	5 to 30	0 to 10	0 to 5
9*	4.75 to 1.18 mm	...	...	...	...	...	...	...	...	...	100	85 to 100	10 to 40	0 to 10	0 to 5

*Size number 9 aggregate is defined in Terminology C 125 as a fine aggregate. It is included as a coarse aggregate when it is combined with a size number 8 material to create a size number 89, which is a coarse aggregate as defined by Terminology C 125.

Table 5.6 gives the aggregate grading requirements for Superpave asphalt mix designs (McGennis et al., 1995). These specifications define the range of allowable gradations for asphalt concrete for mix design purposes. Note that the percentage of material passing the 0.075-mm sieve, the fines or mineral filler, is carefully controlled for asphalt concrete due to its significance to the properties of the mix.

TABLE 5.6 Aggregate Grading Requirements for Superpave Asphalt Mix Designs (AASHTO MP-2)

Sieve Size, mm	Nominal Maximum Size (mm)					
	37.5	25	19	12.5	9.5	4.75
50	100	—	—	—	—	—
37.5	90–100	100	—	—	—	—
25	90 max	90–100	100	—	—	—
19	—	90 max	90–100	100	—	—
12.5	—	—	90 max	90–100	100	100
9.5	—	—	—	90 max	90–100	95–100
4.75	—	—	—	—	90 max	90–100
2.36	15–41	19–45	23–49	28–58	32–67	—
1.18	—	—	—	—	—	30–60
0.075	0.0–6.0	1.0–7.0	2.0–8.0	2.0–10.0	2.0–10.0	6.0–12.0

Once aggregate gradation from asphalt concrete mix design is established for a project, the contractor must produce aggregates that fall within a narrow band around the single gradation line established for developing the mix design. For example, the Arizona Department of Transportation will give the contractor full pay only if the gradation of the aggregates is within the following limits with respect to the accepted mix design gradations:

Sieve Size	Allowable Deviations for Full Pay
9.5 mm and larger	±3%
2.36 to 0.45 mm	±2%
0.075 mm	±0.5%

Fineness Modulus   The *fineness modulus* is a measure of the fine aggregates' gradation and is used primarily for portland cement concrete mix design. It can also be used as a daily quality control check in the production of concrete. The fineness modulus is one-hundredth of the sum of the cumulative percentage weight retained on the 0.15-, 0.3-, 0.6-, 1.18-, 2.36-, 4.75-, 9.5-, 19.0-, 37.5-, 75-, and 150-mm sieves. When the fineness modulus is determined for fine aggregates, sieves larger than 9.5 mm are not used. The fineness modulus for fine aggregates should be in the range of 2.3 to 3.1, with a higher number being a coarser aggregate. Table 5.7 demonstrates the calculation of the fineness modulus.

TABLE 5.7   Sample Calculation of Fineness Modulus

Sieve Size, mm	Percentage of Individual Fraction Retained, by Weight	Cumulative Percentage Retained by Weight	Percentage Passing by Weight
9.5	0	0	100
4.75	2	2	98
2.36	13	15	85
1.18	25	40	60
0.60	15	55	45
0.30	22	77	23
0.15	20	97	3
Pan	3	~~100~~	0
Total	100	$\sum = 286$	

Fineness Modulus = 286/100 = 2.86

Sample Problem 5.5

Calculate the fineness modulus of the sieve analysis results of Sample Problem 5.4.

Solution

According to the definition of fineness modulus, sieves 2.00 and 0.075 mm are not included.

$$\text{Fineness modulus} = \frac{6 + 34 + 64 + 71 + 86}{100} = 2.61$$

Blending Aggregates to Meet Specifications   A single aggregate source is generally unlikely to meet gradation requirements for portland cement or asphalt concrete mixes. Thus, blending of aggregates from two or more sources would be required to satisfy the specifications. Figure 5.16 shows a graphical method for selecting the combination of two aggregates to meet a specification. Table 5.8 presents the data used for Figure 5.16. Determining a satisfactory aggregate blend with the graphical method entails the following steps (The Asphalt Institute, 1995):

1. Plot the percentages passing through each sieve on the right axis for aggregate A and on the left axis for aggregate B, shown as open circles in Figure 5.16.
2. For each sieve size, connect the left and right axes.
3. Plot the specification limits of each sieve on the corresponding sieve lines; that is, a mark is placed on the 9.5-mm sieve line corresponding to 70% and 90% on the vertical axis, shown as closed circles in Figure 5.16.
4. Connect the upper- and lower-limit points on each sieve line.

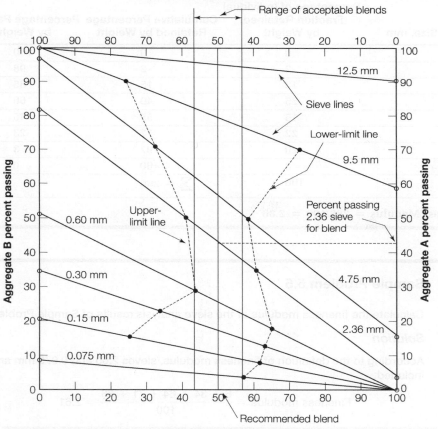

FIGURE 5.16 Graphical method for determining aggregate blend to meet gradation requirements. (See Table 5.8.)

TABLE 5.8   Example of Aggregate Blending Analysis by Graphical Method

Sieve Size, mm	19	12.5	9.5	4.75	2.36	0.60	0.30	0.15	0.075
Specification	100	80–100	70–90	50–70	35–50	18–29	13–23	8–16	4–10
Aggregate A	100	90	59	16	3	0	0	0	0
Aggregate B	100	100	100	96	82	51	36	21	9
Blend	100	95	80	56	43	26	18	11	4.5

Note: Numbers shown are percent passing each sieve.

5. Draw vertical lines through the rightmost point of the upper-limit line and the leftmost point of the lower-limit line. If the upper- and lower-limit lines overlap, no combination of the aggregates will meet specifications.
6. Any vertical line drawn between these two vertical lines identifies an aggregate blend that will meet the specification. The intersection with the upper axis defines the percentage of aggregate B required for the blend. The projection to the lower axis defines the percentage of aggregate A required.
7. Projecting intersections of the blend line and the sieve lines horizontally gives an estimate of the gradation of the blended aggregate. Figure 5.16 shows that a 50-50 blend of aggregates A and B will result in a blend with 43% passing through the 2.36-mm sieve. The gradation of the blend is shown in the last line of Table 5.8.

When more than two aggregates are required, the graphical procedure can be repeated in an iterative manner. However, a trial and error process is generally used to determine the proportions. The basic equation for blending is

$$P_i = A_i a + B_i b + C_i c + \ldots \tag{5.17}$$

where

$$
\begin{aligned}
P_i &= \text{percent blend material passing sieve size } i \\
A_i, B_i, C_i &= \text{percent of aggregates } A, B, C, \ldots \text{ passing sieve } i \\
a, b, c &= \text{decimal fractions by weight of aggregates } A, B, \text{ and } C \text{ used in the} \\
&\quad \text{blend, where the total is 1.00}
\end{aligned}
$$

Table 5.9 demonstrates these calculations for two aggregate sources. The table shows the required specification range and the desired (or target) gradation, usually the midpoint of the specification. A trial percentage of each aggregate source is assumed and is multiplied by the percentage passing each sieve. These gradations are added to get the composite percentage passing each sieve for the blend. The gradation of the blend is compared to the specification range to determine if the blend is acceptable. With practice, blends of four aggregates can readily be resolved. These calculations are easily performed using a spreadsheet program.

TABLE 5.9   Example of Aggregate Blending Analysis by Iterative Method

Sieve Size, mm	12.5	9.5	4.75	2.00	0.425	0.180	0.075
Specification	100	95–100	70–85	55–70	20–40	10–20	4–8
Target gradation	100	98	77.5	62.5	30	15	6
% Passing Aggregate A (A_i)	100	100	98	90	71	42	19
% Passing Aggregate B (B_i)	100	94	70	49	14	2	1
30% A_i $(a.A_i)$	30	30	29.4	27	21.3	12.6	5.7
70% B_i $(b.B_i)$	70	65.8	49	34.3	9.8	1.4	0.7
Blend (P_i)	100	96	78	61	31	14	6.4

Properties of Blended Aggregates   When two or more aggregates from different sources are blended, some of the properties of the blend can be calculated from the properties of the individual components. With the exception of specific gravity and density, the properties of the blend are the simple weighted averages of the properties of the components. This relationship can be expressed as

$$X = P_1 X_1 + P_2 X_2 + P_3 X_3 \ldots \tag{5.18a}$$

where

$X =$ composite property of the blend
$X_1, X_2, X_3 =$ properties of fractions 1, 2, 3
$P_1, P_2, P_3 =$ decimal fractions by weight of aggregates 1, 2, 3 used in the blend, where the total is 1.00

This equation applies to properties such as angularity, absorption, strength, and modulus.

Sample Problem 5.6

Coarse aggregates from two stockpiles having coarse aggregate angularity (crushed faces) of 40% and 90% were blended at a ratio of 30:70 by weight, respectively. What is the percent of crushed faces of the aggregate blend?

Solution

Crushed faces of the blend = (0.3)(40) + (0.7)(90) = 75%

Equation 5.18a is used for properties that apply to the whole aggregate materials in all stockpiles that are blended. However, some properties apply to either coarse aggregate only or fine aggregate only. Therefore, the percentage of coarse or fine aggregate in each stockpile has to be considered. The relationship in this case is expressed as

$$X = \frac{(x_1 P_1 p_1 + x_2 P_2 p_2 + \ldots x_n P_n p_n)}{(P_1 p_1 + P_2 p_2 + \ldots P_n p_n)} \tag{5.18b}$$

where

X = the test value for the aggregate blend
x_i = the test result for stockpile i
P_i = the percent of stockpile i in the blend
p_i = the percent of stockpile i that either passes or is retained on the dividing sieve

Sample Problem 5.7

Aggregates from two stockpiles, A and B having coarse aggregate angularity (crushed faces) of 40% and 90% were blended at a ratio of 30:70 by weight, respectively. The percent material passing the 4.75 mm sieve was 25% and 55% for stockpiles A and B, respectively. What is the percent of crushed faces of the aggregate blend?

Solution
Crushed faces of the blend =

$$X = \frac{(x_1 P_1 p_1 + x_{20} P_2 p_2 + \cdots + x_n P_n p_n)}{(P_1 p_1 + P_2 p_2 + \cdots + P_n p_n)}$$

$$= \frac{(40 \times 30 \times (100 - 25) + 90 \times 70 \times (100 - 55))}{(30 \times (100 - 25) + 70 \times (100 - 55))} = 69\%$$

Note that the percentage of coarse aggregate in each stockpile was calculated by subtracting the percentage passing the 4.75 mm sieve from 100.

Asphalt concrete mix design requires that the engineer knows the composite specific gravity of all aggregates in the mix. The composite specific gravity of a mix of different aggregates is obtained by the formula

$$G = \frac{1}{\dfrac{P_1}{G_1} + \dfrac{P_2}{G_2} + \dfrac{P_3}{G_3} + \ldots} \tag{5.19}$$

where

$$G = \text{composite specific gravity}$$
$$G_1, G_2, G_3 = \text{specific gravities of fractions 1, 2, and 3}$$
$$P_1, P_2, P_3 = \text{decimal fractions by weight of aggregates 1, 2, and 3 used in the}$$
$$\text{blend, where the total is 1.00}$$

Note that Equation 5.19 is used only to obtain the combined specific gravity and density of the blend, whereas Equation 5.18 is used to obtain other combined properties.

Sample Problem 5.8

Aggregates from three sources having bulk specific gravities of 2.753, 2.649, and 2.689 were blended at a ratio of 70:20:10 by weight, respectively. What is the bulk specific gravity of the aggregate blend?

Solution

$$G = \frac{1}{\dfrac{0.7}{2.753} + \dfrac{0.2}{2.649} + \dfrac{0.1}{2.689}} = 2.725$$

5.5.9 ■ Cleanness and Deleterious Materials

Since aggregates are a natural product, there is the potential they can be contaminated by clay, shale, organic matter, and other deleterious materials, such as coal. Owner agencies use specifications and test methods to evaluate and limit the amount of deleterious materials. A deleterious substance is any material that adversely affects the quality of portland cement or asphalt concrete made with the aggregate. Table 5.10 identifies the main deleterious substances in aggregates and their effects on portland cement concrete. In asphalt concrete, deleterious substances are clay lumps, soft or friable particles, and coatings. These substances decrease the adhesion between asphalt and aggregate particles.

The Superpave mix design method requires use of the Sand Equivalency test, AASHTO T176, to limit the clay content of fine aggregates used in asphalt concrete. Figure 5.17 shows the concept of the test and Figure 5.18 shows the apparatus. An 85 ml sample of sand and flocculating agent is poured into a graduated cylinder. The cylinder is agitated to permit all the clay in the sample to be released from the sand and go into suspension in the flocculating agent. The sample is allowed to rest for a specified period of time. Since the sand settles much faster than the

TABLE 5.10   Main Deleterious Substances and Their Effects on Portland Cement Concrete

Substance	Harmful Effect
Organic impurities	Delay settling and hardening, may reduce strength gain, may cause deterioration
Minus 0.075 mm	Weaken bond, may increase water materials requirements
Coal, lignite or other low-density materials	Reduce durability, may cause popouts or stains
Clay lumps and friable particles	Popouts, reduce durability and wear resistance
Soft particles	Reduce durability and wear resistance, popouts

clay, a distinct interface develops between the sand and clay. The heights to the top of the sand (h_{sand}) and the top of the clay (h_{clay}) are measured, and the sand equivalency (*SE*) is computed as:

$$SE = {100h_{sand}}/{h_{clay}} \qquad (5.20)$$

The sand equivalent value is an indication of aggregate cleanness, where a larger SE value indicates cleaner aggregate.

5.5.10 ▮ Alkali–Aggregate Reactivity

Some aggregates react with portland cement, harming the concrete structure. The most common reaction, particularly in humid and warm climates, is between the active silica constituents of an aggregate and the alkalis in cement (sodium oxide,

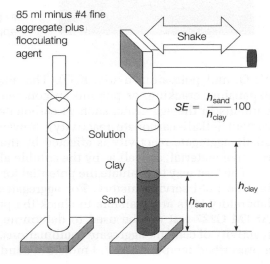

85 ml minus #4 fine aggregate plus flocculating agent

Shake

$$SE = \frac{h_{sand}}{h_{clay}} 100$$

Solution

Clay

Sand

h_{clay}

h_{sand}

FIGURE 5.17   Concept of sand equivalency test.

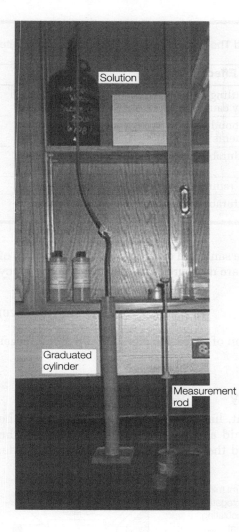

Graduated
cylinder

Measurement
rod

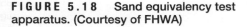

FIGURE 5.18   Sand equivalency test
apparatus. (Courtesy of FHWA)

Na_2O, and potassium oxide, K_2O). The alkali–silica reaction results in excessive expansion, cracking, or popouts in concrete, as shown in Figure 5.19. Other constituents in the aggregate, such as carbonates, can also react with the alkali in the cement (alkali–carbonate reactivity); however, their reaction is less harmful. The alkali–aggregate reactivity is affected by the amount, type, and particle size of the reactive material, as well as by the soluble alkali and water content of the concrete.

The best way to evaluate the potential for alkali–aggregate reactivity is by reviewing the field service history. For aggregates without field service history, several laboratory tests are available to check the potential alkali–aggregate reactivity. The ASTM C227 test can be used to determine the potentially expansive alkali–silica reactivity of cement–aggregate combinations. In this test, a mortar bar is stored under a prescribed temperature and moisture conditions and its expansion is determined.

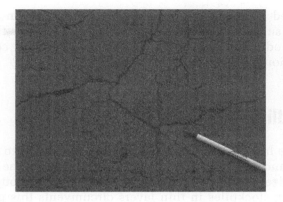

FIGURE 5.19 Example of cracking in concrete due to alkali–silica reactivity.

The quick chemical test (ASTM C289) can be used to identify potentially reactive siliceous aggregates. ASTM C586 is used to determine potentially expansive carbonate rock aggregates (alkali–carbonate reactivity).

If alkali-reactive aggregate must be used, the reactivity can be minimized by limiting the alkali content of the cement. The reactivity can also be reduced by keeping the concrete structure as dry as possible. Fly ash, ground granulated blast furnace slag, silica fume, or natural pozzolans can be used to control the alkali–silica reactivity. Lithium-based admixtures have also been used for the same purpose. Finally, replacing about 30% of a reactive sand–gravel aggregate with crushed limestone (limestone sweetening) can minimize the alkali reactivity (Kosmatka et al., 2011).

5.5.11 ▪ Affinity for Asphalt

Stripping, or moisture-induced damage, is a separation of the asphalt film from the aggregate through the action of water, reducing the durability of the asphalt concrete and resulting in pavement failure. The mechanisms causing stripping are complex and not fully understood. One important factor is the relative affinity of the aggregate for either water or asphalt. *Hydrophilic* (water-loving) aggregates, such as silicates, have a greater affinity for water than for asphalt. They are usually acidic in nature and have a negative surface charge. Conversely, *hydrophobic* (water-repelling) aggregates have a greater affinity for asphalt than for water. These aggregates, such as limestone, are basic in nature and have a positive surface charge. Hydrophilic aggregates are more susceptible to stripping than hydrophobic aggregates. Other stripping factors include porosity, absorption, and the existence of coatings and other deleterious substances.

Since stripping is the result of a compatibility problem between the asphalt and the aggregate, tests for stripping potential are performed on the asphalt concrete mix. Early compatibility tests submerged the sample in either room-temperature water (ASTM D1664) or boiling water (ASTM D3625); after a period of time, the technician observed the percentage of particles stripped from the asphalt. More recent procedures subject asphalt concrete to cycles of freeze–thaw conditioning. The strength or modulus of the specimens is measured and compared with the values of unconditioned specimens (ASTM D1075).

5.6   Handling Aggregates

Aggregates must be handled and stockpiled in such a way as to minimize segregation, degradation, and contamination. If aggregates roll down the slope of the stockpile, the different sizes will segregate, with large stones at the bottom and small ones at the top. Building stockpiles in thin layers circumvents this problem. The drop height should be limited to avoid breakage, especially for large aggregates. Vibration and jiggling on a conveyor belt tends to work fine material downward while coarse particles rise. Segregation can be minimized by moving the material on the belt frequently (up and down, side to side, and in and out) or by installing a baffle plate, rubber sleeve, or paddle wheel at the end of the belt to remix coarse and fine particles. Rounded aggregates segregate more than crushed aggregates. Also, large aggregates segregate more readily than smaller aggregates. Therefore, different sizes should be stockpiled and batched separately. Stockpiles should be separated by dividers or placed in bins to avoid mixing and contamination (Figure 5.20) (Meininger and Nichols, 1990).

5.6.1   Sampling Aggregates

In order for any of the tests described in this chapter to be valid, the sample of material being tested must represent the whole population of materials that is being quantified with the test. This is a particularly difficult problem with aggregates due

FIGURE 5.20   Aggregate bins used to stockpile aggregates with different sizes.

to potential segregation problems. Samples of aggregates can be collected from any location in the production process, that is, from the stockpile, conveyor belts, or from bins within the mixing machinery (ASTM D75). Usually, the best location for sampling the aggregate is on the conveyor belt that feeds the mixing plant. However, since the aggregate segregates on the belt, the entire width of the belt should be sampled at several locations or times throughout the production process. The samples would then be mixed to represent the entire lot of material.

Sampling from stockpiles must be performed carefully to minimize segregation. Typically, aggregate samples are taken from the top, middle, and bottom of the stockpile and then combined. Before taking the samples, discard the 75 to 150 mm material at the surface. A board shoved vertically into the pile just above the sampling point aids in preventing rolling of coarse aggregates during sampling. Samples are collected using a square shovel and are placed in sample bags or containers and labeled.

Sampling tubes 1.8 m long and 30 mm in diameter are used to sample fine aggregate stockpiles. At least five samples should be collected from random locations in the stockpile. These samples are then combined before laboratory testing.

Field sample sizes are governed by the nominal maximum size of aggregate particles (ASTM D75). Larger-sized aggregates require larger samples to minimize segregation errors. Field samples are typically larger than the samples needed for testing. Therefore, field samples must be reduced using sample splitters (Figure 5.21) or by quartering (Figure 5.22) (ASTM C702).

FIGURE 5.21   Aggregate sample splitter.

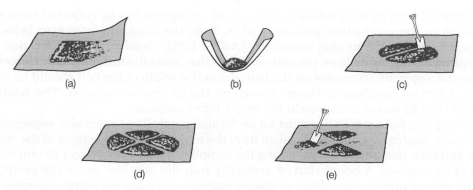

FIGURE 5.22 Steps for reducing the sample size by quartering: (a) mixing by rolling on blanket, (b) forming a cone after mixing, (c) flattening the cone and quartering, (d) finishing quartering, and (e) retaining opposite quarters (the other two quarters are rejected). (ASTM C702. Copyright ASTM. Reprinted with permission)

5.7 Aggregates Sustainability

5.7.1 ▨ LEED Considerations

Potential LEED credits can be earned through the reclamation of aggregate bearing products, for example, concrete, during the demolition phase of the project. This produces both a recyclable material and a reduction of materials placed in construction material landfills. During construction, LEED credits may be earned by specifying the use of either primary or secondary recycled materials. Primary recycled materials include crushed products that were primarily aggregate based, such as crushed concrete. Secondary products are waste products from other industries that are used as aggregate, such as slag from iron and steel production.

5.7.2 ▨ Other Sustainability Considerations

The NSSGA has published guideline principles for sustainable aggregate operations. These principles identify sustainability as a business approach that integrates environmental, social, and economic aspects to provide a long-term supply of aggregates for society. Extraction of aggregates from a quarry involves the disruption of the natural landscape. States have implemented codes for the development, operation, and closure (or reclamation) of quarries (e.g., WVDEP, 2001). Applications for development of a quarry in West Virginia must include a plan for reclamation of the site into a useable function such as a recreational area or wildlife preserve.

Summary

Aggregates are widely used as a base material for foundations and as an ingredient in portland cement concrete and asphalt concrete. While the geological classification of aggregates gives insight into the properties of the material, the suitability of a specific source of aggregates for a particular application requires testing and evaluation. The most significant attributes of aggregates include the gradation, specific gravity, shape and texture, and soundness. When used in concrete, the compatibility of the aggregate and the binder must be evaluated.

Questions and Problems

5.1   What are the three mineralogical or geological classifications of rocks, and how are they formed?

5.2   Discuss five different desirable characteristics of aggregate used in portland cement concrete.

5.3   Discuss five different desirable characteristics of aggregate used in asphalt concrete.

5.4   Using the Internet, find the aggregate specifications for your state and identify the limitations for deleterious materials in aggregates used for asphalt concrete.

5.5   The shape and surface texture of aggregate particles are important for both portland cement concrete and asphalt concrete.
        a. For preparing PCC, would you prefer round and smooth aggregate or rough and angular aggregate? Briefly explain why (no more than two lines).
        b. For preparing HMA, would you prefer round and smooth aggregate or rough and angular aggregate? Briefly explain why (no more than two lines).

5.6   Define the following terms:
        a. Saturated surface-dry condition of aggregates
        b. Absorption of aggregates
        c. Free water on aggregates
        d. Referring to Figure 5.9, which portion of the water in the aggregate reacts with the cement in the PCC mix?

5.7   A sample of wet aggregate weighed 297.2 N. After drying in an oven, this sample weighed 281.5 N. The absorption of this aggregate is 2.5%. Calculate the percent of free water in the original wet sample.

5.8   46.5 kg of fine aggregate is mixed with 72.3 kg of coarse aggregate. The fine aggregate has a moisture content of 2.0% and absorption of 3.4%, whereas the coarse aggregate has a moisture content of 1.3% and absorption of 3.8%. What is the amount of water required to increase the moisture contents of both fine and coarse aggregates to reach absorption? Why is it important to determine the amount of water required to increase the moisture content of aggregate to reach absorption?

5.9 Three samples of fine aggregate have the properties shown in Table P5.9.

TABLE P5.9

Measure	Sample		
	A	B	C
Wet Mass (g)	521.0	522.4	523.4
Dry Mass (g)	491.6	491.7	492.1
Absorption (%)	2.5	2.4	2.3

Determine in percentage: (a) total moisture content and (b) free moisture content for each sample and the average of the three samples.

5.10 Samples of coarse aggregate from a stockpile are brought to the laboratory for determination of specific gravities. The following weights are found:
Mass of moist aggregate sample as brought to the laboratory: 5,298 g
Mass of oven dried aggregate: 5,216 g
Mass of aggregates submerged in water: 3,295 g
Mass of SSD (Saturated Surface Dry) Aggregate: 5,227 g
Find
a. The aggregate bulk dry specific gravity
b. The aggregate apparent specific gravity
c. The moisture content of stockpile aggregate (report as a percent)
d. Absorption (report as percent)

5.11 Base course aggregate has a target dry density of 1917 kg/m^3 in place. It will be laid down and compacted in a rectangular street repair area of 600 m × 15 m × 0.15 m. The aggregate in the stockpile contains 3.1% moisture. If the required compaction is 95% of the target, how many tons of aggregate will be needed?

5.12 Calculate the percent voids between aggregate particles that have been compacted by rodding, if the dry-rodded unit weight is 1410 kg/m^3 and the bulk dry specific gravity is 2.701.

5.13 Calculate the percent voids between aggregate particles that have been compacted by rodding, if the dry-rodded unit weight is 1161 kg/m^3 and the bulk dry specific gravity is 2.639.

5.14 Coarse aggregate is placed in a rigid bucket and rodded with a tamping rod to determine its unit weight. The following data are obtained:
Volume of bucket = 14 L
Weight of empty bucket = 9.21 kg
Weight of bucket filled with dry rodded coarse aggregate:
Trial 1 = 34.75 kg
Trial 2 = 34.06 kg

Trial 3 = 35.74 kg
a. Calculate the average dry-rodded unit weight
b. If the bulk dry specific gravity of the aggregate is 2.620, calculate the percent voids between aggregate particles for each trial.

5.15   The following laboratory tests are performed on aggregate samples:
a. Specific gravity and absorption
b. Soundness
c. Sieve analysis test
What is the significance and use of each of these tests?

5.16   Students in the materials lab performed the specific gravity and absorption test (ASTM C127) on coarse aggregate and they obtained following data:
Dry weight = 3862.1 g
SSD weight = 3923.4 g
Submerged weight = 2452.1 g

Calculate the specific gravity values (dry bulk, SSD, and apparent) and the absorption of the coarse aggregate.

5.17   The specific gravity and absorption test (ASTM C128) was performed on fine aggregate and the following data were obtained:
Mass of SSD sand = 500.0 g
Mass of pycnometer with water only = 623.0 g
Mass of pycnometer with sand and water = 938.2 g
Mass of dry sand = 495.5 g

Calculate the specific gravity values (dry bulk, SSD, and apparent) and the absorption of the fine aggregate.

5.18   Referring to ASTM specification C33 (Table 5.5), what are the maximum sieve size and the nominal maximum sieve size (traditional definition) for each of the standard sizes Numbers 357, 57, and 8?

5.19   Calculate the sieve analysis shown in Table P5.19 and plot on a semilog gradation paper. What is the maximum size? What is the nominal maximum size?

TABLE 5.19

Sieve Size, mm	Amount Retained, g	Cumulative Amount Retained, g	Cumulative Percent Retained	Percent Passing
25	0			
9.5	47.1			
4.75	239.4			
2.00	176.5			
0.425	92.7			
0.075	73.5			
Pan	9.6			

5.20   Calculate the sieve analysis shown in Table P5.20, and plot on a 0.45 power gradation chart. What is the maximum size? What is the nominal maximum size?

TABLE P5.20

Sieve Size, mm	Amount Retained, g	Cumulative Amount Retained, g	Cumulative Percent Retained	Percent Passing
37.5	0			
25	315			
19	782			
9.5	1493			
4.75	677			
0.60	1046			
0.075	1502			
Pan	45			

5.21   A sieve analysis test was performed on a sample of aggregate and produced the results shown in Table P5.21.

TABLE P5.21

Sieve Size, mm	Amount Retained, g	Sieve Size, mm	Amount Retained, g
25	0	1.18	891.5
19	376.7	0.60	712.6
12.5	888.4	0.30	625.2
9.5	506.2	0.15	581.5
4.75	1038.4	0.075	242.9
2.36	900.1	Pan	44.9

Calculate the percent passing through each sieve. Plot the percent passing versus sieve size on:
a. a semilog gradation chart, and
b. a 0.45 gradation chart (Figure A.25).

What is the maximum size? What is the nominal maximum size?

5.22   A sieve analysis test was performed on a sample of coarse aggregate and produced the results in Table P5.22.
a. Calculate the percent passing through each sieve.
b. What is the maximum size?
c. What is the nominal maximum size?
d. Plot the percent passing versus sieve size on a semilog gradation chart.

e. Plot the percent passing versus sieve size on a 0.45 gradation chart (Figure A.25).

f. Referring to Table 5.5 (ASTM C33), what is the closest size number and does it meet the gradation for that standard size?

TABLE P5.22

Sieve Size, mm	Amount Retained, g
75.0	0
50.0	0
37.5	1678
25.0	7212
19.0	5443
12.5	6124
9.5	12111
4.75	4581
Pan	590

5.23 Draw a graph to show the cumulative percent passing through the sieve versus sieve size for well-graded, gap-graded, open-graded, and one-sized aggregates. What is the format name of this graph?

5.24 Three aggregates are to be mixed together in the following ratio:
   Aggregate A: 35%
   Aggregate B: 40%
   Aggregate C: 25%

For each aggregate, the percent passing a set of five sieves is shown in Table P5.24.

TABLE P5.24

Sieve Size, mm	% Passing Aggregate A	% Passing Aggregate B	% Passing Aggregate C
9.5	85	50	40
4.75	70	35	30
0.6	35	20	5
0.3	25	13	1
0.15	17	7	0

5.25 Referring to Table 5.6, plot the specification limits of Superpave gradation with a nominal maximum size of 19 mm on a 0.45 gradation chart (Figure A.25). What is the maximum aggregate size of this gradation? Is this a dense, open, or gap graded gradation? Why?

5.26   Referring to the aggregate gradations A, B, and C in Figure P5.26, answer the
following questions:
   a. What is the maximum size of each gradation?
   b. What is the nominal maximum size of each gradation?
   c. Classify each gradation as dense, open, or gap indicating the reason for
each classification.

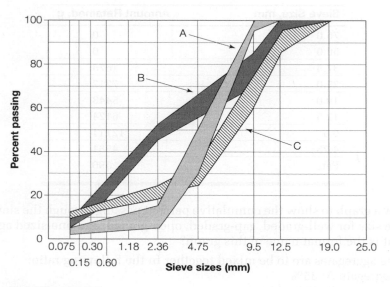

FIGURE P5.26

5.27   Table P5.27 shows the grain size distributions of aggregates A, B, and C. The
three aggregates must be blended at a ratio of 15:25:60 by weight, respectively.
Using a spreadsheet program, determine the grain size distribution of the blend.

TABLE P5.27

	Percent Passing								
	Sieve Size, mm								
	25	19	12.5	9.5	4.75	1.18	0.60	0.30	0.15
Aggregate A	100	100	100	77	70	42	34	28	20
Aggregate B	100	85	62	43	24	13	7	0	0
Aggregate C	100	100	84	51	29	19	18	14	9

5.28   Table P5.28 shows the grain size distributions of two aggregates A and B.

TABLE P5.28

Sieve Size, mm	25	19	12.5	9.5	4.750	2.36	1.18	0.600	0.300	0.150	0.075
% Passing Agg. A	100	92	76	71	53	38	32	17	10	5	3.0
% Passing Agg. B	100	100	92	65	37	31	30	29	28	21	15.4

Answer the following questions and show all calculations:
   a. What are the maximum sizes of aggregates A and B?
   b. Is aggregate A well graded? Why?
   c. Is aggregate B well graded? Why?

Note: In order to answer questions b and c, you need to plot the gradations on 0.45 gradation charts. You may create 0.45 power gradation charts using Excel or you may manually plot the data on a copy of Figure A.25.

Determine the percent passing each sieve for the blended aggregate.

5.29   Table P5.29 shows the grain size distribution for two aggregates and the specification limits for an asphalt concrete. Determine the blend proportion required to meet the specification and the gradations of the blend. On a semilog gradation graph, plot the gradations of aggregate A, aggregate B, the selected blend, and the specification limits.

TABLE P5.29

	Percent Passing								
	Sieve Size, mm								
	19	12.5	9.5	4.75	2.36	0.60	0.30	0.15	0.075
Specification limits	100	80–100	70–90	50–70	35–50	18–29	13–23	8–16	4–10
Aggregate A	100	85	55	20	2	0	0	0	0
Aggregate B	100	100	100	85	67	45	32	19	11

5.30 Laboratory specific gravity and absorption tests are run on two coarse aggregate sizes, which have to be blended. The results are as follows:
    Aggregate A: Bulk specific gravity = 2.814; absorption = 0.4%
    Aggregate B: Bulk specific gravity = 2.441; absorption = 5.2%
  a. What is the specific gravity of a mixture of 35% aggregate A and 65% aggregate B by weight?
  b. What is the absorption of the mixture?

5.31 Table P5.31 shows the grain size distribution for two aggregates and the specification limits for an asphalt concrete. Determine the blend proportion required to meet the specification and the gradations of the blend. On a semilog gradation graph, plot the gradations of aggregate A, aggregate B, the selected blend, and the specification limits.

TABLE P5.31

	Percent Passing								
	Sieve Size, mm								
	19	**12.5**	**9.5**	**4.75**	**2.36**	**0.60**	**0.30**	**0.15**	**0.075**
Spec. limits	100	80–100	70–90	50–70	35–50	18–29	13–23	8–16	4–10
Aggregate A	100	100	100	79	66	41	38	21	12
Aggregate B	100	92	54	24	3	1	0	0	0

5.32 Make a spread sheet blend template program to perform a blend gradation analysis for up to four stockpiles with the ability to produce automatically both semilog and power 0.45 gradation charts. Use Table P5.32 to demonstrate the template.

5.33 Laboratory specific gravity and absorption tests are run on two coarse aggregate sizes, which have to be blended. The results are as follows:
    Aggregate A: Bulk specific gravity = 2.491; absorption = 0.8%
    Aggregate B: Bulk specific gravity = 2.773; absorption = 4.6%
  a. What is the specific gravity of a mixture of 60% aggregate A and 40% aggregate B by weight?
  b. What is the absorption of the mixture?

TABLE P5.32

Blend	Aggregate				Blend
	A	**B**	**C**	**D**	**Blend**
Blend	20%	15%	25%	40%	
Sieve Size, mm	Percent Passing				
37.5	100	100	100	100	
25	100	100	100	100	
19	95	100	100	100	
12.5	89	100	100	100	
9.5	50	85	100	100	
4.75	10	55	100	100	
2.36	2	15	88	98	
1.18	2	5	55	45	
0.6	2	3	35	34	
0.3	2	2	22	25	
0.15	2	2	15	14	
0.075	2	1	6	5	

5.34 The mix design for an asphalt concrete mixture requires 2 to 6% minus 0.075 mm. The three aggregates shown in Table P.5.34 are available.

TABLE P5.34

	Minus 0.075 mm
Coarse	0.5%
Intermediate	1.5%
Fine Aggregate	11.5%

Considering that approximately equal amounts of coarse and intermediate aggregate will be used in the mix, what is the percentage of fine aggregate that will give a resulting minus 0.075 mm in the mixture in the middle of the range, about 4%?

5.35 Define the fineness modulus of aggregate. What is it used for?

5.36 Calculate the fineness modulus of aggregate A in Problem 5.28. Is your answer within the typical range for fineness modulus? If not, why not?

5.37 Calculate the fineness modulus of aggregate B in Problem 5.29. Is your answer within the typical range for fineness modulus? If not, why not? (Note that the percent passing the 1.18-mm sieve is not given and must be estimated.)

5.38 A portland cement concrete mix requires mixing sand having a gradation following the midpoint of the ASTM gradation band (Table 5.4) and gravel having a gradation following the midpoint of size number 467 of the ASTM gradation band (Table 5.5) at a ratio of 2:3 by weight. On a 0.45 power gradation chart, plot the gradations of the sand, gravel, and the blend. Is the gradation of the blend well graded? If not, what would you call it?

5.39 Discuss the effect of the amount of material passing the 0.075-mm sieve on the stability, drainage, and frost susceptibility of aggregate base courses.

5.40 Aggregates from three sources having the properties shown in Table P5.40 were blended at a ratio of 55:25:20 by weight. Determine the properties of the aggregate blend.

TABLE P5.40

Property	Aggregate 1	Aggregate 2	Sand
Coarse aggregate angularity, percent crushed faces	100	87	N/A
Bulk specific gravity	2.631	2.711	2.614
Apparent specific gravity	2.732	2.765	2.712

5.41 Aggregates from three sources having the properties shown in Table P5.41 were blended at a ratio of 25:60:15 by weight. Determine the properties of the aggregate blend.

TABLE P5.41

Property	Aggregate 1	Aggregate 2	Sand
Coarse aggregate angularity, percent crushed faces	73	95	N/A
Bulk specific gravity	2.774	2.390	2.552
Apparent specific gravity	2.810	2.427	2.684

5.42   A contractor is considering using three stockpiles for a Superpave mix design that requires evaluation of three different blend percentages. Determine the blended specific gravity for the three blends shown in Table P5.42.

TABLE P5.42

Material	G_{sb}	Stockpile 1	Stockpile 2	Stockpile 3
Crushed limestone	2.702	45%	55%	50%
Blast furnace slag	2.331	35%	20%	30%
Sand	2.609	20%	25%	20%

What is the bulk specific gravity of the blended aggregates?

5.43   What is alkali–silica reactivity? What kinds of problems are caused by ASR? Mention two ways to minimize ASR.

5.44   What are the typical deleterious substances in aggregates that affect portland cement concrete? Discuss these effects.

5.45   Review ASTM D75 and summarize the following:
a. Sampling aggregates from conveyer belts.
b. Sampling aggregates from stockpiles.
c. Sampling aggregates from roadway bases and subbases.
d. If the nominal maximum aggregate size is 19 mm, what is the minimum sample size that needs to be obtained?

5.46   Prepare a 2-page single-spaced report on the local aggregates used in your state using online materials, state specifications, and personal contacts. Topics covered in the report should include aggregate types and sources, typical gradations for concrete and asphalt mixes, need for crushing and blending, and laboratory tests used for quality control and quality assurance. Show the list of references used and cite them.

5.8   References

Arizona Department of Transportation. *Standard Specifications for Roads and Bridge Construction.* Phoenix, AZ: Arizona Department of Transportation, 2008.

The Asphalt Institute. *Mix Design Methods for Asphalt Concrete and Other Hot-Mix Types.* 6th ed. Manual Series No. 2 (MS-2). Lexington, KY: The Asphalt Institute, 1995.

Federal Highway Administration. *Asphalt Concrete Mix Design and Field Control.* Technical Advisory T 5040.27. Washington, DC: Federal Highway Administration, 1988.

Goetz, W. H. and L. E. Wood. Bituminous Materials and Mixtures. *Highway Engineering Handbook,* Section 18. New York: McGraw-Hill, 1960.

Kosmatka, S. H., and M. L. Wilson. *Design and Control of Concrete Mixtures.* 15th ed. (Revised). Skokie, IL: Portland Cement Association, 2011.

McGennis, R. B., et al. *Background of Superpave Asphalt Mixture Design and Analysis.* Publication No. FHWA-SA-95-003. Washington, DC: Federal Highway Administration, 1995.

Meininger, R. C. and F. P. Nichols. *Highway Materials Engineering, Aggregates and Unbound Bases.* Publication No. FHWA-HI-90-007, NHI Course No. 13123. Washington, DC: Federal Highway Administration, 1990.

NSSGA, National Stone, Sand and Gravel Association, web page http://www.nssga.org/nssga-recognized-as-leader-in-sustainability/, 2015.

The Aggregate Handbook, 2nd ed., National Stone, Sand & Gravel Association, Washington, DC, 2012.

WVDEP, *Quarry Handbook,* West Virginia Department of Environmental Protection, Division of Mining and Reclamation, Charleston, WV 2001.

CHAPTER

6

PORTLAND CEMENT, MIXING WATER, AND ADMIXTURES

Portland cement concrete is the most widely used manufactured construction material in the world. The importance of concrete in our daily lives cannot be overstated. It is used in structures such as buildings, bridges, tunnels, dams, factories, pavements, and playgrounds. Portland cement concrete consists of portland cement, aggregates, water, air voids, and, in many cases, admixtures. This chapter covers the topics of portland cement, mixing water, and admixtures; Chapter 7 describes portland cement concrete. The main use of portland cement is to make portland cement concrete, but it can be used for other purposes, such as stabilizing soils and aggregate bases for highway construction.

There are many types of concrete, based on different cements. However, portland cement concrete is so prevalent that, unless otherwise identified, the term concrete is always assumed to mean portland cement concrete. Portland cement was patented by Joseph Aspdin in 1824 and was named after the limestone cliffs on the Isle of Portland in England (Kosmatka et al., 2011).

Portland cement is an instant glue (just add water) that bonds aggregates together to make portland cement concrete. Materials specialists concerned with the selection, specification, and quality control of civil engineering projects should understand the production, chemical composition, hydration rates, and physical properties of portland cement.

6.1 Portland Cement Production

Production of portland cement starts with two basic raw ingredients: a calcareous material and an argillaceous material. The calcareous material is a calcium oxide, such as limestone, chalk, or oyster shells. The argillaceous material is a combination of silica and alumina that can be obtained from clay, shale, and blast furnace slag. As shown in Figure 6.1, these materials are crushed and then stored in silos. The raw materials, in the desired proportions, are passed through a grinding mill, using either a wet or dry process. The ground material is stored until it can be sent

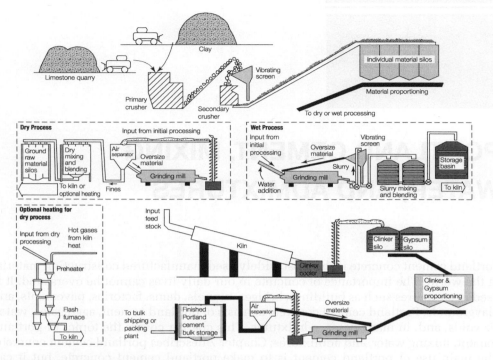

FIGURE 6.1 Steps in the manufacture of portland cement.

to the kiln. Modern dry process cement plants use a heat recovery cycle to preheat the ground material, or feed stock, with the exhaust gas from the kiln. In addition, some plants use a flash furnace to further heat the feed stock. Both the preheater and flash furnace improve the energy efficiency of cement production. In the kiln, the raw materials are melted at temperatures of 1400 to 1650°C, changing the raw materials into cement *clinker*. The clinker is cooled and stored. The final process involves grinding the clinker into a fine powder. During grinding, a small amount of gypsum is added to regulate the setting time of the cement in the concrete.

The finished product may be stored and transported in either bulk or sacks. The cement can be stored for long periods of time, provided it is kept dry.

6.2 Chemical Composition of Portland Cement

The raw materials used to manufacture portland cement are lime, silica, alumina, and iron oxide. These raw materials interact in the kiln, forming complex chemical compounds. *Calcination* in the kiln restructures the molecular composition, producing four main compounds, as shown in Table 6.1.

TABLE 6.1   Main Compounds of Portland Cement

Compound	Chemical Formula	Common Formula*	Usual Range by Weight (%)
Tricalcium silicate	$3\,CaO \cdot SiO_2$	C_3S	45–60
Dicalcium silicate	$2\,CaO \cdot SiO_2$	C_2S	15–30
Tricalcium aluminate	$3\,CaO \cdot Al_2O_3$	C_3A	6–12
Tetracalcium aluminoferrite	$4\,CaO \cdot Al_2O_3 \cdot Fe_2O_3$	C_4AF	6–8

*The cement industry commonly uses shorthand notation for chemical formulas:
C = Calcium oxide, S = silicon dioxide, A = Aluminum oxide, and F = Iron oxide.

C_3S and C_2S, when hydrated, provide the desired characteristics of the concrete. Alumina and iron, which produce C_3A and C_4AF, are included with the other raw materials to reduce the temperature required to produce C_3S from 2000°C to 1350°C. This saves energy and reduces the cost of producing the portland cement.

In addition to these main compounds, there are minor compounds, such as magnesium oxide, titanium oxide, manganese oxide, sodium oxide, and potassium oxide. These minor compounds represent a few percent by weight of cement. The term minor compounds refers to their quantity and not to their importance. In fact, two of the minor compounds, sodium oxide (Na_2O) and potassium oxide (K_2O), are known as alkalis.

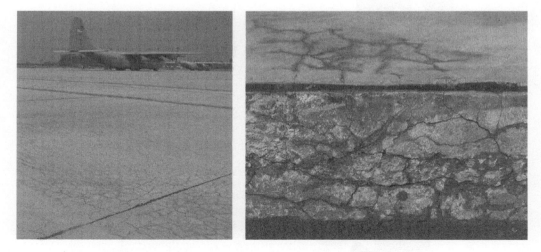

FIGURE 6.2   Cracks in concrete pavement due to alkali–silica reactivity.

6.3 Fineness of Portland Cement

Fineness of cement particles is an important property that must be carefully controlled. Since hydration starts at the surface of cement particles, the finer the cement particles, the larger the surface area and the faster the hydration. Therefore, finer material results in faster strength development and a greater initial heat of hydration. Increasing fineness beyond the requirements for a type of cement increases production costs and can be detrimental to the quality of the concrete.

The maximum size of the cement particles is 0.09 mm; 85% to 95% of the particles are smaller than 0.045 mm, and the average diameter is 0.01 mm. A kilogram of portland cement has approximately 7 trillion particles with a total surface area of about 300 to 400 m^2. The total surface area per unit weight is a function of the size of the particles and is more readily measured. Thus, particle size specifications are defined in terms of the surface area per unit weight.

Fineness of cement is usually measured indirectly by measuring the surface area with the Blaine air permeability apparatus (ASTM C204) or the Wagner turbidimeter apparatus (ASTM C115). In the Blaine test (Figure 6.3), the surface area of the cement particles in cm^2/g is determined by measuring the air permeability of a cement sample and relating it to the air permeability of a standard material. The Wagner turbidimeter determines the surface area by measuring the rate of sedimentation of cement suspended in kerosene. The finer the cement particles,

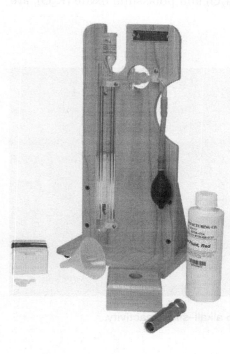

FIGURE 6.3   Blaine air permeability apparatus. (Courtesy of Humboldt Mfg. Co.)

the slower the sedimentation. Both the Blaine and Wagner tests are indirect measurements of surface area and use somewhat different measurement principles. Therefore, tests on a single sample of cement will produce different results. Fineness can also be measured by determining the percent passing the 0.045 mm sieve (ASTM C430).

6.4 Specific Gravity of Portland Cement

The specific gravity of cement is needed for mixture-proportioning calculations. The specific gravity of portland cement (without voids between particles) is about 3.15 and can be determined according to ASTM C188. The density of the bulk cement (including voids between particles) varies considerably, depending on how it is handled and stored. For example, vibration during transportation of bulk cement consolidates the cement and increases its bulk density. Thus, cement quantities are specified and measured by weight rather than volume.

6.5 Hydration of Portland Cement

Hydration is the chemical reaction between the cement particles and water. The features of this reaction are the change in matter, the change in energy level, and the rate of reaction. The primary chemical reactions are shown in Table 6.2. Since portland cement is composed of several compounds, many reactions are occurring concurrently.

The hydration process occurs through two mechanisms: through-solution and topochemical. The through-solution process involves the following steps (Mehta and Monteiro, 2013):

1. dissolution of anhydrous compounds into constituents
2. formation of hydrates in solution
3. precipitation of hydrates from the supersaturated solution

The through-solution mechanism dominates the early stages of hydration. Topochemical hydration is a solid-state chemical reaction occurring at the surface of the cement particles.

The aluminates hydrate much faster than the silicates. The reaction of tricalcium aluminate with water is immediate and liberates large amounts of heat. Gypsum is used to slow down the rate of aluminate hydration. The gypsum goes into the solution quickly, producing sulfate ions that suppress the solubility of the aluminates. The balance of aluminate to sulfate determines the rate of *setting* (solidification). Cement paste that sets at a normal rate requires low concentrations of both aluminate and sulfate ions. The cement paste will remain workable for about 45 minutes; thereafter, the paste starts to stiffen as crystals displace the water in the pores. The paste begins to solidify within 2 to 4 hours after the water is added to the cement. If there

TABLE 6.2 Primary Chemical Reactions During Cement Hydration

$2 (3\ CaO \cdot SiO_2)$ + 6 H_2O = $3\ CaO \cdot 2\ SiO_2 \cdot 3\ H_2O$ + $3\ Ca(OH)_2$
Tricalcium silicate   Water   Calcium silicate hydrates   Calcium hydroxide

$2 (2\ CaO \cdot SiO_2)$ + 4 H_2O = $3\ CaO \cdot 2\ SiO_2 \cdot 3\ H_2O$ + $Ca(OH)_2$
Dicalcium silicate   Water   Calcium silicate hydrates   Calcium hydroxide

$3\ CaO \cdot Al_2O_3$ + 12 H_2O + $Ca(OH)_2$ = $3\ CaO \cdot Al_2O_3 \cdot Ca(OH)_2 \cdot 12\ H_2O$
Tricalcium aluminate   Water   Calcium hydroxide   Calcium aluminate hydrate

$4\ CaO \cdot Al_2O_3 \cdot Fe_2O_3$ + 10 H_2O + $2\ Ca(OH)_2$ = $6\ CaO \cdot Al_2O_3 \cdot Fe_2O_3 \cdot 12\ H_2O$
Tetracalcium aluminoferrite   Water   Calcium Hydroxide   Calcium aluminoferrite hydrate

$3\ CaO \cdot Al_2O_3$ + 10 H_2O + $CaSO_4 \cdot 2\ H_2O$ = $3\ CaO \cdot Al_2O_3 \cdot CaSO_4 \cdot 12\ H_2O$
Tricalcium aluminate   Water   Gypsum   Calcium monosulfoaluminate hydrate

is an excess of both aluminate and sulfate ions, the workability stage may only last for 10 minutes and setting may occur in 1 to 2 hours. If the availability of aluminate ions is high, and sulfates are low, either a *quick set* (10 to 45 minutes) or *flash set* (less than 10 minutes) can occur. Finally, if the aluminate ions availability is low and the sulfate ions availability is high, the gypsum can recrystalize in the pores within 10 minutes, producing a flash set. Flash set is associated with large heat evolution and poor ultimate strength (Mehta and Monteiro, 2013).

Calcium silicates combine with water to form calcium-silicate-hydrate, C-S-H. The crystals begin to form a few hours after the water and cement are mixed and can be developed continuously as long as there are unreacted cement particles and free water. C-S-H is not a well-defined compound. The calcium-to-silicate ratio varies between 1.5 and 2.0, and the structurally combined water content is more variable.

As shown in Table 6.2, the silicate hydration produces both C-S-H and calcium hydroxide. Complete hydration of C_3S produces 61% C-S-H and 39% calcium hydroxide; hydration of C_2S results in 82% C-S-H and 18% calcium hydroxide. Since C-S-H is what makes the hydrated cement paste strong, the ultimate strength of the concrete is enhanced by increasing the content of C_2S relative to the amount of C_3S. Furthermore, calcium hydroxide is susceptible to attack by sulfate and acidic waters.

C_3S hydrates more rapidly than C_2S, contributing to the final set time and early strength gain of the cement paste. The rate of hydration is accelerated by sulfate ions in solution. Thus, a secondary effect of the addition of gypsum to cement is to increase the rate of development of the C-S-H.

6.5.1 ▨ Structure Development in Cement Paste

The sequential development of the structure in a cement paste is summarized in Figure 6.4. The process begins immediately after water is added to the cement [Figure 6.4(a)]. In less than 10 minutes, the water becomes highly alkaline. As the cement particles hydrate, the volume of the cement particle reduces, increasing the space between the particles. During the early stages of hydration, weak bonds can form, particularly from the hydrated C_3A [Figure 6.4(b)]. Further hydration stiffens the mix and begins locking the structure of the material in place [Figure 6.4(c)]. Final set occurs when the C-S-H phase has developed a rigid structure, all components of the paste lock into place and the spacing between grains increases as the grains are consumed by hydration [Figure 6.4(d)]. The cement paste continues hardening and gains strength as hydration continues [Figure 6.4(e)]. Hardening develops rapidly at early ages and continues, as long as unhydrated cement particles and free water exist. However, the rate of hardening decreases with time.

6.5.2 ▨ Evaluation of Hydration Progress

Several methods are available to evaluate the progress of cement hydration in hardened concrete. These include measuring the following properties (Neville, 1996):

1. the heat of hydration
2. the amount of calcium hydroxide in the paste developed due to hydration

The C-S-H phase is initially formed. C_3A forms a gel fastest.

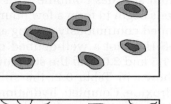

(a)

The volume of cement grain decreases as a gel forms at the surface. Cement grains are still able to move independently, but as hydration grows, weak interlocking begins. Part of the cement is in a thixotropic state; vibration can break the weak bonds.

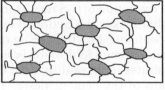

(b)

The initial set occurs with the development of a weak skeleton in which cement grains are held in place.

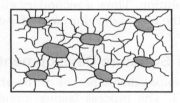

(c)

Final set occurs as the skeleton becomes rigid, cement particles are locked in place, and spacing between cement grains increases due to the volume reduction of the grains.

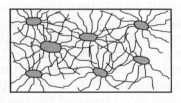

(d)

Spaces between the cement grains are filled with hydration products as cement paste develops strength and durability.

(e)

FIGURE 6.4   Development of structure in the cement paste: (a) initial C-S-H phase, (b) forming of gels, (c) initial set—development of weak skeleton, (d) final set—development of rigid skeleton, and (e) hardening (Hover and Phillco, 1990).

3. the specific gravity of the paste
4. the amount of chemically combined water
5. the amount of unhydrated cement paste using X-ray quantitative analysis
6. the strength of the hydrated paste, an indirect measurement

6.6   Voids in Hydrated Cement

Due to the random growth of the crystals and the different types of crystals, voids are left in the paste structure as the cement hydrates. Concrete strength, durability, and volume stability are greatly influenced by voids. Two types of voids are formed during hydration: the interlayer hydration space and capillary voids.

Interlayer hydration space occurs between the layers in the C-S-H. The space thickness is between 0.5 and 2.5 nm, which is too small to affect the strength. It can, however, contribute 28% to the porosity of the paste. Water in the interparticle space is strongly held by hydrogen bonds but can be removed when humidity is less than 11%, resulting in considerable shrinkage.

Capillary voids are the result of the hydrated cement paste having a lower bulk specific gravity than the cement particles. The amount and size of capillary voids depends on the initial separation of the cement particles, which is largely controlled by the ratio of water to cement paste. For a highly hydrated cement paste in which a minimum amount of water was used, the capillary voids will be on the order of 10 to 50 nm. A poorly hydrated cement produced with excess water can have capillary voids on the order of 3 to 5 μm. Capillary voids greater than 50 nm decrease strength and increase permeability. Removal of water from capillary voids greater than 50 nm does not cause shrinkage, whereas removal of water from the smaller voids causes shrinkage.

In addition to the interlayer space and capillary voids, air can be trapped in the cement paste during mixing. The *trapped air* reduces strength and increases permeability. However, well-distributed, minute air bubbles can greatly increase the durability of the cement paste. Hence, as described later in this chapter, admixtures are widely used to *entrain air* into the cement paste.

6.7   Properties of Hydrated Cement

The proper hydration of portland cement is a fundamental quality control issue for cement producers. While specifications control the quality of the portland cement, they do not guarantee the quality of the concrete made with the cement. Mix design, quality control, and the characteristics of the mixing water and aggregates also influence the quality of the concrete. Properties of the hydrated cement are evaluated with either *cement paste* (water and cement) or *mortar* (paste and sand).

6.7.1   Setting

Setting refers to the stiffening of the cement paste or the change from a plastic state to a solid state. Although with setting comes some strength, it should be distinguished from *hardening*, which refers to the strength gain in a set cement paste. Setting is usually described by two levels: initial set and final set. The definitions of the initial and final sets are arbitrary, based on measurements by either the Vicat apparatus (ASTM C191) or the Gillmore needles (ASTM C266).

The Vicat test (Figure 6.5) requires that a sample of cement paste be prepared, using the amount of water required for normal consistency according to a specified procedure. The 1-mm diameter needle is allowed to penetrate the paste for 30 seconds and the amount of penetration is measured. The penetration process is repeated every 15 minutes (every 10 minutes for Type III cement) until a penetration of 25 mm or less is obtained. By interpolation, the time when a penetration of 25 mm occurs is determined and recorded as the initial set time. The final set time is when the needle does not penetrate visibly into the paste.

Similar to the Vicat test, the Gillmore test (Figure 6.6) requires that a sample of cement paste of normal consistency be prepared. A pat with a flat top is molded, and the initial Gillmore needle is applied lightly to its surface. The application process is repeated until the pat bears the force of the needle without appreciable indentation, and the elapsed time is recorded as the initial set time. This process is then repeated with the final Gillmore needle and the final set time is recorded. Due to the differences in the test apparatuses and procedures, the Vicat and Gillmore tests produce different results for a single sample of material.

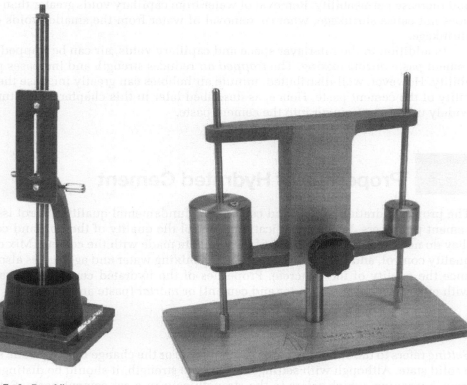

FIGURE 6.5 Vicat set time apparatus. (Courtesy of Humboldt Mfg. Co.)

FIGURE 6.6 Gillmore set time apparatus. (Courtesy of Humboldt Mfg. Co.)

The initial set time must allow for handling and placing the concrete before stiffening. The maximum final set time is specified and measured to ensure normal hydration. During cement manufacturing, gypsum is added to regulate the setting time. Other factors that affect the set time include the fineness of the cement, the water–cement ratio, and the use of admixtures.

If the cement is exposed to humidity during storage, a *false set* might occur in which the cement stiffens within a few minutes of being mixed, without the evolution of much heat. To resolve this problem, the cement paste can be vigorously remixed, without adding water, in order to restore plasticity of the paste and to allow it to set in a normal manner without losing strength. A false set is different than a quick set and a flash set mentioned earlier; a false set can be remedied by remixing, whereas a quick set and a flash set cannot be remedied.

6.7.2 ▓ Soundness

Soundness of the cement paste refers to its ability to retain its volume after setting. Expansion after setting, caused by delayed or slow hydration or other reactions, could result if the cement is unsound. The autoclave expansion test (Figure 6.7) (ASTM C151) is used to check the soundness of the cement paste. In this test, cement

FIGURE 6.7   Cement autoclave expansion apparatus. (Courtesy of Humboldt Mfg. Co.)

paste bars are subjected to heat and high pressure, and the amount of expansion is measured. ASTM C150 limits autoclave expansion to 0.8%.

6.7.3 ■ Compressive Strength of Mortar

Compressive strength of mortar is measured by preparing 50-mm cubes and subjecting them to compression according to ASTM C109. The mortar is prepared with cement, water, and standard sand (ASTM C778). Minimum compressive strength values are specified by ASTM C150 for different cement types at different ages. The compressive strength of mortar cubes is proportional to the compressive strength of concrete cylinders. However, the compressive strength of the concrete cannot be predicted accurately from mortar cube strength, since the concrete strength is also affected by the aggregate characteristics, the concrete mixing, and the construction procedures.

6.8   Water–Cement Ratio

In 1918, Abrams found that the ratio of the weight of water to the weight of cement, *water–cement ratio,* influences all the desirable qualities of concrete. For fully compacted concrete made with sound and clean aggregates, strength and other desirable properties are improved by reducing the weight of water used per unit weight of cement. This concept is frequently identified as Abrams' law.

Supplementary cementitious materials, such as fly ash, slag, silica fume, and natural *pozzolans*, have been used as admixtures in recent years to alter some of the properties of portland cement concrete. Therefore, the term water–cement ratio has been expanded to *water–cementitious materials ratio* to include these cementitious materials. Usually, the terms "water–cement ratio" and "water–cementitous materials ratio" are used interchangeably.

The amount of water added to concrete must be sufficient for hydration, water absorbed by the aggregate, water lost through evaporation and absorption into the forms, and additional water needed for the workability of the plastic concrete. Hydration is the chemical reaction between cement and water. There are two components to hydration: chemical and physical. Chemical bonding requires approximately 0.22 to 0.25 kg of water per 1 kg of cement. Historically, this was considered the minimum water to cement ratio required for hydration. More recently, the role of physical bonding of water to the cement gel, termed gel-water has been recognized. Cement cannot fully hydrate without gel-water. Each kilogram of cement requires 0.19 kg of gel-water (Aitcin, 2008). Hence, the PCA now recommends a minimum of 0.40.

The water added for workability is in excess of the water needed for hydration and causes the development of capillary voids in the concrete. These voids increase the porosity and permeability of the concrete and reduce strength. Figure 6.8 shows a typical relationship between the water–cement ratio and compressive strength. It is easy to see that increasing the water–cement ratio decreases the compressive strength

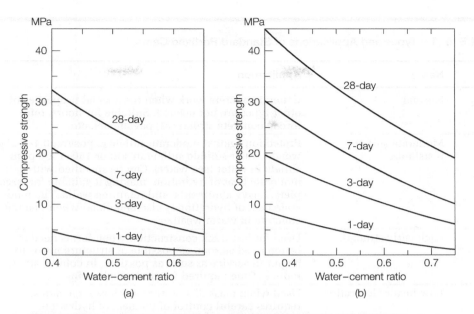

FIGURE 6.8 Typical age-strength relationships of concrete based on compression tests of 0.15 × 0.30 m cylinders, using Type I portland cement and moist-curing at 21°C: (a) air entrained concrete and (b) non-air entrained concrete. (Courtesy of Portland Cement Association)

of the concrete for various curing times. A low water–cement ratio also increases resistance to weathering, provides a good bond between successive concrete layers, provides a good bond between concrete and steel reinforcement, and limits volume change due to wetting and drying. Air-entrained concrete includes an air entraining agent, an admixture, which is used to increase the concrete's resistance to freezing and thawing, as will be discussed later in this chapter. Curing maintains satisfactory moisture content and temperature in the hardened concrete for a definite period of time to allow for hydration (see Chapter 7).

6.9 Types of Portland Cement

Different concrete applications require cements with different properties. Some applications require rapid strength gain to expedite construction. Other applications require a low heat of hydration to control volume change and associated shrinkage cracking. In some cases, the concrete is exposed to sulfates (SO_4), which can deteriorate normal portland cement concrete. Fortunately, these situations can be accommodated by varying the raw materials used to produce the cement, thereby altering the ratios of the four main compounds of portland cement listed in Table 6.1. The

TABLE 6.3   Types and Applications of Standard Portland Cement

Type	Name	Application
I	Normal	General concrete work when the special properties of other types are not needed. Suitable for floors, reinforced concrete structures, pavements, etc.
II	Moderate sulfate resistance	Protection against moderate sulfate exposure, 0.1–0.2% weight water-soluble sulfate in soil or 150–1500 ppm sulfate in water (sea water). Can be specified with a moderate heat of hydration, making it suitable for large piers, heavy abutments, and retaining walls. The moderate heat of hydration is also beneficial when placing concrete in warm weather.
III	High early strength	Used for fast-track construction when forms need to be removed as soon as possible or structure needs to be put in service as soon as possible. In cold weather, reduces time required for controlled curing.
IV	Low heat of hydration	Used when mass of structure, such as large dams, requires careful control of the heat of hydration.
V	High sulfate resistance	Protection from severe sulfate exposure, 0.2–2.0% weight water-soluble sulfate in soils or 1500–10,800 ppm sulfate in water.

rate of hydration can also be altered by varying the fineness of the cement produced in the final grinding mill. Cement is classified into five standard types, as well as other special types.

6.9.1 ■ Standard Portland Cement Types

Table 6.3 describes the five standard types of portland cement (Types I through V) specified by ASTM C150. In addition to these five types, air entrainers can be added to Type I, II, and III cements during manufacturing, producing Types IA, IIA, and IIIA, which provide better resistance to freeze and thaw than do non–air-entrained cements. The use of air-entrained cements (Types IA, IIA, and IIIA) has diminished, due to improved availability and reliability of the air entrainer admixtures that can be added during concrete mixing. The uses and effects of air entrainers will be described in the section on admixtures. The ASTM physical property specifications of the standard cement types are shown in Table 6.4 and the chemical limits are shown in Table 6.5.

The existence of an ASTM specification for a type of cement does not guarantee that cement's availability. Type I cement is widely available and represents most of the United States' cement production. Type II is the second most available type. Cements can be manufactured that meet all the requirements of both Types I and II; these are labeled Type I/II. Type III cement represents about 4% of U.S. production.

TABLE 6.4 Standard Physical Requirements of Portland Cement (Reprinted, with permission, from ASTM C150, Table 3, copyright ASTM International, 100 Barr Harbor Drive, West Conshohocken, PA 19428)

Cement Type[A]	Applicable Test Method	I	IA	II	IIA	II(MH)	II(MH)A	III	IIIA	IV	V
Air content of mortar,[B] volume %:	C185										
max		12	22	12	22	12	22	12	22	12	12
min		...	16	...	16	...	16	...	16	...	...
Fineness, specific surface, m²/kg Air permeability test	C204										
min		260	260	260	260	260	260	...	...	260	260
max		...	...	...	...	430[C]	430[C]	...	...	430	...
Autoclave expansion, max, %	C151	0.80	0.80	0.80	0.80	0.80	0.80	0.80	0.80	0.80	0.80
Strength, not less than the values shown for the ages Indicated as follows:[D]											
Compressive strength, MPa:	C109										
1 day		...	...	...	...	...	...	12.0	10.0	...	...
3 days		12.0	10.0	10.0	8.0	10.0 7.0[E]	8.0 6.0[E]	24.0	19.0	...	8.0
7 days		19.0	16.0	17.0	14.0	17.0 12.0[E]	14.0 9.0[E]	...	...	7.0	15.0
28 days		...	...	...	...	...	...	...	...	17.0	21.0
Time of setting: *Vi test*[F]	C191										
Time of setting, min, not less than		45	45	45	45	45	45	45	45	45	45
Time of setting, min, not more than		375	375	375	375	375	375	375	375	375	375

[A] With the exception of Type I, not all types are carried in stock in some areas.
[B] Compliance with the requirements of this specification does not necessarily ensure that the desired air content will be obtained in concrete.
[C] Maximum fineness limits do not apply if the sum of $C_3S + 4.75C_3A$ is less than or equal to 90.
[D] The strength at any specified test age shall be not less than that attained at any previous specified test age.
[E] When the optional heat of hydration in Table 4 (ASTM C150) is specified.
[F] The time of setting is that described as initial setting time in Test Method C191.

TABLE 6.5 Standard Composition Requirements for Portland Cement (Reprinted, with permission, from ASTM C150, Table 1, copyright ASTM International, 100 Barr Harbor Drive, West Conshohocken, PA 19428.)

Cement Type[A]	Applicable Test Method	I and IA	II and IIA	II(MH) and II(MH)A	III and IIIA	IV	V
Aluminum oxide (Al_2O_3), max, %	C114	...	6.0	6.0	...	...	...
Ferric oxide (Fe_2O_3), max, %	C114	...	6.0[B]	6.0[B, C]	...	6.5	...
Magnesium oxide (MgO), max, %	C114	6.0	6.0	6.0	6.0	6.0	6.0
Sulfur trioxide (SO_3),[D] max, %	C114						
When (C_3A)[E] is 8% or less		3.0	3.0	3.0	3.5	2.3	2.3
When (C_3A)[E] is more than 8 %		3.5	F	F	4.5	F	F
Loss on ignition, max, %	C114	3.0	3.0	3.0	3.0	2.5	3.0
Insoluble residue, max, %	C114	0.75	0.75	0.75	0.75	0.75	0.75
Tricalcium silicate (C_3S)[E], max, %	See Annex A1	...	...	...	...	35[C]	...
Dicalcium silicate (C_2S)[E], min, %	See Annex A1	...	...	...	...	40[C]	...
Tricalcium aluminate (C_3A)[E], max, %	See Annex A1	...	8	8	15	7[C]	5[B]
Sum of $C_3S + 4.75C_3A$[G], max, %	See Annex A1	...	...	100[C, H]	...	...	...
Tetracalcium aluminoferrite plus twice the tricalcium aluminate ($C_4AF + 2(C_3A)$), or solid solution ($C_4AF + C_2F$), as applicable, max, %	See Annex A1	...	...	...	...	...	25[B]

[A] With the exception of Type I, not all types are carried in stock in some areas.

[B] Does not apply when the sulfate resistance limit in Table 4 of ASTM C150 is specified.

[C] Does not apply when the heat of hydration limit in Table 4 of ASTM C150 is specified.

[D] It is permissible to exceed the values in the table for SO_3 content, provided it has been demonstrated by Test Method C1038 that the cement with the increased SO_3 will not develop expansion exceeding 0.020% at 14 days. When the manufacturer supplies cement under this provision, supporting data shall be supplied to the purchaser.

[E] See Annex A1 (ASTM C150) for calculation.

[F] Not applicable.

[G] The limit on the sum, $C_3S + 4.75C_3A$ provides control on the heat of hydration of the cement and is consistent with Test Method C186 7-day heat of hydration limit of 335 kJ/kg [80 cal/g].

[H] In addition, 7-day heat of hydration testing by Test Method C186 shall be conducted at least once every six months. Such testing shall not be used for acceptance or rejection of the cement, but results shall be reported for informational purposes.

Due to the stricter grinding requirements for Type III, it is more expensive than Type I. The strength gain of Type I cement can be accelerated by increasing the cement content per unit volume of concrete, so the selection of Type III becomes a question of economics and availability. Type IV can be manufactured on demand. As discussed later, adding fly ash to Type I or II portland cement reduces the heat of hydration, producing the benefits of Type IV, but at a lower cost. Type V cement is produced only in locations with a severe sulfate problem.

6.9.2  Other Cement Types

Other than the five standard types of portland cement, several hydraulic cements are manufactured in the United States, including:

> White portland cement
> Blended hydraulic cements (ASTM C595)
> Type IS (Portland blast furnace slag cement)
> Type IP (Portland-pozzolan cement)
> Hydraulic cements (ASTM C1157)
>> Type GU (General use)
>> Type HE (High early strength)
>> Type MS (Moderate sulfate resistance)
>> Type HS (High sulfate resistance)
>> Type MH (Moderate heat of hydration)
>> Type LH (Low heat of hydration)
> Masonry and mortar cements
> Plastic cements
> Finely-ground cements (ultrafine cements)
> Expansive cements
> Specialty cements

In general, these cements have limited applications. Civil and construction engineers should be aware of their existence but should study them further before using them.

6.10 Mixing Water

Any potable water is suitable for making concrete. However, some nonpotable water may also be suitable. Sometimes, concrete suppliers will use unprocessed surface or well water if it can be obtained at a lower cost than processed water. However, impurities in the mixing water can affect concrete set time, strength, and long-term durability. In addition, chloride ions in the mixing water can accelerate corrosion of reinforcing steel.

6.10.1 ■ Acceptable Criteria

The acceptance criteria for questionable water are specified in ASTM C94. After 7 days, the average compressive strength of mortar cubes made with the questionable water should not be less than 90% of the average strength of cubes made with potable or distilled water (ASTM C109). Also, the set time of cement paste made with the questionable water should, as measured using the Vicat apparatus (ASTM C191), not be 1 hour less than or $1\text{-}\frac{1}{2}$ hours more than the set time of paste made with potable or distilled water.

Sample Problem 6.1

Three standard mortar cubes were made using questionable water available at the job site, and three others were made using potable water. The cubes were tested for compressive strength after 7 days of curing and produced the failure loads in kN shown in the following table:

Questionable Water	Potable Water
56.19	63.17
53.13	65.01
52.56	60.06

The Vicat test was conducted on the cement paste made with the questionable water and showed that the set time is 1 hour more than the set time of paste made with potable water. Based on these results, would you accept that water for mixing concrete according to ASTM standards? Explain why.

Solution

Average failure load of mortar cubes with the questionable water = 53.96 kN.

Average failure load of mortar cubes with potable = 62.75 kN.

Ratio = 53.96/62.75 = 0.86 = 86%

The average strength of the cubes made with questionable water is less than 90% of the strength of the cubes made with potable water. Therefore, the strength requirement is not satisfied.

The set time of mortar made with the questionable water is not more than $1\frac{1}{2}$ hours of the set time of paste made with potable water. Therefore, the set time requirement is satisfied.

Based on these results, the questionable water is not acceptable according to the ASTM standards.

Other adverse effects caused by excessive impurities in mixing water include efflorescence (white stains forming on the concrete surface due to the formation of calcium carbonate), staining, corrosion of reinforcing steel, volume instability, and reduced durability. Therefore, in addition to the compressive strength and set time, there are maximum chemical limits that should not be exceeded in the mixing water, as shown in Table 6.6. Several tests are available to evaluate the chemical impurities of questionable water. Over 100 different compounds and ions can exist in the mixing water and can affect concrete quality; the more important effects are described in the Table 6.7.

TABLE 6.6 Chemical Limits for Wash Water Used as Mixing Water (Reprinted, with permission, from ASTM C1602, Table 2, copyright ASTM International, I00 Barr Harbor Drive, West Conshohocken, PA 19428)

	Maximum Concentration Chemical (ppm)	ASTM Test Method
Chloride, as Cl		
Prestressed concrete or concrete in bridge decks	500	C114
Other reinforced concrete in moist environments or containing aluminum embedments or dissimilar metals or with stay-in-place galvanized metal forms	1000	C114
Sulfate, as SO_4	3000	C114
Alkalis, as ($Na_2O + 0.658 K_2O$)	600	C114
Total solids	50,000	C1603

TABLE 6.7 Summary of Effects of Water Impurities on Concrete Quality

Impurity	Effect
Alkali carbonate and bicarbonate	Can retard or accelerate strength test setting and 28-day strength when total dissolved salts exceed 1000 ppm. Can also aggravate alkali–aggregate reaction.
Chloride	Corrosion of reinforcing steel is primary concern. Chloride can enter the mix through admixtures, aggregates, cement, and mixing water, so limits are expressed in terms of total free chloride ions. ACl limits water-soluble ion content based on the type of reinforcement: Prestressed concrete 0.06% Reinforced concrete exposed to chloride in service 0.15% Reinforced concrete protected from moisture 1.00% Other reinforced concrete 0.30%

(Continued)

TABLE 6.7 (*Continued*)

Impurity	Effect
Sulfate	Can cause expansive reaction and deterioration.
Other salts	Not harmful when concentrations limited to: Calcium bicarbonate 400 ppm Magnesium bicarbonate 400 ppm Magnesium sulfate 25000 ppm Magnesium chloride 40000 ppm Iron salts 40000 ppm Sodium sulfide 100 ppm
Sea water	Do not use for reinforced concrete. Can accelerate strength gain but reduces ultimate strength. Can aggravate alkali reactions.
Acid water	Limit concentrations of hydrochloric, sulfuric, and other inorganic acids to less than 10000 ppm.
Alkaline water	Possible increase in alkali–aggregate reactivity. Sodium hydroxide may introduce quick set at concentrations higher than 0.5%. Strength may be lowered. Potassium hydroxide in concentrations over 1.2% may reduce 28-day strength of some cements.
Sanitary sewage water	Dilute to reduce organic matter to less than 20 ppm.
Sugar	Concentrations over 500 ppm can retard setting time and alter strength development. Sucrose in the range of 0.03 to approximately 0.15% usually retards setting. Concentrations over 0.25% by weight of cement can accelerate strength gain but substantially reduce 28-day strength.
Oils	Mineral oil (petroleum) in excess of 2.5% by weight of mix may reduce strength by 20%.
Algae	Can reduce hydration and entrain air. Do not use water containing algae.

6.10.2 ▧ Disposal and Reuse of Concrete Wash Water

Disposal of waste water from ready-mixed concrete operations is a great concern of the ready-mixed concrete producers (Chini and Mbwambo, 1996). This waste water is usually generated from truck wash systems, washing of central mixing plant, storm water runoff from the ready-mix plant yard, waste water generated from water sprayed dust control, and conveyor washdown. According to the Water Quality Act (part 116), waste water from ready-mixed concrete operations is a hazardous substance (it contains caustic soda and potash), and its disposal is regulated by the Environmental Protection Agency (EPA). In addition, the high pH makes concrete wash water hazardous under the EPA definition of corrosivity.

The conventional practices for disposing of concrete wash water include dumping at the job site, dumping at a landfill, or dumping into a concrete wash water pit at the ready-mix plant. The availability of landfill sites for the disposal of concrete wash water has been drastically reduced in the past few decades. In addition, the current environmental restrictions either prevent or limit these conventional disposal practices. As a result, most ready-mix batch plants have developed a variety of operational configurations to manage their own wash water. The alternatives include settling ponds, storm water detention/retention facilities, and water reuse systems. Chemical stabilizing admixture systems have been used to circumvent the necessity to remove any wash water from concrete truck drums and allow wash water to be reused for mixing more concrete. Studies have concluded that concrete properties are not significantly affected by the use of stabilized wash water (Borger et al., 1994).

Currently, concrete wash water is being used as mixing water for concrete in many places throughout the United States.

6.11　Admixtures for Concrete

Admixtures are ingredients other than portland cement, water, and aggregates that may be added to concrete to impart a specific quality to either the plastic (fresh) mix or the hardened concrete (ASTM C494). Some admixtures are charged into the mix as solutions. In such cases, the liquid should be considered part of the mixing water. If admixtures cannot be added in solution, they are either weighed or measured by volume as recommended by the manufacturer. Admixtures are classified by the following chemical and functional physical characteristics:

1. air entrainers
2. water reducers
3. retarders
4. hydration controller admixtures
5. accelerators
6. specialty admixtures

The Portland Cement Association (PCA) identifies four major reasons for using admixtures (Kosmatka et al., 2011):

1. to reduce the cost of concrete construction
2. to achieve certain properties in concrete more effectively than by other means
3. to ensure quality of concrete during the stages of mixing, transporting, placing, and curing in adverse weather conditions
4. to overcome certain emergencies during concrete operations

6.11.1　Air Entrainers

Air entrainers produce tiny air bubbles in the hardened concrete to provide space for water to expand upon freezing (Figure 6.9). As moisture within the concrete

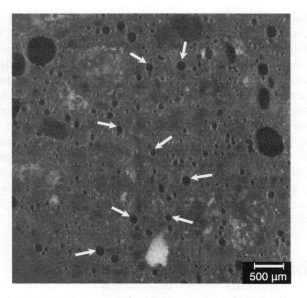

FIGURE 6.9 Magnified photo of concrete showing entrained air bubbles. (Courtesy of DRP Consulting, Inc.)

pore structure freezes, three mechanisms contribute to the development of internal stresses in the concrete:

1. Critical saturation—Upon freezing, water expands in volume by 9%. If the percent saturation exceeds 91.7%, the volume increase generates stress in the concrete.
2. Hydraulic pressure—Freezing water draws unfrozen water to it. The unfrozen water moving throughout the concrete pores generates stress, depending on length of flow path, rate of freezing, permeability, and concentration of salt in pores.
3. Osmotic pressure—Water moves from the gel to capillaries to satisfy thermodynamic equilibrium and equalize alkali concentrations. Voids permit water to flow from the interlayer hydration space and capillaries into the air voids, where it has room to freeze without damaging the parts.

Internal stresses reduce the durability of hardened concrete, especially when cycles of freeze and thaw are repeated many times. The impact of each of these mechanisms is mitigated by providing a network of tiny air voids in the hardened concrete using air entrainers. In the late 1930s, the introduction of air entrainment in concrete represented a major advance in concrete technology. Air entrainment is recommended for all concrete exposed to freezing.

All concrete contains entrapped air voids, which have diameters of 1 mm or larger and which represent approximately 0.2% to 3% of the concrete volume. Entrained air voids have diameters that range from 0.01 to 1 mm, with the majority being less than 0.1 mm. The entrained air voids are not connected and have a total volume between 1% and 7.5% of the concrete volume. Concrete mixed for severe frost conditions should contain approximately 14 billion bubbles per cubic meter. Frost resistance improves with decreasing void size, and small voids reduce strength less than large ones. The fineness of air voids is measured by the specific surface index, equal to the total surface area of voids in a unit volume of paste. The specific surface index should exceed 23,600 m^2/m^3 for frost resistance.

In addition to improving durability, air entrainment provides other important benefits to both freshly mixed and hardened concrete. Air entrainment improves concrete's resistance to several destructive factors, including freeze–thaw cycles, deicers and salts, sulfates, and alkali–silica reactivity. Air entrainment also increases the workability of fresh concrete. Air entrainment decreases the strength of concrete, as shown in Figure 6.8; however, this effect can be reduced for moderate-strength concrete by lowering the water–cement ratio and increasing the cement factor. High strength is difficult to attain with air-entrained concrete.

Air-entraining admixtures are available from several manufacturers and can be composed of a variety of materials, such as:

salts of wood resins (Vinsol resin)
synthetic detergents
salts of sulfonated lignin (by-product of paper production)
salts of petroleum acids
salts of proteinaceous material
fatty and resinous acids
alkylbenzene sulfonates
salts of sulfonated hydrocarbons

Air entrainers are usually liquid and should meet the specifications of ASTM C260. The agents enhance air entrainment by lowering the surface tension of the mixing water. Anionic air entrainers are hydrophobic (water hating). The negative charge of the agent is attracted to the positive charge of the cement particle. The hydrophobic agent forms tough, elastic, air-filled bubbles. Mixing disperses the air bubbles throughout the paste, and the sand particles form a grid that holds the air bubbles in place. Other types of air entrainers have different mechanisms but produce similar results.

6.11.2 ■ Water Reducers

Workability of fresh or plastic concrete requires more water than is needed for hydration. The excess water, beyond the hydration requirements, is detrimental to all desirable properties of hardened concrete. Therefore, water-reducing admixtures have been developed to gain workability and, at the same time, maintain quality. Water reducers increase the mobility of the cement particles in the plastic mix, allowing workability to be achieved at lower water contents. Water reducers are produced with different levels of effectiveness: conventional, mid-range, and high-range. Figure 6.10 shows concrete without the addition of admixture and with the addition of conventional, mid-range, and high-range water reducers. As shown in

FIGURE 6.10   Slumps of concretes with the same water–cement ratio: (a) no water reducer, (b) conventional water reducer, (c) mid-range water reducer, and (d) high-range water reducer (superplasticizer).

the figure, the slump of the concrete increases, indicating an increase in workability. The high-range water reducer is typically called plasticizer or superplasticizer.

Water Reducers Mechanism   Cement grains develop static electric charge on their surface as a result of the cement-grinding process. Unlike charges attract, causing the cement grains to cluster or "flocculate" [Figure 6.11(a)], which in turn limits the workability. The chemicals in the water-reducing admixtures reduce the static attraction among cement particles. The molecules of water-reducing admixtures have both positive and negative charges at one end, and a single charge (usually negative) on the other end, as illustrated in Figure 6.11(b). These molecules are attracted by the charged surface of the cement grains. The water reducers neutralize the static attraction on the cement surfaces. As a result, the clusters of cement grains are broken apart. Mutual repulsion of like charges pushes the cement grains apart, achieving a better distribution of particles [see Figure 6.11(c)], more uniform hydration, and a less-viscous paste.

Water reducing admixtures can be used indirectly to gain strength. Since the water-reducing admixture increases workability, we can take advantage of this phenomenon to decrease the mixing water, which in turn reduces the water–cement ratio and increases strength. Hewlett (1988) demonstrated that water reducers can actually be used to accomplish three different objectives, as shown in Table 6.8.

TABLE 6.8   Effects of Water Reducer

	Cement Content kg/m^3	Water–Cement Ratio	Slump (mm)	Compressive Strength (MPa)	
				7 day	28 day
Base mix	300	0.62	50	25	37
Improve consistency	300	0.62	100	26	38
Increase strength	300	0.56	50	34	46
Reduce cost	270	0.62	50	2	37.5

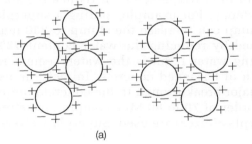

(a)

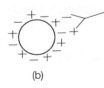

(b)

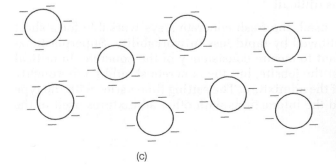

(c)

FIGURE 6.11   Water reducer mechanism: (a) clustering of cement grains without water reducer, (b) molecule of water reducer, and (c) better distribution of cement grains due to the use of water reducer.

1. Adding a water reducer without altering the other quantities in the mix increases the slump, which is a measure of concrete consistency and an indicator of workability, as discussed in Chapter 7.
2. The strength of the mix can be increased by using the water reducer by lowering the quantity of water and keeping the cement content constant.
3. The cost of the mix, which is primarily determined by the amount of cement, can be reduced. In this case, the water reducer allows a decrease in the amount of water. The amount of cement is then reduced to keep the water–cement ratio equal to the original mix. Thus, the quality of the mix, as measured by compressive strength, is kept constant, although the amount of cement is decreased.

The name "water reducer" may imply the admixture reduces the water in the concrete mix; this is not the case. A water reducer *allows* the use of a lower amount of mixing water while maintaining the same workability level. Used in this manner, the water reducer allows a lower water–cement ratio and therefore increases the strength and other desirable properties of the concrete.

Superplasticizers   Superplasticizers (plasticizers), or high-range water reducers, can either greatly increase the flow of the fresh concrete or reduce the amount of water required for a given consistency. For example, adding a superplasticizer to a concrete with a 75-mm slump can increase the slump to 230 mm, or the original slump can be maintained by reducing the water content 12% to 30%. Reducing the amount of mixing water reduces the water–cement ratio, which in turn, increases the strength of hardened concrete. In fact, the use of superplasticizers has resulted in a major breakthrough in the concrete industry. Now, high-strength concrete in the order of 70 to 80 MPa compressive strength or more can be produced when superplasticizers are used. Superplasticizers can be used when:

1. a low water–cement ratio is beneficial (e.g., high-strength concrete, early strength gain, and reduced porosity)
2. placing thin sections
3. placing concrete around tightly spaced reinforcing steel
4. placing cement underwater
5. placing concrete by pumping
6. consolidating the concrete is difficult

When superplasticizers are used, the fresh concrete stays workable for a short time, 30–60 minutes, and is followed by rapid loss in workability. Superplasticizers are usually added at the plant to ensure consistency of the concrete. In critical situations, they can be added at the jobsite, but the concrete should be thoroughly mixed following the addition of the admixture. The setting time varies with the type of agents, the amount used, and the interactions with other admixtures used in the concrete.

Sample Problem 6.2

The results of an experiment to evaluate the effects of a water reducer are shown in the following table:

	Without Water Reducer	With Water Reducer		
		Case 1	Case 2	Case 3
Cement Content, kg/m³	380	380	380	342
Water Content,* kg/m³	209	171	209	187
Slump, mm	50	50	100	50
28-Day Compressive Strength, MPa	24.1	29.6	24.2	24.2

* Assume that the water content is above the saturated surface-dry (SSD) condition of the aggregate.

Referring to the appropriate cases in the table

    a. Calculate the water/cement ratio in each case.
    b. Using water reducer, how can we increase the compressive strength of concrete without changing workability?
    c. Using water reducer, how can we improve workability without changing the compressive strength?
    d. Using water reducer, how can we reduce cost without changing workability or strength? (Assume that the cost of the small amount of water reducer added is less than the cost of cement.)

Solution

    a. Water/cement ratio for the case without water reducer = 209/380 = 0.55
       Water/cement ratio for case 1 = 171/380 = 0.45
       Water/cement ratio for case 2 = 209/380 = 0.55
       Water/cement ratio for case 3 = 187/342 = 0.55
    b. Add water reducer, use the same amount of cement, and decrease the amount of water (Case 1)
    c. Add water reducer and use the same amounts of cement and water (Case 2)
    d. Add water reducer and decrease both amounts of cement and water to maintain the same water to cement ratio (Case 3)

6.11.3 ■ Retarders

Some construction conditions require that the time between mixing and placing or finishing the concrete be increased. In such cases, retarders can be used to delay the initial set of concrete. Retarders are used for several reasons, such as the following:

1. offsetting the effect of hot weather
2. allowing for unusual placement or long haul distances
3. providing time for special finishes (e.g., exposed aggregate)

Retarders can reduce the strength of concrete at early ages (e.g., 1 to 3 days). In addition, some retarders entrain air and improve workability. Other retarders increase the time required for the initial set but reduce the time between the initial and final set. The properties of retarders vary with the materials used in the mix and with job conditions. Thus, the use and effect of retarders must be evaluated experimentally during the mix design process.

6.11.4 ▒ Hydration-Control Admixtures

These admixtures have the ability to stop and reactivate the hydration process of concrete. They consist of two parts: a stabilizer and an activator. Adding the stabilizer completely stops the hydration of the cementing materials for up to 72 hours, while adding the activator to the stabilized concrete reestablishes normal hydration and setting. These admixtures are very useful in extending the use of ready-mixed concrete when the work at the jobsite is stopped for various reasons. They are also useful when concrete is being hauled for a long time.

6.11.5 ▒ Accelerators

Accelerators are used to develop early strength of concrete at a faster rate than that developed in normal concrete. The ultimate strength, however, of high early strength concrete is about the same as that of normal concrete. Accelerators are used to

1. increase rate of strength gain
2. reduce the amount of time before finishing operations begin
3. reduce curing time
4. plug leaks under hydraulic pressure efficiently

The first three reasons are particularly applicable to concrete work placed during cold temperatures. The increased strength gained helps to protect the concrete from freezing and the rapid rate of hydration generates heat that can reduce the risk of freezing.

Calcium chloride, $CaCl_2$, is the most widely used accelerator (ASTM D98). Both initial and final set times are reduced with calcium chloride. The initial set time of 3 hours for a typical concrete can be reduced to 1.5 hours by adding an amount of calcium chloride equal to 1% of the cement weight; 2% reduces the initial set time to 1 hour. Typical final set times are 6 hours, 3 hours, and 2 hours for 0%, 1%, and 2% calcium chloride. Figure 6.12 shows that strength development is also affected by $CaCl_2$ for plain portland cement concrete (PCC) and portland cement concrete with 2% calcium chloride. Concrete with $CaCl_2$ develops higher early strength compared with plain concrete cured at the same temperature (Hewlett, 1988).

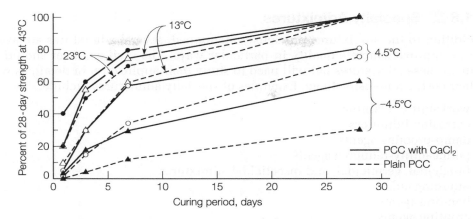

FIGURE 6.12   Effect of CaCl$_2$ on strength development at different curing temperatures.

The PCA recommends against using calcium chloride under the following conditions:

1. concrete is prestressed
2. concrete contains embedded aluminum such as conduits, especially if the aluminum is in contact with steel
3. concrete is subjected to alkali–aggregate reaction
4. concrete is in contact with water or soils containing sulfates
5. concrete is placed during hot weather
6. mass applications of concrete

The American Concrete Institute (ACI) recommends the following limits to water-soluble chloride ion content based on percent weight of cement (American Concrete Institute, 1999):

Member Type	Chloride Ion Limit, %
Prestressed concrete	0.06
Reinforced concrete subjected to chloride in service	0.15
Reinforced concrete protected from moisture	1.00
Other reinforced concrete	0.30

Several alternatives to the use of calcium chloride are available. These include the following:

1. using high early strength (Type III) cement
2. increasing cement content
3. curing at higher temperatures
4. using non–calcium chloride accelerators such as triethanolamine, sodium thiocyanate, calcium formate, or calcium nitrate

6.11.6 ■ Specialty Admixtures

In addition to the admixtures previously mentioned, several admixtures are available to improve concrete quality in particular ways. The civil engineer should be aware of these admixtures but will need to study their application in detail, as well as their cost, before using them. Examples of specialty admixtures include:

workability retaining
corrosion inhibitors
damp-proofing agents
permeability-reducing agents
fungicidal, germicidal, and insecticidal admixtures
pumping aids
bonding agents
grouting agents
gas-forming agents
coloring agents
shrinkage reducing

6.12 Supplementary Cementitious Materials

Several by-products of other industries have been used in concrete as supplementary cementitious materials, SCM, since the 1970s, especially in North America. These materials have been used to improve some properties of concrete and reduce the problem of discarding them. Most of today's high-performance concrete mixtures have one or more SCMs to enhance workability, durability, strength, etc. (CJSI, 2015a). Since these materials are cementitious, they can be used in addition to or as a partial replacement for portland cement. In fact, two or more of these supplementary cementitious additives have been used together to enhance concrete properties. These supplementary cementitious materials include fly ash, ground granulated blast furnace slag, silica fume, and natural pozzolans.

Fly Ash   Fly ash is the most commonly used pozzolan in civil engineering structures. Fly ash is a by-product of the coal-fired electricity production. Combusting pulverized coal in an electric power plant burns off the carbon and most volatile materials (Figure 6.13). However, depending on the source and type of coal, a significant amount of impurities passes through the combustion chamber. The carbon contents of common coals ranges from 70 to 100%. The noncarbon percentages are impurities (e.g., clay, feldspar, quartz, and shale), which fuse as they pass through the combustion chamber. Exhaust gas carries the fused material, fly ash, out of the combustion chamber. In the 1990s processes were implemented to control the carbon content of fly ash to consistently low levels resulting in a specified manufactured product with desired properties (CJSI, 2015a). The fly ash cools into spheres, which may be solid, hollow (cenospheres), or hollow and filled with other spheres (plerospheres). Particle diameters range from 1 μm to more than 0.1 mm, with an average of 0.015

(a)   (b)

FIGURE 6.13   Fly ash resulting from the combustion of ground or powdered coal (a) being discharged from a coal fired electric plant before environmental regulations were put into effect (Fotolia/manfredxy) and (b) collected into a bag house.

to 0.020 mm, and are 70 to 90% smaller than 0.045 mm. Fly ash is primarily a silica glass composed of silica (SiO_2), alumina (Al_2O_3), iron oxide (Fe_2O_3), and lime (CaO). Fly ash is classified (ASTM C618) as follows:

Class N—Raw or calcined natural pozzolans, including diatomaceous earths, opaline cherts and shales, ruffs and volcanic ashes or pumicites, and some calcined clays and shales

Class F—Fly ash with pozzolan properties

Class C—Fly ash with pozzolan and cementitious properties

Class F fly ash usually has less than 5% CaO but may contain up to 10%. Class C fly ash has 15% to 30% CaO.

Fly ash increases the workability of the fresh concrete. In addition, fly ash extends the hydration process, allowing a greater strength development and reduced porosity. Studies have shown that concrete containing more than 20% fly ash by weight of cement has a much smaller pore size distribution than portland cement concrete without fly ash. The lower heat of hydration reduces the early strength of the concrete so the curing period may need to be extended. The extended reaction permits a continuous gaining of strength beyond what can be accomplished with plain portland cement.

Slag Cement   Slag cement is made from iron blast furnace slag. It is a nonmetallic hydraulic cement consisting basically of silicates and aluminosilicates of calcium, which is developed in a molten condition simultaneously with iron in a blast furnace. The molten slag is rapidly chilled by quenching in water to form a glassy, sandlike granulated material. The material is then ground to less than 45 μm. The specific gravity of slag cement is in the range of 2.85 to 2.95.

The rough and angular-shaped ground slag in the presence of water and an activator, NaOH or CaOH, both supplied by portland cement, hydrates and sets in a manner similar to portland cement.

Slag cement has been used as a cementitious material in concrete since the beginning of the 1900s. Slag cement commonly constitutes between 30% and 45% of the cementing material in the mix. Some slag concretes have a slag component of 70% or more of the cementitious material. ASTM C 989 (AASHTO M 302) classifies slag by its increasing level of reactivity as Grade 80, 100, or 120.

Silica Fume   Silica fume is a by-product of the production of silicon metal or ferrosilicon alloys. One of the most beneficial uses for silica fume is as a mineral admixture in concrete. Because of its chemical and physical properties, it is a very reactive pozzolan. Concrete containing silica fume can have very high strength and can be very durable. Silica fume is available from suppliers of concrete admixtures and, when specified, is simply added during concrete production either in wet or dry forms. Placing, finishing, and curing silica fume concrete require special attention on the part of the concrete contractor.

Silicon metal and alloys are produced in electric furnaces. The raw materials are quartz, coal, and woodchips. The smoke that results from furnace operation is collected and sold as silica fume.

Silica fume consists primarily of amorphous (noncrystalline) silicon dioxide (SiO_2). The individual particles are extremely small, approximately 1/100th the size of an average cement particle. Because of its fine particles, large surface area, and the high SiO_2 content, silica fume is a very reactive pozzolan when used in concrete. The quality of silica fume is specified by ASTM C 1240 and AASHTO M 307.

In addition to producing high-strength concrete, silica fume can reduce concrete corrosion induced by deicing or marine salts. Silica fume concrete with a low water content is highly resistant to penetration by chloride ions. More information is available at the www.silicafume.org website.

Natural Pozzolans   A pozzolan is a siliceous and aluminous material which, in itself, possesses little or no cementitious value but will, in finely divided form and in the presence of moisture, react chemically with calcium hydroxide at ordinary temperatures to form compounds possessing cementitious properties (ASTM C595). Naturally occurring pozzolans, such as fine volcanic ash, combined with burned lime, were used about 2000 years ago for building construction, and pozzolan continues to be used today. As shown in Table 6.2, calcium hydroxide is one of the products generated by the hydration of C_3S and C_2S. In fact, up to 15% of the weight of portland cement is hydrated lime. Adding a pozzolan to portland cement generates an opportunity to convert this free lime into a cementitious material.

Tables 6.9 and 6.10 summarize the effects of supplementary cementitious admixtures on fresh and hardened concrete. These summaries are based on general trends and should be verified experimentally for specific materials and construction conditions.

T A B L E   6 . 9   Effect of Supplementary Cementitious Materials on Freshly Mixed Concrete

Quality Measure	Effect
Water requirements	Fly ash reduces water requirements. Silica fume increases water requirements.
Air content	Fly ash and silica fume reduce air content, compensate by increasing air entrainer.
Workability	Fly ash, ground slag, and inert minerals generally increase workability. Silica fume reduces workability; use superplasticizer to compensate.
Hydration	Fly ash reduces heat of hydration. Silica fume may not affect heat, but superplasticizer used with silica fume can increase heat.
Set time	Fly ash, natural pozzolans and blast furnace slag increase set time. Can use accelerator to compensate.

T A B L E   6 . 1 0   Effect of Supplementary Cementitious Admixtures on Hardened Concrete

Quality Measure	Effect
Strength	Fly ash increases ultimate strength but reduces rate of strength gain. Silica fume has less effect on rate of strength gain than pozzolans.
Drying shrinkage and creep	Low concentrations usually have little effect. High concentrations of ground slag or fly ash may increase shrinkage. Silica fume may reduce shrinkage.
Permeability and absorption	Generally reduce permeability and absorption. Silica fume is especially effective.
Alkali–aggregate reactivity	Generally reduced reactivity. Extent of improvement depends on type of admixture.
Sulfate resistance	Improved due to reduced permeability.

6.13   Cement Sustainability

6.13.1   LEED Considerations

The sustainability considerations of cement and concrete are closely tied together. LEED credit would be claimed for the sustainability of concrete, which is addressed in Chapter 7.

6.13.2 ■ Other Sustainability Considerations

The production of portland cement is very energy intensive and releases a considerable amount of greenhouse gasses, CO_2 in particular. According to the CJSI (2015b), the concrete industry produces 5% of all man-made CO_2 on a global basis. In the United States, the concrete industry produces 1.5% of all man-made CO_2 due to other sectors that produce CO_2. Of that amount 96% comes from the production of the cement. In addition to the carbon footprint, cement production uses 85% of the energy used to construct concrete facilities. The heat required to fire of the raw materials in the kiln accounts for 35% of the CO_2; calcination (the chemical transformation of the limestone into clinker) accounts for the other 65% of the CO_2. The cement industry is responding to the sustainability challenges the industry faces through voluntary programs to reduce CO_2 emissions by 10% and a 20% energy efficiency improvement per ton of cement produced by 2020 relative to a 1990 baseline (CJSI, 2015b). In 2008, 54% of the cement plants had auditable environmental management systems; the industry target is 90% by 2020.

Coal is a primary fuel used in the cement kilns. The industry is turning to tire-derived fuel, TDF, as an alternative to coal (PCA, 2008). Potential benefits of TDF include removal of the tires from the waste stream and a reduction in CO_2 emissions with either a reduction or no significant effect other emissions.

The use of SCM is also a sustainability advantage as it both offsets the use of cement and is a beneficial use of a secondary product.

Summary

The development of portland cement as the binder material for concrete is one of the most important innovations of civil engineering. It is extremely difficult to find civil engineering projects that do not include some component constructed with portland cement concrete. The properties of portland cement are governed by the chemical composition and the fineness of the particles. These control the rate of hydration and the ultimate strength of the concrete. Abrams's discovery of the importance of the water-to-cement ratio as the factor that controls the quality of concrete is perhaps the single most important advance in concrete technology. Second to this development was the introduction of air entrainment. The subsequent development of additional admixtures for concrete has improved the workability, set time, strength, and economy of concrete construction.

Questions and Problems

6.1   What ingredients are used for the production of portland cement?

6.2   What is the role of gypsum in the production of portland cement?

6.3   What is a typical value for the fineness of portland cement?

6.4   What are the primary chemical reactions during the hydration of portland cement?

6.5   Define the C-S-H phase of cement paste.

6.6   What are the four main chemical compounds in portland cement?

6.7   What chemical compounds contribute to early strength gain?

6.8   Define
      a. interlayer hydration space
      b. capillary voids
      c. entrained air
      d. entrapped air

6.9   Define *initial set* and *final set*. Briefly discuss one method used to determine them.

6.10 The following laboratory tests are performed:
      a. Setting time test of cement paste samples
      b. Compressive strength of mortar cubes

      What are the significance and use of each of these tests?

6.11 What is a false set of portland cement? State one reason for false set. If false set is encountered at the job site, what would you do?

6.12 You are an engineer in charge of mixing concrete in an undeveloped area where no potable water is available for mixing concrete. A source of water is available that has some impurities. What tests would you run to evaluate the suitability of this water for concrete mixing? What criteria would you use?

6.13 The water–cement ratio is important because it influences all of the desirable qualities of concrete.
      a. Referring to Figure 5.9, which portion of water is used to calculate the water–cement ratio?
      b. What is a typical water–cement ratio for normal strength concrete?
      c. What is the water–cement ratio needed for hydration only when concrete is in the plastic state?
      d. Why is the extra water necessary?
      e. Briefly describe how super high strength concrete ($f_c' = 100$ MPa) can be made.

6.14 Discuss the effect of water–cement ratio on the quality of hardened concrete. Explain why this effect happens.

6.15 Draw a graph to show the general relationship between the compressive strength of the concrete and the water–cement ratio for different curing times. Label all axes and curves.

6.16 Students in the materials class prepared three mortar mixes with water to cement ratios of 0.50, 0.55, and 0.60. Three 50-mm mortar cubes were prepared for each mix. The cubes were cured for 7 days and then tested for compressive strength. The test results were as shown in Table P6.16.

TABLE P6.16

Mix No.	w/c Ratio	Cube No.	Maximum Load (kN)	Compressive Strength (MPa)	Average Compressive Strength (MPa)
1	0.50	1	79.4		
		2	80.1		
		3	81.9		
2	0.55	1	74.7		
		2	74.5		
		3	72.5		
3	0.60	1	65.8		
		2	69.3		
		3	71.2		

Determine the following:

a. The compressive strength of each cube.
b. The average compressive strength for each mix.
c. Plot the average compressive strength versus w/c ratios for all mixes.
d. Comment on the effect of increasing w/c ratio on the compressive strength of the cubes.

6.17 Two batches of cement mortar with properties as shown in Table P6.17. If mortar cubes are made from these batches, which batch do you expect to have larger compressive strength? Why?

TABLE P6.17

Batch Number	Water–Cement Ratio	Superplasticizer Content
1	0.5	0
2	0.5	39 mL/m^3

6.18 What are the five primary types and functions of portland cement? Describe an application for each type.

6.19 Why isn't pozzolan used with Type III cement?

6.20 What type of cement would you use in each of the following cases? Why?
a. Construction of a large pier
b. Construction in cold weather
c. Construction in a warm climate region such as the Phoenix area
d. Concrete structure without any specific exposure condition
e. Building foundation in a soil with severe sulfate exposure

6.21 In order to evaluate the suitability of nonpotable water available at the job site for mixing concrete, six standard mortar cubes were made using that water and six others using potable water. The cubes were tested for compressive strength after 7 days of curing and produced the loads to failure (in kN) shown in Table P6.21.

   a. Based on these results only, would you accept that water for mixing concrete according to ASTM C94?

   b. According to ASTM C94, are there other tests to be performed to evaluate the suitability of that water? Discuss briefly.

TABLE P6.21

Cubes Made with Nonpotable Water	Cubes Made with Potable Water
64.6	73.5
70.4	74.8
63.0	79.7
71.6	69.8
70.6	83.7
64.0	74.4

6.22 Three standard mortar cubes were made using nonpotable water available at the job site, and three others were made using potable water. The cubes were tested for compressive strength after 7 days of curing and produced the failure loads in kN shown in Table P6.22.

   The Vicat test was conducted on the cement paste made with the questionable water and showed that the set time was 45 minutes more than the set time of paste made with potable water. Based on these results, would you accept that water for mixing concrete according to ASTM standards? Explain why.

TABLE P6.22

Nonpotable Water	Potable Water
70.3	75.2
73.0	83.2
73.8	77.4

6.23   Four standard mortar cubes were made using nonpotable water available at the job site, and four others were made using potable water. The cubes were tested for compressive strength after 7 days of curing and produced the failure stresses in MPa shown in Table P6.23.

Also, the Vicat test was conducted on the cement paste made with the questionable water and showed that the set time is $1/2$ hour less than the set time of paste made with potable water. Based on these results, would you accept that water for mixing concrete according to ASTM standards? Explain why.

TABLE P6.23

Nonpotable Water	Potable Water
27.6	27.4
26.9	28.1
25.5	29.3
27.7	27.9

6.24   Discuss the problem of disposal of waste water from ready-mixed concrete operations. State three alternate methods that can be used to alleviate this problem.

6.25   State five types of admixtures and discuss their applications.

6.26   If you were a materials engineer in Minnesota (cold climate) and could use only one type of admixture, which would you select? Explain.

6.27   Under what condition is an air-entraining agent needed? Why? Discuss how the air-entraining agent performs its function. What is the typical diameter of air-entrained voids?

6.28   If a water reducer is added to the concrete mix without changing other ingredients, what will happen to the properties of the concrete? If the intention of adding the water reducer is to increase the compressive strength of hardened concrete, how can this be achieved?

6.29   How can the water reducer be used to achieve each of the following functions?
a. improve strength
b. improve workability
c. improve economy

6.30   A concrete mix includes the following ingredients per cubic meter:
a. Cement = 400 kg
b. Water = 176 kg
c. No admixture

Table P6.30 shows possible changes that can be made to the mix ingredients. Indicate in the appropriate boxes in the table what will happen in each case for the workability and the ultimate compressive strength as *increase, decrease,* or *approximately the same.*

TABLE P6.30

			What Will Happen?	
Cement (kg)	Water (kg)	Admixture	Workability	Ultimate Compressive Strength
400	240	None		
449	176	None		
400	176	Water reducer		
400	128	Water reducer		
400	176	Superplasticizer		
400	176	Air entrainer		
400	176	Accelerator		

6.31 The results of an experiment to evaluate the effects of a water reducer are shown in Table P6.31.
   a. Calculate the water–cement ratio in each of the three cases.
   b. Using water reducer, how can we increase the compressive strength of concrete without changing workability? Refer to the appropriate case in the table.
   c. Using water reducer, how can we improve workability without changing the compressive strength? Refer to the appropriate case in the table.
   d. Using water reducer, how can we reduce cost without changing workability or strength? (Assume that the cost of the small amount of water reducer added is less than the cost of cement.) Refer to the appropriate case in the table.
   e. Summarize all possible effects of water reducers on concrete.

TABLE P6.31

		With Water Reducer		
	Without Water Reducer	Case 1	Case 2	Case 3
Cement content, kg/m³	436	436	436	393
Water content,* kg/m³	218	218	174	196
Slump, mm	50	100	50	50
28-Day compressive strength, MPa	28.27	28.30	34.85	28.29

*Assume that the water content is above the saturated surface-dry (SSD) condition of the aggregate.

6.32 The results of a laboratory experiment to evaluate the effects of a plasticizer are shown below.
   a. Calculate the water–cement ratio in each of the three cases.
   b. Using water reducer, how can we increase the compressive strength of concrete without changing workability? Refer to the appropriate case in Table P6.32.
   c. Using water reducer, how can we improve workability without changing the compressive strength? Refer to the appropriate case in the table.

TABLE P6.32

| | Without Water Reducer | With Water Reducer | | |
		Case 1	Case 2	Case 3
Cement content, kg/m^3	815	815	815	734
Water content,* kg/m^3	446	446	355	402
Slump, mm	50	100	50	50
Compressive strength, MPa	37.8	38.0	46	37.9

*Assume that the water content is above the saturated surface-dry (SSD) condition of the aggregate.

   d. Using water reducer, how can we reduce cost without changing workability or strength? (Assume that the cost of the small amount of water reducer added is less than the cost of cement.) Refer to the appropriate case in the table.
   e. Summarize all possible effects of water reducers on concrete.

6.33 Referring to Table P6.33, Mix No. 1 was designed to produce a 25 mm slump without water reducer. Mix Nos. 2–4 are trial mixes with water reducer. All mixes have the same amounts of aggregate and cement, different amounts of mixing water.

If the compressive strength at 28 days (f_c') follows the results shown in Figure 6.8, determine the following items:
   a. Water–cement ratio of each mix.
   b. f_c' of each mix
   c. The weight of mixing water required for a mix with water reducer having a 25 mm slump using the same amounts of aggregate and cement shown in the table.
   d. f_c' of the mix in Item c.
   e. Comment on the effect of adding water reducer to the mix and discuss how to make use of it to increase the compressive strength of concrete.

Assume that Figure 6.8 can be extrapolated. Use one system of units only as specified by the instructor.

TABLE P6.33

Mix No.	Aggregate,* kg	Cement, kg	Water, kg	Water Reducer?	Slump, mm
1	45.5	10.5	5.23	No	50
2	45.5	10.5	5.23	Yes	175
3	45.5	10.5	4.32	Yes	100
4	45.5	10.5	3.18	Yes	25

* Assume all aggregates are in the SSD moisture state.

6.34 Two batches of concrete cylinders were made with and without a calcium chloride ($CaCl_2$) accelerator admixture. Six cylinders were made of each batch and submerged in water for different times before testing for compressive strength and the results are shown in Table P6.34.

It is required to do the following:
a. Using an Excel sheet, plot the relationship between curing time and compressive strength of the two mixes on the same graph. Label all axes and curves.
b. Comment on the effect of the accelerator on the compressive strength of PCC at early and late ages.

TABLE P6.34

Cylinder no.	1	2	3	4	5	6
Water submergence (days)	1	3	7	14	21	28
Compressive strength of plain concrete (MPa)	0.8	7.8	15.8	21.1	24.5	26.8
Compressive strength of concrete with accelerator (MPa)	6.2	14.8	20.1	24.0	25.9	26.8

6.35 As a materials engineer in charge of designing concrete mix and facing the following problems, which admixture would you use in each case during mixing or at the jobsite?
a. There is a large quantity of freshly mixed concrete and the work at the jobsite had to stop because of rain.
b. More time is expected to be needed for finishing concrete.
c. The structure is subject to freezing.
d. Concrete is to be around tightly spaced reinforcing steel.
e. Concrete mix will be hauled a long distance.
f. High early strength concrete is required, but not necessarily high ultimate strength.

6.36 What is the source of fly ash? Why is fly ash sometimes added to concrete? What are the different classes of fly ash?

6.37 A materials engineer is working in a research project to evaluate the effect of one type of admixture on the compressive strength of concrete. She tested 10 mortar cubes made without admixture and 10 others with admixture after 28 days of curing. The compressive strengths of cubes without admixture were 25.1, 24.4, 25.8, 25.2, 23.9, 24.7, 24.3, 26.0, 23.8, and 24.6 MPa. The compressive strengths of cubes with admixture were 25.3, 26.8, 26.5, 24.5, 27.2, 24.8, 24.1, 25.9, 25.3, and 25.0 MPa. Using the statistical t-test, does this admixture show an increase in the compressive strength of the cement mortar at a level of significance of 0.05?

6.38 A materials engineer is working in a research project to evaluate the effect of one type of admixture on the compressive strength of concrete. He tested eight mortar cubes made with admixture and eight others without admixture after 28 days of curing. The compressive strengths of cubes in MPa with and without admixture are shown in Table P6.38.

Using the statistical t-test, is there a significant difference between the means of the compressive strengths of the two cement mortars at a level of significance of 0.10?

TABLE P6.38

	Compressive Strength, MPa	
Cube No.	With Admixture	Without Admixture
1	24.2	24.4
2	25.1	26.0
3	25.6	25.6
4	24.1	25.1
5	23.5	24.5
6	25.4	25.1
7	24.4	23.4
8	24.2	24.5

6.14 References

Aitcin, J., *Binders for Durable and Sustainable Concrete.* Taylor and Francis, New York, NY, 2008.

American Concrete Institute. *Building Code Requirements for Reinforced Concrete.* ACI Committee 318 Report, ACI 318-99. Farmington Hills, MI: American Concrete Institute, 1999.

Borger, J., R. L. Carrasquillo, and D. W. Fowler. "Use of Recycled Wash Water and Returned Plastic Concrete in the Production of Fresh Concrete." *Advanced Cement Based Materials,* Nov. 1 (6), 1994.

Chini, S. A. and W. J. Mbwambo. "Environmentally Friendly Solutions for the Disposal of Concrete Wash Water from Ready Mixed Concrete Operations." *Proceedings, Beijing International Conference.* Beijing, 1996.

CJSI. Concrete Joint Sustainability Initiative, web source http://www.sustainable concrete.org/?q=node/142, 2015a

CJSI. Concrete Joint Sustainability Initiative, web source http://www.sustainable concrete.org/?q=node/141, 2015b

Hewlett, P. C. *Concrete Admixtures: Use and Applications.* Longman Scientific & Technical, 1988.

Hover, K. and R. E. Phillco. *Highway Materials Engineering, Concrete.* Publication No. FHWA-H1-90-009, NHI Course No. 13123. Washington, D.C.: Federal Highway Administration, 1990.

Kosmatka, S. H. and W. C. Panarese. *Design and Control of Concrete Mixtures.* 13th ed. Skokie, IL: Portland Cement Association, 1988.

Kosmatka, S. H., and M. L. Wilson. *Design and Control of Concrete Mixtures.* 15th ed. (Revised). Skokie, IL: Portland Cement Association, 2011.

Mehta, P. K. and P. J. M. Monteiro. *Concrete: Microstructure, Properties, and Materials.* 4th ed. New York: McGraw-Hill Professional, 2013.

Neville, A. M. *Properties of Concrete.* 4th ed. New York: John Wiley and Sons, 1996.

PCA, *Tire-Derived Fuel.* Portland Cement Association, Washington DC, 2008.

C H A P T E R

PORTLAND CEMENT
CONCRETE

Civil and construction engineers are directly responsible for the quality control of portland cement concrete and the proportions of the components used in it. The quality of the concrete is governed by the chemical composition of the portland cement, hydration and development of the microstructure, admixtures, and aggregate characteristics. The quality is strongly affected by placement, consolidation, and curing, as well.

How a concrete structure performs throughout its service life is largely determined by the methods of mixing, transporting, placing, and curing the concrete in the field. In fact, the ingredients of a "good" concrete may be the same as those of a "bad" concrete. The difference, however, depends on the expertise of the engineer and technicians who are handling the concrete during construction.

Because of the advances made in concrete technology in the past few decades, concrete can be used in many more applications. Civil and construction engineers should be aware of the alternatives to conventional concrete, such as lightweight concrete, high-strength concrete, polymer concrete, fiber-reinforced concrete, and roller-compacted concrete. Before using these alternatives to conventional concrete, the engineer needs to study them, and their costs, in detail. This chapter covers basic principles of conventional portland cement concrete, its proportioning, mixing and handling, curing, and testing. Alternatives to conventional concrete that increase the applications and improve the performance of concrete are also introduced. Figure 7.1 shows order of activities involved in the construction process of concrete structures.

7.1   Proportioning of Concrete Mixes

The properties of concrete depend on the mix proportions and the placing and curing methods. Designers generally specify or assume a certain strength or modulus of elasticity of the concrete when determining structural dimensions. The materials engineer is responsible for assuring that the concrete is properly proportioned, mixed, placed, and cured so as to have the properties specified by the designer.

I.   Mix Design (Proportioning)
II.   Trial Mixes & Testing
III.   Batching

--Start the Clock

IV.   Mixing
V.   Transporting
VI.   Pouring (Placing)

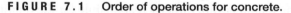

| Sampling & Testing |

VII.   Vibrating (Consolidating)

--Initial Set

VIII.   Finishing

--Final Set

IX.   Curing
X.   Maintenance

FIGURE 7.1   Order of operations for concrete.

The proportioning of the concrete mix affects its properties in both the plastic and solid states. During the plastic state, the materials engineer is concerned with the workability and finishing characteristics of the concrete. Properties of the hardened concrete important to the materials engineer are the strength, modulus of elasticity, durability, and porosity. Strength is generally the controlling design factor. Unless otherwise specified, concrete strength f'_c refers to the average compressive strength of three tests. Each test is the average result of two 0.15-m $\times$ 0.30-m cylinders tested in compression after curing for 28 days.

The PCA specifies three qualities required of properly proportioned concrete mixtures (Kosmatka et al., 2011):

1. acceptable workability of freshly mixed concrete
2. durability, strength, and uniform appearance of hardened concrete
3. economy

In order to achieve these characteristics, the materials engineer must determine the proportions of cement, water, fine and coarse aggregates, and the use of admixtures. Several mix design methods have been developed over the years, ranging from an arbitrary volume method (1:2:3 cement: sand: coarse aggregate) to the weight and absolute volume methods prescribed by the American Concrete Institute's Committee 211. The weight method provides relatively simple techniques for estimating mix proportions, using an assumed or known unit weight of concrete. The absolute volume method uses the specific gravity of each ingredient to calculate the unit volume each will occupy in a unit volume of concrete. The absolute volume method is more accurate than the weight method. The mix design process for the weight and absolute volume methods differs only in how the amount of fine aggregates is determined.

7.1.1 ▦ Basic Steps for Weight and Absolute Volume Methods

The basic steps required for determining mix design proportions for both weight and absolute volume methods are as follows (Kosmatka et al., 2011):

1. Evaluate strength requirements.
2. Determine the water–cement (water–cementitious materials) ratio required.
3. Evaluate coarse aggregate requirements.

   ■ maximum aggregate size of the coarse aggregate

   ■ quantity of the coarse aggregate

4. Determine air entrainment requirements.
5. Evaluate workability requirements of the plastic concrete.
6. Estimate the water content requirements of the mix.
7. Determine cementing materials content and type needed.
8. Evaluate the need and application rate of admixtures.
9. Evaluate fine aggregate requirements.
10. Determine moisture corrections.
11. Make and test trial mixes.

Most concrete supply companies have a wealth of experience about how their materials perform in a variety of applications. This experience, accompanied by reliable test data on the relationship between strength and water–cementitious materials ratio, is the most dependable method for selecting mix proportions. However, understanding the basic principles of mixture design and the proper selection of materials and mixture characteristics is as important as the actual calculation. Therefore, the PCA procedure provides guidelines and can be adjusted to match the experience obtained from local conditions. The PCA mix design steps are discussed in the following.

1. Strength Requirements   Variations in materials, batching, and mixing of concrete results in deviations in the strength of the concrete produced by a plant. Generally, the structural design engineer does not consider this variability when determining the size of the structural members. If the materials engineer provides a material with an average strength equal to the strength specified by the designer, then half of the concrete will be weaker than the specified strength. Obviously, this is undesirable. To compensate for the variance in concrete strength, the materials engineer designs the concrete to have an average strength greater than the strength specified by the structural engineer.

In order to compute the strength requirements for concrete mix design, three quantities must be known:

1. the specified compressive strength f'_c
2. the variability or standard deviation s of the concrete
3. the allowable risk of making concrete with an unacceptable strength

The standard deviation in the strength is determined for a plant by making batches of concrete, testing the strength for many samples, and computing the

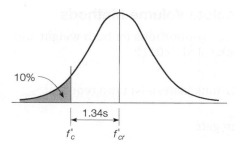

10%

1.34s

f'_c f'_{cr}

FIGURE 7.2   Use of normal distribution and risk criteria to estimate average required concrete strength.

standard deviation using Equation 1.16 in Chapter 1. The allowable risk has been established by the American Concrete Institute (ACI). One of the risk rules states that there should be less than 10% chance that the strength of a concrete mix is less than the specified strength. Assuming that the concrete strength has a normal distribution, the implication of the ACI rule is that 10% of the area of the distribution must be to the left of f'_c, as shown in Figure 7.2. Using a table of standard z values for a normal distribution curve, we can determine that 90% of the area under the curve will be to the right of f'_c if the average strength is 1.34 standard deviations from f'_c. In other words, the required average strength f'_{cr} for this criterion can be calculated as

$$f'_{cr} = f'_c + 1.34s \tag{7.1}$$

where

f'_{cr} = required average compressive strength, MPa
f'_c = specified compressive strength, MPa
s = standard deviation, MPa

For mixes with a large standard deviation in strength, the ACI has another risk criterion that requires

$$f'_{cr} = f'_c + 2.33s - 3.45 \tag{7.2}$$

The required average compressive strength f'_{cr} is determined as the larger value obtained from Equations 7.1 and 7.2.

The standard deviation should be determined from at least 30 strength tests. If the standard deviation is computed from 15 to 30 samples, then the standard deviation is multiplied by the following factor, F, to determine the modified standard deviation s'.

Number of Tests	Modification Factor F
15	1.16
20	1.08
25	1.03
30 or more	1.00

Linear interpolation is used for an intermediate number of tests, and s' is used in place of s in Equations 7.1 and 7.2.

If fewer than 15 tests are available, the following adjustments are made to the specified strength, instead of using Equations 7.1 and 7.2:

Specified Compressive Strength f'_c, MPa	Required Average Compressive Strength f'_{cr}, MPa
< 21	f'_c + 7.0
21 to 35	f'_c + 8.5
> 35	f'_c + 10.0

These estimates are very conservative and should not be used for large projects, since the concrete will be overdesigned, and therefore not economical.

Sample Problem 7.1

The design engineer specifies a concrete strength of 31.0 MPa. Determine the required average compressive strength for:

a. a new plant for which s is unknown
b. a plant for which $s = 3.6$ MPa for 17 test results
c. a plant with extensive history of producing concrete with $s = 2.4$ MPa
d. a plant with extensive history of producing concrete with $s = 3.8$ MPa

Solution

a. $f'_{cr} = f'_c + 8.5 = 31.0 + 8.5 = 39.5$ MPa
b. Need to interpolate modification factor:

$$F = 1.16 - \left(\frac{1.16 - 1.08}{20 - 15}\right)(17 - 15) \cong 1.13$$

Multiply standard deviation by the modification factor

$$s' = (s)(F) = 3.6(1.13) = 4.1 \text{ MPa}$$

Determine maximum from Equations 7.1 and 7.2

$$f'_{cr} = 31.0 + 1.34(4.1) = 36.5 \text{ MPa}$$

$$f'_{cr} = 31.0 + 2.33(4.1) - 3.45 = 37.1 \text{ MPa}$$

Use $f'_{cr} = 37.1$ MPa

    c. Determine maximum from Equations 7.1 and 7.2

$$f'_{cr} = 31.0 + 1.34(2.4) = 34.2 \text{ MPa}$$

$$f'_{cr} = 31.0 + 2.33(2.4) - 3.45 = 33.1 \text{ MPa}$$

Use $f'_{cr} = 34.2$ MPa

    d. Determine maximum from Equations 7.1 and 7.2

$$f'_{cr} = 31.0 + 1.34(3.8) = 36.1 \text{ MPa}$$

$$f'_{cr} = 31.0 + 2.33(3.8) - 3.45 = 36.4 \text{ MPa}$$

Use $f'_{cr} = 36.4$ MPa

2. Water–Cement Ratio Requirements The next step is to determine the water–cement ratio needed to produce the required strength, f'_{cr}. Historical records are used to plot a strength-versus–water–cement ratio curve, such as that seen in Figure 7.3. If historical data are not available, three trial batches are made at different water–cement ratios to establish a curve similar to Figure 7.3. Table 7.1 can be used for estimating the water–cement ratios for the trial mixes when no other data are available. The required average compressive strength is used with the strength versus water–cement relationship to determine the water–cement ratio required for the strength requirements of the project.

    For small projects of noncritical applications, Table 7.2 can be used in lieu of trial mixes, with the permission of the project engineer. Table 7.2 is conservative with respect to the strength versus water–cement ratio relationship, which results in higher cement factors and greater average strengths than would be required if a mix design is performed. This table is not intended for use in designing trial batches; use Table 7.1 for trial batch design.

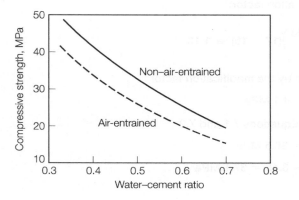

FIGURE 7.3 Example trial mixture or field data strength curves. (Kosmatka et al., 2011)

TABLE 7.1   Typical Relationship Between Water–Cement Ratio and Compressive Strength of Concrete*

Compressive Strength at 28 days, f'_{cr} MPa**	Water–Cement Ratio by Weight	
	Non-Air-Entrained Concrete	Air-Entrained Concrete
48	0.33	—
41	0.41	0.32
35	0.48	0.40
28	0.57	0.48
21	0.68	0.59
14	0.82	0.74

* American Concrete Institute (ACI 211.1 and ACI 211.3)
** Strength is based on cylinders moist-cured 28 days in accordance with ASTM C31 (AASHTO T23). Relationship assumes nominal maximum size of aggregate about 19 to 25 mm.

TABLE 7.2   Maximum Permissible Water–Cement Ratios for Concrete when Strength Data from Field Experience or Trial Mixtures Are Not Available*

Specified 28-day compressive Strength, f'_{cr} MPa	Water–Cement Ratio by Weight	
	Non-Air-Entrained Concrete	Air-Entrained Concrete
17	0.67	0.54
21	0.58	0.46
24	0.51	0.40
28	0.44	0.35
31	0.38	**
35	**	**

* American Concrete Institute (ACI 318), 1999.
** For strength above 31.0 MPa (non-air-entrained concrete) and 27.6 MPa (air-entrained concrete), concrete proportions shall be established from field data or trial mixtures.

The water–cement ratio required for strength is checked against the maximum allowable water–cement ratio for the exposure conditions. Tables 7.3 and 7.4 provide guidance on the maximum allowable water–cement ratio and the minimum design compressive strength for exposure conditions. Generally, more severe exposure conditions require lower water–cement ratios. The minimum of the water–cement ratio for strength and exposure is selected for proportioning the concrete.

If a pozzolan is used in the concrete, the water–cement plus pozzolan ratio (water–cementitious materials ratio) by weight may be used instead of the traditional water–cement ratio. In other words, the weight of the water is divided by the sum of the weights of cement plus pozzolan.

3. Coarse Aggregate Requirements The next step is to determine the suitable aggregate characteristics for the project. In general, large dense graded aggregates provide the most economical mix. Large aggregates minimize the amount of water required and, therefore, reduce the amount of cement required per cubic meter of mix. Round aggregates require less water than angular aggregates for an equal workability.

TABLE 7.3 Maximum Water–Cement Material Ratios and Minimum Design Strengths for Various Exposure Conditions*

Exposure Condition	Maximum Water–Cement Ratio by Mass for Concrete	Minimum Design Compressive Strength, f'_c, MPa
Concrete protected from exposure to freezing and thawing, application of deicing chemicals, or aggressive substances	Select water–cement ratio on basis of strength, workability, and finishing needs	Select strength based on structural requirements
Concrete intended to have low permeability when exposed to water	0.50	28
Concrete exposed to freezing and thawing in a moist condition or deicers	0.45	31
Reinforced concrete exposed to chlorides from deicing salts, salt water, brackish water, seawater, or spray from these sources	0.40	35

* American Concrete Institute (ACI 318), 2008.

TABLE 7.4   Requirements for Concrete Exposed to Sulfates in Soil or Water*

Sulfate Exposure	Water-Soluble Sulfate (SO₄) in Soil, Percent by Weight**	Sulfate (SO₄) in Water, ppm**	Cement Type***	Maximum Water–Cement Ratio By Weight
Negligible	Less than 0.10	Less than 150	No special type required	—
Moderate****	0.10–0.20	150–1500	II, II(MH), IP(MS), IS(<70)(MS), IT(P≥S)(MS), IT(P<S<70)(MS), MS	0.50
Severe	0.20–2.00	1500–10,000	V, IP(HS), IS(<70)(HS), IT(P≥S)(HS), IT(P<S<70)(HS), HS	0.45
Very Severe	Over 2.00	Over 10,000	V, IP(HS), IS(<70)(HS), IT(P≥S)(HS), IT(P<S<70)(HS), HS	0.40

*Adopted from American Concrete Institute (ACI 318), 2008.
** Tested in accordance with the Method for Determining the Quantity of Soluble Sulfate in Solid (Soil and Rock) and Water Samples, Bureau of Reclamation, Denver, 1977.
*** Cement Types II, II(MH) and V are in ASTM C150 (AASHTO M85), Types MS and HS in ASTM C1157, and the remaining types are in ASTM C595 (AASHTO M240). Pozzolans or slags that have been determined by test or severe record to improve sulfate resistance may also be used.
**** Includes sea water.

The maximum allowable aggregate size is limited by the dimensions of the structure and the capabilities of the construction equipment. The largest maximum aggregate size practical under job conditions that satisfies the size limits in the table should be used. Once the maximum aggregate size is determined, the nominal maximum aggregate size, which is generally one sieve size smaller than the maximum aggregate size, is used for the remainder of the proportioning analysis.

Situation	Maximum Aggregate Size
Form dimensions	1/5 of minimum clear distance
Clear space between reinforcement or prestressing tendons	3/4 of minimum clear space
Clear space between reinforcement and form	3/4 of minimum clear space
Unreinforced slab	1/3 of thickness

Sample Problem 7.2

A structure is to be built with concrete with a minimum dimension of 0.2 m, minimum space between rebars of 40 mm, and minimum cover over rebars of 40 mm. Two types of aggregate are locally available, with maximum sizes of 19 and 25 mm, respectively. If both types of aggregate have essentially the same cost, which one is more suitable for this structure?

Solution:

25 mm < (1/5)(200 mm) minimum dimensions.

25 mm < (3/4)(40 mm) rebar spacing.

25 mm < (3/4)(40 mm) rebar cover.

Therefore, both sizes satisfy the dimension requirements. However, 25 mm aggregate is more suitable, because it will produce a more economical concrete mix. Note that the 25 mm maximum aggregate size would correspond to a nominal maximum aggregate size of 19 mm.

The gradation of the fine aggregates is defined by the fineness modulus. The desirable fineness modulus depends on the coarse aggregate size and the quantity of cement paste. A low fineness modulus is desired for mixes with low cement content to promote workability.

Once the fineness modulus of the fine aggregate and the nominal maximum size of the coarse aggregate are determined, the volume of coarse aggregate per unit volume of concrete is determined using Table 7.5. For example, if the fineness modulus of the fine aggregate is 2.60 and the nominal maximum aggregate size is 19 mm, the coarse aggregate will have a volume of 0.64 m^3/m^3 of concrete. Table 7.5 is based on the unit weight of aggregates in a dry-rodded condition (ASTM C29). The values given are based on experience in producing an average degree of workability. The volume of coarse aggregate can be increased by 10% when less workability is required, such as in pavement construction. The volume of coarse aggregate should be reduced by 10% to increase workability, for example, to allow placement by pumping.

4. Air Entrainment Requirements   Next, the need for air entrainment is evaluated. Air entrainment is required whenever concrete is exposed to freeze–thaw conditions and deicing salts. Air entrainment is also used for workability in some situations. The amount of air required varies based on exposure conditions and is affected by the size of the aggregates. The exposure levels are defined as follows:

Mild exposure—Indoor or outdoor service in which concrete is not exposed to freezing and deicing salts. Air entrainment may be used to improve workability.

Moderate exposure—Some freezing exposure occurs, but concrete is not exposed

TABLE 7.5   Bulk Volume of Coarse Aggregate per Unit Volume of Concrete*

Nominal Maximum Size of Aggregate, mm	Bulk Volume of Dry-Rodded Coarse Aggregate Per Unit Volume of Concrete for Different Fineness Moduli of Fine Aggregate**			
	Fineness Modulus			
	2.40	2.60	2.80	3.00
9.5	0.50	0.48	0.46	0.44
12.5	0.59	0.57	0.55	0.53
19	0.66	0.64	0.62	0.60
25	0.71	0.69	0.67	0.65
37.5	0.75	0.73	0.71	0.69
50	0.78	0.76	0.74	0.72
75	0.82	0.80	0.78	0.76
150	0.87	0.85	0.83	0.81

* American Concrete Institute (ACI 211.1).
** Bulk volumes are based on aggregates in a dry-rodded condition as described in ASTM C29 (AASHTO T19).

to moisture or free water for long periods prior to freezing. Concrete is not exposed to deicing salts. Examples include exterior beams, columns, walls, etc., not exposed to wet soil.

Severe exposure—Concrete is exposed to deicing salts, saturation, or free water. Examples include pavements, bridge decks, curbs, gutters, canal linings, etc.

Table 7.6 presents the recommended air contents for different combinations of exposure conditions and nominal maximum aggregate sizes. The values shown in Table 7.6 are the interlayer hydration space (see Section 6.6) for non–air-entrained concrete and the interlayer hydration space plus entrained air in case of air-entrained concrete. The recommended air content decreases with increasing the nominal maximum aggregate size.

5. Workability Requirements   The next step in the mix design is to determine the workability requirements for the project. Workability is defined as the ease of placing, consolidating, and finishing freshly mixed concrete. Concrete should be workable but should not segregate or excessively bleed (migration of water to the top surface of concrete). The slump test (Figure 7.4) is an indicator of workability when evaluating similar mixtures. This test consists of filling a truncated cone with concrete, removing the cone, then measuring the distance the concrete slumps (ASTM C143). The slump is increased by adding water, air entrainer, water reducer, superplasticizer,

TABLE 7.6   Approximate Target Percent Air Content Requirements for Different Nominal Maximum Sizes of Aggregates*

	Nominal Maximum Aggregate Size, mm							
	9.5	12.5	19	25	37.5	50	75	150
Non-air-entrained concrete	3	2.5	2	1.5	1	0.5	0.3	0.2
Air-entrained concrete**								
Mild exposure	4.5	4.0	3.5	3.0	2.5	2.0	1.5	1.0
Moderate exposure	6.0	5.5	5.0	4.5	4.5	4.0	3.5	3.0
Severe exposure	7.5	7.0	6.0	6.0	5.5	5.0	4.5	4.0

* American Concrete Institute (ACI 211.1 and ACI 318).
** The air content in job specifications should be specified to be delivered within −1 to +2 percentage points of the table target value for moderate and severe exposures.

FIGURE 7.4   Slump test apparatus. (Fotolia/Bradlee Mauer)

or by using round aggregates. Table 7.7 provides recommendations for the slump of concrete used in different types of projects. For batch adjustments, slump increases about 25 mm for each 6 kg of water added per m^3 of concrete.

6. Water Content Requirements   The water content required for a given slump depends on the nominal maximum size and shape of the aggregates and whether an air entrainer is used. Table 7.8 gives the approximate mixing water requirements for angular coarse aggregates (crushed stone). The recommendations in Table 7.8 are reduced for other aggregate shape as shown in this table.

Aggregate Shape	Reduction in Water Content, kg/m^3
Subangular	12
Gravel with crushed particles	21
Round gravel	27

These recommendations are approximate and should be verified with trial batches for local materials.

7. Cementing Materials Content Requirements   With the water–cement ratio (water–cementitious materials ratio) and the required amount of water estimated, the amount of cementing materials required for the mix is determined by dividing the weight of the water by the water–cement ratio. PCA recommends

TABLE 7.7   Recommended Slumps for Various Types of Construction*

Concrete Construction	Slump, mm	
	Maximum**	Minimum
Reinforced foundation walls and footings	75	25
Plain footings, caissons, and substructure walls	75	25
Beams and reinforced walls	100	25
Building columns	100	25
Pavements and slabs	75	25
Mass concrete	75	25

* American Concrete Institute (ACI 211.1).

** May be increased 25 mm for consolidation by hand methods such as rodding and spading. Plasticizers can safely provide higher slumps.

TABLE 7.8 Approximate Mixing Water in kg/m³ for Different Slumps and Nominal Maximum Aggregate Sizes*

Slump, mm	Nominal Maximum Aggregate Size in mm**								
	9.5	12.5	19	25	37.5	50***	75***	150***
	Non-air-entrained concrete								
25 to 50	207	199	190	179	166	154	130	113
75 to 100	228	216	205	193	181	169	145	124
150 to 175	243	228	216	202	190	178	160	—
	Air-entrained concrete								
25 to 50	181	175	168	160	150	142	122	107
75 to 100	202	193	184	175	165	157	133	119
150 to 175	216	205	197	184	174	166	154	—

* American Concrete Institute (ACI 211.1 and ACI 318).
** These quantities of mixing water are for use in computing cementitious material contents for trial batches. They are maximums for reasonably well-shaped angular coarse aggregates graded within limits of accepted specifications.
*** The slump values for concrete containing aggregates larger than 37.5 mm are based on slump tests made after removal of particles larger than 37.5 mm by wet screening.

TABLE 7.9　Minimum Requirements of Cementing Materials for
Concrete Used in Flatwork*

Nominal Maximum Size of Aggregate, mm	Cementing Materials, kg/m^{3}**
37.5	280
25.0	310
19.0	320
12.5	350
9.5	360

* American Concrete Institute (ACI 302).
** Cementing materials quantities may need to be greater for severe exposure. For
example, for deicer exposures, concrete should contain at least 335 kg/m^3 of
cementing materials.

a minimum cement content of 334 kg/m^3 for concrete exposed to severe freeze–
thaw, deicers, and sulfate exposures, and not less than 385 kg/m^3 for concrete
placed under water. In addition, Table 7.9 shows the minimum cement requirements for proper placing, finishing, abrasion resistance, and durability in flatwork, such as slabs.

8. Admixture Requirements　If one or more admixtures are used to add a specific quality in the concrete (as discussed in Chapter 6), their quantities should be considered
in the mix proportioning. Admixture manufacturers provide specific information on
the quantity of admixture required to achieve the desired results.

9. Fine Aggregate Requirements　At this point, water, cement, and dry coarse aggregate weights per cubic meter are known and the volume of air is estimated. The only
remaining factor is the amount of dry fine aggregates needed. The weight mix design
method uses Table 7.10 to estimate the total weight of a "typical" freshly mixed concrete for different nominal maximum aggregate sizes. The weight of the fine aggregates is determined by subtracting the weight of the other ingredients from the total
weight. Since Table 7.10 is based on a "typical" mix, the weight-based mix design
method is only approximate.

　　In the absolute volume method of mix design, the component weight and the
specific gravity are used to determine the volumes of the water, coarse aggregate,
and cement. These volumes, along with the volume of the air, are subtracted from a
unit volume of concrete to determine the volume of the fine aggregate required. The
volume of the fine aggregate is then converted into a weight using the unit weight.
Generally, the bulk SSD specific gravity of aggregates is used for the weight–volume
conversions of both fine and coarse aggregates.

TABLE 7.10 Estimate of Weight of Freshly Mixed Concrete

Nominal Maximum Aggregate Size, mm	Non-Air-Entrained Concrete, kg/m^3	Air-Entrained Concrete, kg/m^3
9.5	2276	2187
12.5	2305	2228
19.0	2347	2276
25.0	2376	2311
37.5	2412	2347
50.0	2441	2370
75.0	2465	2394
150	2507	2441

10. Moisture Corrections Mix designs assume that water used to hydrate the cement is the free water in excess of the moisture content of the aggregates at the SSD condition (absorption), as discussed in Section 5.5.4. Therefore, the final step in the mix design process is to adjust the weight of water and aggregates to account for the existing moisture content of the aggregates. If the moisture content of the aggregates is more than the SSD moisture content, the weight of mixing water is reduced by an amount equal to the free weight of the moisture on the aggregates. Similarly, if the moisture content is below the SSD moisture content, the mixing water must be increased. Also, since the weights of coarse and fine aggregates estimated in steps 3 and 9 assume dry condition, they need to be adjusted to account for the absorbed moisture in the aggregates.

11. Trial Mixes After computing the required amount of each ingredient, a trial batch is mixed to check the mix design. Three 0.15 m × 0.30 m cylinders are made, cured for 28 days, and tested for compressive strength. In addition, the air content and slump of fresh concrete are measured. If the slump, air content, or compressive strength does not meet the requirements, the mixture is adjusted and other trial mixes are made until the design requirements are satisfied.

Additional trial batches could be made by slightly varying the material quantities in order to determine the most workable and economical mix.

Sample Problem 7.3

You are working on a concrete mix design that requires each cubic yard of concrete to have a 0.43 water–cement ratio, 1232 kg/m^3 of dry gravel, 145 kg/m^3 of water, and 4% air content. The available gravel has a specific gravity of G_{gravel} = 2.60, a moisture content of 2.3%, and absorption of 4.5%. The available sand has a specific gravity of G_{sand} = 2.40, a moisture content of 2.2%, and absorption of 1.7%. Air entrainer is to be included using the manufacturers specification of 6.3 mL/1% air/100 kg cement.

For each cubic meter of concrete needed on the job, calculate the weight of cement, moist gravel, moist sand, and water that should be added to the batch. Summarize and total the mix design when finished.

Solution

Step 7: W_{cement} = 145/0.43 = 337 kg/m^3

Step 8: air entrainer = 6.3 mL (4)(337/100) = 84.9 mL

Step 9: γ_w = 1000 kg/m^3

$$V_{cement} = 337/(3.15 \times 1000) = 0.107 \text{ m}^3$$
$$V_{water} = 145/1000 \qquad\quad = 0.145 \text{ m}^3$$
$$V_{gravel} = 1232/(2.6 \times 1000) = 0.474 \text{ m}^3$$
$$V_{air} = 4\% \qquad\qquad\quad = \underline{0.040 \text{ m}^3}$$
$$\text{subtotal} = 0.766 \text{ m}^3$$

$$V_{sand} = 1 - 0.766 = 0.234 \text{ m}^3$$

$$W_{sand} = (0.234)(2.4)(1000) = 562 \text{ kg/m}^3$$

Step 10: mix water = 145 − 1232(0.023 − 0.045) − 562(0.022 − 0.017)
$$= 169 \text{ kg/m3}$$

moist gravel = 1232(1.023) = 1261 kg/m^3

moist sand = 562(1.022)   = 574 kg/m^3

cement   = 337 kg/m^3

air entrainer   = 84.9 mL

Sample Problem 7.4

Design a concrete mix for the following conditions and constraints using the absolute volume method:

Design Environment

Bridge pier exposed to freezing and subjected to deicing chemicals

Required design strength = 24.1 MPa

Minimum dimension = 0.3 m

Minimum space between rebars = 64 mm

Minimum cover over rebars = 64 mm

Standard deviation of compressive strength of 2.4 MPa is expected (more than 30 samples)

Only air entrainer is allowed

Available Materials
Cement

Select Type V due to exposure

Air Entrainer

Manufacturer specification 6.3 mL/1% air/100 kg cement

Coarse aggregate

25 mm nominal maximum size, river gravel (round)

Bulk oven dry specific gravity = 2.621, absorption = 2.4%

Oven dry-rodded density = 1681 kg/m³

Moisture content = 1.5%

Fine aggregate

Natural sand

Bulk oven-dry specific gravity = 2.572, absorption = 0.8%

Moisture content = 4%

Fineness modulus = 2.60

Solution

1. Strength Requirements

   s = 2.4 MPa (enough samples so that no correction is needed)
   $f'_{cr} = f'_c + 1.34s = 24.1 + 1.34(2.4) = 27.3$ MPa
   $f'_{cr} = f'_c + 2.33s - 3.45 = 24.1 + 2.33(2.4) - 3.45 = 26.2$ MPa
   $$f'_{cr} = 27.3 \text{ MPa}$$

2. Water–Cement Ratio

   Strength requirement (Table 7.1), water–cement ratio = 0.48 by interpolation
   Exposure requirement (Tables 7.3 and 7.4), maximum water–cement ratio = 0.45
   $$\text{Water} - \text{cementitious materials ratio} = 0.45$$

3. Coarse Aggregate Requirements

   The 25 mm nominal maximum size corresponds to 37.5 mm maximum size.

   37.5 mm < 1/5 (300 mm) minimum dimensions

   37.5 mm < 3/4 (64 mm) rebar spacing

   37.5 mm < 3/4 (64 mm) rebar cover

   Aggregate size okay for dimensions

   (Table 7.5) 25 mm nominal maximum size coarse aggregate and 2.60 FM fine aggregate

   Coarse aggregate factor = 0.69

   Dry mass of coarse aggregate = (1681)(0.69) = 1160 kg/m³

   Coarse aggregate = 1160 kg/m³

4. **Air Content**
   (Table 7.6) Severe exposure, target air content = 6.0%
   Job range = 5% to 8% base

   Design using 7%

5. **Workability**
   (Table 7.7) Pier best fits the column requirement in the table
   Slump range = 25 to 100 mm

   Use 75 mm

6. **Water Content**
   (Table 7.8) 25 mm aggregate with air entrainment and 75 mm slump
   Water = 175 kg/m^3 for angular aggregates. Since we have round coarse aggregates, reduce by 27 kg/m^3

   Required water = 148 kg/m^3

7. **Cementing Materials Content**
   Water–cement ratio = 0.45, water = 148 kg/m^3

   Cement = 148/0.45 = 329 kg/m^3

   Increase for minimum criterion of 334 kg/m^3 for exposure

   Note that increasing the cement content without increasing water will slightly reduce water–cement ratio, producing a less permeable mix that is good for severe exposure.

   Cement = 334 kg/m^3

8. **Admixture**
   7% air, cement = 334 kg/m^3

   Admixture = (6.3)(7)(334/100) = 147 mL/m^3

   Admixture = 147 mL/m^3

9. **Fine Aggregate Requirements**
   Find fine aggregate content; use the absolute volume method.

   Water volume = 148/(1 × 1000) = 0.148 m^3/m^3

   Cement volume = 334/(3.15 × 1000) = 0.106 m^3/m^3

   Air volume = 0.07 m^3/m^3

   Coarse aggregate volume = 1160/(2.621 × 1000) = 0.443 m^3/m^3

   Subtotal volume = 0.767 m^3/m^3

   Fine aggregate volume = 1 - 0.767 = 0.233 m^3/m^3

   Fine aggregate dry mass = (0.233)(2.572)(1000) = 599 kg/m^3

   Fine aggregate = 599 kg/m^3

10. **Moisture Corrections**
    Coarse aggregate: Need 1160 kg/m^3 in dry condition, so increase by 1.5% for moisture

    Moist coarse aggregate = (1160)(1.015) = 1177 kg/m^3

Fine aggregate: Need 599 kg/m^3 in dry condition, so increase 4% for moisture

Fine aggregate in moist condition $= (599)(1.04) = 623$ kg/m^3

Water: Since the moisture content of coarse aggregate is less than the absorption level (SSD moisture content), the mixing water needs to be increased by 1160 $(0.024 - 0.015) = 10.4$ kg/m^3. On the contrary, since the moisture content of the fine aggregate is higher the absorption level, the mixing water needs to be reduced by 599 $(0.040 - 0.008) = 19.2$ kg/m^3.

Thus, adjusted water content $= 148 + 10.4 - 19.2 = 139$ kg/m^3

Summary	
Batch Ingredients Required for 1 m^3 PCC	
Water	139 kg
Cement	334 kg
Fine aggregate	623 kg
Coarse aggregate	1177 kg
Admixture	147 mL

7.1.2 ■ Mixing Concrete for Small Jobs

The mix design process applies to large jobs. For small jobs, for which a large design effort is not economical (e.g., jobs requiring less than one cubic meter of concrete), Tables 7.11 and 7.12 can be used as guides. The values in these tables may need to be adjusted to obtain a workable mix, using the locally available aggregates. Recommendations related to exposure conditions discussed earlier should be followed.

The combined volume is approximately 2/3 of the sum of the original bulk volumes.

Tables 7.11 and 7.12 are used for proportioning concrete mixes by weight and volume, respectively. The tables provide ratios of components, with a sum of one unit. Therefore, the required total weight or volume of the concrete mix can be multiplied by the given ratios to obtain the weight or volume of each component. Note that for proportioning by volume, the combined volume is approximately two-thirds of the sum of the original bulk volumes of the components, since water and fine materials fill the voids between coarse materials.

TABLE 7.11 Relative Components of Concrete for Small Jobs, by Weight*

Nominal Maximum Size of Coarse Aggregate, mm	Air-Entrained Concrete				Non-Air-Entrained Concrete			
	Cement	Wet Fine Aggregate	Wet Coarse Aggregate **	Water	Cement	Wet Fine Aggregate	Wet Coarse Aggregate **	Water
9.5	0.210	0.384	0.333	0.073	0.200	0.407	0.317	0.076
12.5	0.195	0.333	0.399	0.073	0.185	0.363	0.377	0.075
19	0.176	0.296	0.458	0.070	0.170	0.320	0.442	0.068
25	0.169	0.275	0.493	0.063	0.161	0.302	0.470	0.067
37.5	0.159	0.262	0.517	0.062	0.153	0.287	0.500	0.060

* Source: Portland Cement Association, 2011.
** If crushed stone is used, decrease coarse aggregate by 50 kg and increase fine aggregate by 50 kg for each cubic meter of concrete.

TABLE 7.12 Relative Components of Concrete for Small Jobs, by Volume*

Nominal Maximum Size of Coarse Aggregate, mm	Air-Entrained Concrete				Non-Air-Entrained Concrete			
	Cement	Wet Fine Aggregate	Wet Coarse Aggregate	Water	Cement	Wet Fine Aggregate	Wet Coarse Aggregate	Water
9.5	0.190	0.429	0.286	0.095	0.182	0.455	0.272	0.091
12.5	0.174	0.391	0.348	0.087	0.167	0.417	0.333	0.083
19	0.160	0.360	0.400	0.080	0.153	0.385	0.385	0.077
25	0.154	0.346	0.423	0.077	0.148	0.370	0.408	0.074
37.5	0.148	0.333	0.445	0.074	0.143	0.357	0.429	0.071

* Source: Portland Cement Association, 2011.
The combined volume is approximately 2/3 of the sum of the original bulk volumes.

Sample Problem 7.5

Determine the required mass of ingredients to make a 1587.6 kg batch of non-air-entrained concrete mix with a nominal maximum gravel size of 12.5 mm.

Solution

From Table 7.11:

Mass of cement = 1587.6 × 0.185 = 293.7 kg
Mass of wet fine aggregate = 1587.6 × 0.363 = 576.3 kg
Mass of wet coarse aggregate = 1587.6 × 0.377 = 598.5 kg
Mass of water = 1587.6 × 0.075 = 119.1 kg

Sample Problem 7.6

Determine the required volumes of ingredients to make a 0.5-m^3 batch of air-entrained concrete mix with a nominal maximum gravel size of 19 mm.

Solution

Sum of the original bulk volumes of the components = 0.5 × 1.5 = 0.75 m^3. From Table 7.12:

Volume of cement = 0.75 × 0.160 = 0.12 m^3
Volume of wet fine aggregate = 0.75 × 0.360 = 0.27 m^3
Volume of wet coarse aggregate = 0.75 × 0.400 = 0.3 m^3
Volume of water = 0.75 × 0.080 = 0.06 m^3

Sample Problem 7.7

Concrete was mixed with the following ingredients: 20.7 kg of cement, 39.2 kg of sand, 60.9 kg of gravel, and 14.6 kg of water. The sand has a moisture content of 3.3% and an absorption of 4.7%. The gravel has a moisture content of 3.6% and an absorption of 4.5%. Since water absorbed in the aggregate does not react with the cement or improve the workability of the plastic concrete, what is the water–cement ratio of this mix according to the American Concrete Institute's weight mix design method? If a water–cement ratio of 0.5 is required using the same materials and ingredients but different amount of mixing water, what is the mass of the mixing water to use?

Solution

Since moisture content and absorption are related to the aggregate dry mass, the first step of the solution is determine the dry mass of gravel and sand.

Dry mass of sand $= 39.2/1.033 = 37.948$ kg

Dry mass of gravel $= 60.9/1.036 = 58.784$ kg

Water required for sand to reach
absorption $= 37.948 (0.047 - 0.033) = 0.531$ kg

Water required for gravel to reach
absorption $= 58.784 (0.045 - 0.036) = 0.529$ kg

Water available for hydration $= 14.6 - 0.531 - 0.529 = 13.540$ kg

w/c ratio $= 13.540/20.7 = 0.65$

For a w/c ratio of 0.5, amount of free water $= 20.7 \times 0.5 = 10.35$ kg

Required total mixing water $= 10.35 + 0.531 + 0.529 = 11.41$ kg

7.2   Mixing, Placing, and Handling Fresh Concrete

The proper batching, mixing, and handling of fresh concrete are important prerequisites for strong and durable concrete structures. There are several steps and precautions that must be followed in mixing and handling fresh concrete in order to ensure a quality material with the desired characteristics (Mehta and Monteiro, 2013; American Concrete Institute, 1982; American Concrete Institute, 1983).

Batching is measuring and introducing the concrete ingredients into the mixer. Batching by weight is more accurate than batching by volume, since weight batching avoids the problem created by bulking of damp sand. Water and liquid admixtures, however, can be measured accurately by either weight or volume. On the other hand, batching by volume is commonly used with continuous mixers and when hand mixing. Continuous mixers are specifically designed to overcome the bulking problem.

Concrete should be mixed thoroughly, either in a mixer or by hand, until it becomes uniform in appearance. Hand mixing is usually limited to small jobs or situations in which mechanical mixers are not available. Mechanical mixers include on-site mixers and central mixers in ready-mix plants. The capacity of these mixers varies from 1.5 to 9 m³. Mixers also vary in type, such as tilting, nontilting, and pan-type mixers. Most of the mixers are batch mixers, although some mixers are continuous.

Mixing time and number of revolutions vary with the size and type of the mixer. Specifications usually require a minimum of 1 minute of mixing for stationary mixers of up to 0.75 m³ of capacity, with an increase of 15 seconds for each additional 0.75 m³ of capacity. Mixers are usually charged with 10% of the water, followed by uniform additions of solids and 80% of the water. Finally, the remainder of the water is added to the mixer.

7.2.1   Ready-Mixed Concrete

Ready-mixed concrete is mixed in a central plant and delivered to the job site in mixing trucks ready for placing (Figure 7.5). Three mixing methods can be used for ready-mixed concrete:

FIGURE 7.5 Concrete ready mix plant.

1. Central-mixed concrete is mixed completely in a stationary mixer and delivered in an agitator truck (2 rpm to 6 rpm).
2. Shrink-mixed concrete is partially mixed in a stationary mixer and completed in a mixer truck (4 to 16 rpm).
3. Truck-mixed concrete is mixed completely in a mixer truck (4 to 16 rpm).

Truck manufacturers usually specify the speed of rotation for their equipment. Also, specifications limit the number of revolutions in a truck mixer in order to avoid segregation. Furthermore, the concrete should be discharged at the job site within 90 minutes from the start of mixing, even if retarders are used (ASTM C94).

7.2.2 ▨ Mobile Batcher Mixed Concrete

Concrete can be mixed in a mobile batcher mixer at the job site (Figure 7.6). Aggregate, cement, water, and admixtures are fed continuously by volume, and the concrete is usually pumped into the forms.

7.2.3 ▨ Depositing Concrete

Several methods are available to deposit concrete at the jobsite. Concrete should be deposited continuously as close as possible to its final position. Advance planning and good workmanship are essential to reduce delay, early stiffening and drying out, and segregation. Figures 7.7—7.11 show different methods used to deposit concrete at the jobsite.

FIGURE 7.6   Mobile batcher mixer at the job site.

FIGURE 7.7   Loading concrete in a wheelbarrow. (Fotolia/Laure F)

FIGURE 7.8 Placing concrete slab.

FIGURE 7.9 Placing concrete with Chute. (Fotolia/Hoda Bogdan)

F I G U R E   7 . 1 0   Placing concrete pavement with a slip-form paver.

F I G U R E   7 . 1 1   Depositing concrete using a 2m³ bucket.

7.2.4 ■ Pumped Concrete

Pumped concrete is frequently used for large construction projects. Special pumps deliver the concrete directly into the forms (see Figure 7.12). Careful attention must be exercised to ensure well-mixed concrete with proper workability. The slump should be between 40 and 100 mm before pumping. During pumping, the slump decreases by about 12 to 25 mm, due to partial compaction. Blockage could happen during pumping, due to either the escape of water through the voids in the mix or due to friction if fines content is too high (Neville, 1996).

7.2.5 ■ Vibration of Concrete

Quality concrete requires thorough consolidation to reduce the entrapped air in the mix. On small jobs, consolidation can be accomplished manually by ramming and tamping the concrete. For large jobs, vibrators are used to consolidate the concrete. Several types of vibrators are available, depending on the application. *Internal vibrators* are the most common type used on construction projects (see Figure 7.13). These consist of an eccentric weight housed in a spud. The weight is rotated at high speed to produce vibration. The spud is slowly lowered into and through the entire layer of concrete, penetrating into the underlying layer if it is still plastic. The spud is left in place for 5 seconds to 2 minutes, depending on the type of vibrator and the consistency of the concrete. The operator judges the total

FIGURE 7.12   Pumping concrete in a retaining wall.

FIGURE 7.13   Consolidating concrete with an internal vibrator. (Shutterstock/Dmitry Kalinovsky)

vibration time required. Over-vibration causes segregation as the mortar migrates to the surface.

Several specialty types of vibrators are used in the production of precast concrete. These include *external vibrators, vibrating tables, surface vibrators, electric hammers,* and *vibratory rollers* (Neville, 1996).

7.2.6 ▩ Pitfalls and Precautions for Mixing Water

Since the water–cement ratio plays an important role in concrete quality, the water content must be carefully controlled in the field. Water should not be added to the concrete during transportation. Crews frequently want to increase the amount of water in order to improve workability. If water is added, the hardened concrete will suffer serious loss in quality and strength. The engineer in the field must prevent any attempt to increase the amount of mixing water in the concrete beyond that which is specified in the mix design.

7.2.7 ▩ Measuring Air Content in Fresh Concrete

Mixing and handling can significantly alter the air content of fresh concrete. Thus, field tests are used to ensure that the concrete has the proper air content prior to placing. Air content can be measured with the pressure, volumetric, gravimetric, or Chace air indicator methods.

The pressure method (ASTM C231) is widely used, since it takes less time than the volumetric method. The pressure method is based on Boyle's law, which relates pressure to volume. A calibrated cylinder (Figure 7.14) is filled with fresh concrete. The vessel is capped and air pressure is applied. The applied pressure compresses the air in the voids of the concrete. The volume of air voids is determined by measuring the amount of volume reduced by the pressure applied. This method is not valid for concrete made with lightweight aggregates, since air in the aggregate voids is also compressed, confounding the measurement of the air content of the cement paste.

The volumetric method for determining air content (ASTM C173) can be used for concrete made with any type of aggregate. The basic process involves placing concrete in a fixed volume cylinder, as shown in Figure 7.15. An equal volume of water is added to the container. Agitation of the container allows the excess water to displace the air in the cement paste voids. The water level in the container falls as the air rises to the top of the container. Thus, the volume of air in the cement paste is directly measured. The accuracy of the method depends on agitating the sample enough to remove all the air from it.

The gravimetric method (ASTM C138) compares the unit weight of freshly mixed concrete with the theoretical maximum unit weight of the mix. The theoretical unit weight is computed from the mix proportions and the specific gravity of each ingredient. This method requires very accurate specific gravity measurements and thus is more suited to the laboratory rather than the field.

The Chace air indicator test (AASHTO T199) is a quick method used to determine the air content of freshly mixed concrete. The device consists of a small glass tube with a stem, a rubber stopper, and a metal cup mounted on the stopper, as shown in Figure 7.16. The metal cup is filled with cement mortar from the concrete to be tested. The indicator is filled with alcohol to a specified level, and the stopper

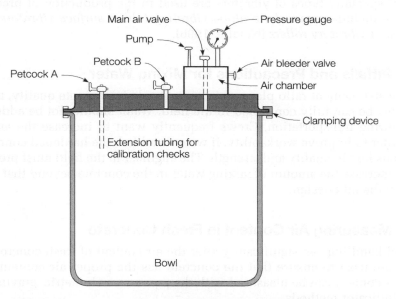

FIGURE 7.14 Pressure method apparatus for determining air voids in fresh concrete–Type B Meter. (reprinted, with permission, from ASTM C231, Figure 2, copyright ASTM International, 100 Barr Harbor Drive, West Conshohocken, PA 19428)

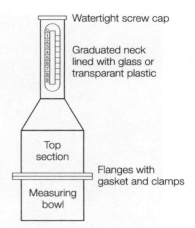

Watertight screw cap

Graduated neck lined with glass or transparant plastic

Top section

Flanges with gasket and clamps

Measuring bowl

FIGURE 7.15 Volumetric method (Roll-A-Meter) apparatus for determining air voids in fresh concrete. (Reprinted, with permission, from ASTM C173, Figure 1, copyright ASTM International, 100 Barr Harbor Drive, West Conshohocken, PA 19428)

FIGURE 7.16   Chace air indicator.

is inserted into the indicator. The indicator is then closed with a finger and gently rolled and tapped until all of the mortar is dispersed in the alcohol and all of the air is displaced with alcohol. With the indicator held in a vertical position, the alcohol level in the stem is read. This reading is then adjusted using calibration tables or figures to determine the air content. The Chace air indicator test can be used to rapidly monitor air content, but it is not accurate, nor does it have the precision required for specification control. It is especially useful for measuring the air content of small areas near the surface that may have lost air content by improper finishing.

These methods of measuring air content determine the total amount of air, including entrapped air and entrained air, as well as air voids in aggregate particles. Only minute bubbles produced by air-entraining agents impart durability to the concrete. However, the current state of the art is unable to distinguish between the types of air voids in fresh concrete.

7.2.8 ▓ Spreading and Finishing Concrete

Different methods are available to spread and finish concrete, depending on the nature of the structure and the available equipment. Tools and equipment used for spreading and finishing concrete include hand floats, power floats, darbies, bullfloats, straightedges, trowels, vibratory screed, and slip forms. (See Figures 7.10 and 7.17–7.23).

(a)

(b)

FIGURE 7.17   Spreading concrete with a straightedge: (a) manually and (b) mechanically.

FIGURE 7.18   Finishing concrete with a trowel. (Fotolia/alisonhancock)

FIGURE 7.19   Finishing concrete manually with a straightedge. (Fotolia/alisonhancock)

FIGURE 7.20 Finishing concrete with a power float.

FIGURE 7.21 Finishing concrete with a vibratory screed. (Courtesy of FHWA)

FIGURE 7.22   Finishing concrete pavement with a mechanical straightedge. (Courtesy of FHWA)

FIGURE 7.23   Using laser level to establish elevation for finishing concrete. (Courtesy of FHWA)

7.3 Curing Concrete

Curing is the process of maintaining satisfactory moisture content and temperature in the concrete for a definite period of time. Hydration of cement is a long-term process and requires water and proper temperature. Therefore, curing allows continued hydration and, consequently, continued gains in concrete strength. In fact, once curing stops, the concrete dries out, and the strength gain stops, as indicated in Figure 7.24. If the concrete is not cured and is allowed to dry in air, it will gain only about 50% of the strength of continuously cured concrete. If concrete is cured for only 3 days, it will reach about 60% of the strength of continuously cured concrete; if it is cured for 7 days, it will reach 80% of the strength of continuously cured concrete. If curing stops for some time and then resumes again, the strength gain will also stop and reactivate.

Increasing temperature increases the rate of hydration and, consequently, the rate of strength development. Temperatures below 10°C are unfavorable for hydration and should be avoided, if possible, especially at early ages.

Although concrete of high strength may not be needed for a particular structure, strength is usually emphasized and controlled since it is an indication of the concrete quality. Thus, proper curing not only increases strength but also provides other desirable properties such as durability, water tightness, abrasion resistance, volume stability, resistance to freeze and thaw, and resistance to deicing chemicals.

Curing should start after the final set of the cement. If concrete is not cured after setting, concrete will shrink, causing cracks. Drying shrinkage can be prevented if ample water is provided for a long period of time. An example of improper curing would be a concrete floor built directly over the subgrade, not cured at the surface, with the moisture in the soil curing it from the bottom. In this case, the concrete slab may curl due to the relative difference in shrinkage.

Curing can be performed by any of the following approaches:

1. maintaining the presence of water in the concrete during early ages. Methods to maintain the water pressure include ponding or immersion, spraying or fogging, and wet coverings.

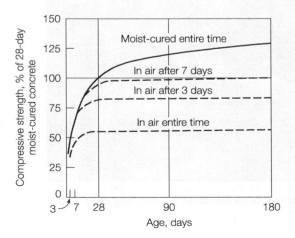

FIGURE 7.24 Compressive strength of concrete at different ages and curing levels.

2. preventing loss of mixing water from the concrete by sealing the surface. Methods to prevent water loss include impervious papers or plastic sheets, membrane-forming compounds, and leaving the forms in place.
3. accelerating the strength gain by supplying heat and additional moisture to the concrete. Accelerated curing methods include steam curing, insulating blankets or covers, and various heating techniques.

Note that preventing loss of mixing water from the concrete by sealing the surface is not as effective as maintaining the presence of water in the concrete during early ages. The choice of the specific curing method or combination of methods depends on the availability of curing materials, size and shape of the structure, in-place versus plant production, economics, and aesthetics (Kosmatka et al., 2011; American Concrete Institute, 1986a).

7.3.1 ▧ Ponding or Immersion

Ponding involves covering the exposed surface of the concrete structure with water. Ponding can be achieved by forming earth dikes around the concrete surface to retain water. This method is suitable for flat surfaces such as floors and pavements, especially for small jobs. The method requires intensive labor and supervision. Immersion is used to cure test specimens in the laboratory, as well as other concrete members, as appropriate.

7.3.2 ▧ Spraying or Fogging

A system of nozzles or sprayers can be used to provide continuous spraying or fogging (see Figures 7.25 and 7.26). This method requires a large amount of water and could be expensive. It is most suitable in high temperature and low humidity environments. Commercial test laboratories generally have a controlled temperature and humidity booth for curing specimens.

FIGURE 7.25   Curing concrete by spraying. (Courtesy of FHWA)

FIGURE 7.26   Curing concrete by fogging. (Courtesy of FHWA)

7.3.3 ▨ Wet Coverings

Moisture-retaining fabric coverings saturated with water, such as burlap, cotton mats, and rugs, are used in many applications (see Figure 7.27). The fabric can be kept wet, either by periodic watering or covering the fabric with polyethylene film to retain moisture. On small jobs, wet coverings of earth, sand, saw dust, hay, or straw can be used. Stains or discoloring of concrete could occur with some types of wet coverings.

7.3.4 ▨ Impervious Papers or Plastic Sheets

Evaporation of moisture from concrete can be reduced using impervious papers, such as kraft papers, or plastic sheets, such as polyethylene film (see Figures 7.28 and 7.29). Impervious papers are suitable for horizontal surfaces and simply shaped concrete structures, while plastic sheets are effective and easily applied to various shapes. Periodic watering is not required when impervious papers or plastic sheets are used. Discoloration, however, can occur on the concrete surface.

7.3.5 ▨ Membrane-Forming Compounds

Various types of liquid membrane-forming compounds can be applied to the concrete surface to reduce or retard moisture loss. These can be used to cure fresh concrete, as well as hardened concrete, after removal of forms or after moist curing. Curing compounds can be applied by hand or by using spray equipment (see Figures 7.30 and 7.31). Either one coat or two coats (applied perpendicular to each other) are used. Normally, the concrete surface should be damp when the curing compound is applied. Curing compounds should not be used when subsequent concrete layers are to be placed, since the compound hinders the bond between successive layers. Also, some compounds affect the bond between the concrete surface and paint.

FIGURE 7.27 Curing concrete by wet covering. (Courtesy of FHWA)

FIGURE 7.28 Curing concrete with impervious fabrics.

FIGURE 7.29   Curing concrete with plastic sheets.

FIGURE 7.30   Curing concrete by manually applying membrane forming compound.

FIGURE 7.31   Curing concrete by machine applying membrane forming compound.

7.3.6 ▧ Forms Left in Place

Loss of moisture can be reduced by leaving the forms in place as long as practical, provided that the top concrete exposed surface is kept wet. If wood forms are used, the forms should also be kept wet. After removing the forms, another curing method can be used.

7.3.7 ▧ Steam Curing

Steam curing is used when early strength gain in concrete is required or additional heat is needed during cold weather. Steam curing can be attained either with or without pressure. Steam at atmospheric pressure is used for enclosed cast-in-place structures and large precast members. High-pressure steam in autoclaves can be used at small manufactured plants.

7.3.8 ▧ Insulating Blankets or Covers

When the temperature falls below freezing, concrete should be insulated using layers of dry, porous material such as hay or straw. Insulating blankets manufactured of fiberglass, cellulose fibers, sponge rubber, mineral wool, vinyl foam, or open-cell polyurethane foam can be used to insulate formwork. Moisture proof commercial blankets can also be used.

7.3.9 ▧ Electrical, Hot Oil, and Infrared Curing

Precast concrete sections can be cured using electrical, oil, or infrared curing techniques. Electrical curing includes electrically heated steel forms, and electrically heated blankets. Reinforcing steel can be used as a heating element, and concrete can be used as the electrical conductor. Steel forms can also be heated by circulating hot oil around the outside of the structure. Infrared rays have been used for concrete curing on a limited basis.

7.3.10 ▉ Curing Period

The curing period should be as long as is practical. The minimum time depends on several factors, such as type of cement, mixture proportions, required strength, ambient weather, size and shape of the structure, future exposure conditions, and method of curing. For most concrete structures, the curing period at temperatures above 5°C should be a minimum of 7 days or until 70% of specified compressive or flexure strength is attained. The curing period can be reduced to 3 days if high early strength concrete is used and the temperature is above 10°C.

7.4 Properties of Hardened Concrete

It is important for the engineer to understand the basic properties of hardened portland cement concrete and to be able to evaluate these properties. The main properties of hardened concrete that are of interest to civil and construction engineers include the early volume change, creep, permeability, and stress–strain relationship.

7.4.1 ▉ Early Volume Change

When the cement paste is still plastic, it undergoes a slight decrease in volume of about 1%. This shrinkage is known as *plastic shrinkage* and is due to the loss of water from the cement paste, either from evaporation or from suction by dry concrete

FIGURE 7.32 Plastic shrinkage cracking.

below the fresh concrete. Plastic shrinkage may cause cracking (Figure 7.32); it can be prevented or reduced by controlling water loss.

In addition to the possible decrease in volume when the concrete is still plastic, another form of volume change may occur after setting, especially at early ages. If concrete is not properly cured and is allowed to dry, it will shrink. This shrinkage is referred to as *drying shrinkage,* and it also causes cracks. Shrinkage takes place over a long period of time, although the rate of shrinkage is high early and then decreases rapidly with time. In fact, about 15% to 30% of the shrinkage occurs in the first 2 weeks, while 65% to 85% occurs in the first year. Shrinkage and shrinkage-induced cracking are increased by several factors, including lack of curing, high water–cement ratio, high cement content, low coarse aggregate content, existence of steel reinforcement, and aging. On the other hand, if concrete is cured continuously in water after setting, concrete will swell very slightly due to the absorption of water. Since swelling, if it happens, is very small, it does not cause significant problems. *Swelling* is accompanied by a slight increase in weight (Neville, 1996).

How much drying shrinkage occurs depends on the size and shape of the concrete structure. Also, nonuniform shrinkage could happen due to the nonuniform loss of water. This may happen in mass concrete structures, where more water is lost at the surface than at the interior. In cases such as this, cracks may develop at the surface. In other cases, curling might develop due to the nonuniform curing throughout the structure and, consequently, nonuniform shrinkage.

In addition, the heat of hydration can produce two mechanisms that will limit the performance of concrete. These are generally associated with mass concrete, which is defined in ACI 301 as "any concrete with dimensions large enough to require that measures be taken to cope with the heat of hydration". There are two issues where the heat of hydration may create a problem:

1. If the internal temperature of the hydrating concrete exceeds 70°C which can result in a delayed ettringite formation, causing long-term durability issues.
2. If the thermal differential in the concrete exceeds 1.67°C thermal cracking can develop.

The PCA indicates thermal control plan should be developed when the minimum dimension of greater than 0.9 m (http://www.nrmca.org/aboutconcrete/cips/42p.pdf). However, the excessive thermal gradients can also develop in a structure with large surface to volume ratio, such as pavements and walls. The Texas Department of Transportation has developed software for the thermal analysis concrete structures (ConcreteWorks at http://www.txdot.gov/inside-txdot/division/information-technology/engineering-software.html) and Transtec has developed software that includes the thermal analysis of concrete pavements (http://www.hiperpav.com/). These programs can be used for the development of a thermal control plan.

Measures that can be taken to minimize the potential for thermally induced stresses include:

1. Using a mix design with a low initial heat of hydration by the substituting fly ash and/or silica fume for cement.
2. Reducing the temperature during mixing by substituting ice for water.

3. Placing the concrete in thin lifts
4. Insulate forms
5. Embed cooling pipes into the structure.
6. Scheduling concrete placement so that the development of the heat of hydration does not correspond with increasing environmental temperatures.

7.4.2 ▨ Creep Properties

Creep is defined as the gradual increase in strain, with time, under sustained load. Creep of concrete is a long-term process, and it takes place over many years. Although the amount of creep in concrete is relatively small, it could affect the performance of structures. The effect of creep varies with the type of structure. In simply supported reinforced concrete beams, creep increases the deflection and, therefore, increases the stress in the steel. In reinforced concrete columns, creep results in a gradual transfer of load from the concrete to the steel. Creep also could result in losing some of the prestress in prestressed concrete structures, although the use of high-tensile stress steel reduces this effect. Rheological models, discussed in Chapter 1, have been used to analyze the creep response of concrete (Neville, 1996).

7.4.3 ▨ Permeability

Permeability is an important factor that largely affects the durability of hardened concrete. Permeable concrete allows water and chemicals to penetrate, which, in turn, reduces the resistance of the concrete structure to frost, alkali–silica reactivity, and other chemical attacks. Water that permeates into reinforced concrete causes corrosion of steel rebars. Furthermore, impervious concrete is a prerequisite in watertight structures, such as tanks and dams.

Typically, the air voids in the cement paste and aggregates are small and do not affect permeability. However, the air voids that do affect permeability of hardened concrete are obtained from two main sources: incomplete consolidation of fresh

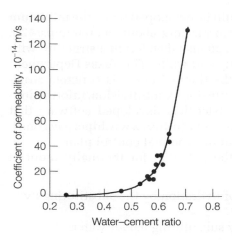

FIGURE 7.33 Relation between water–cement ratio and permeability of mature cement paste.

concrete and voids resulting from evaporation of mixing water that is not used for hydration of cement.

Therefore, increasing the water–cement ratio in fresh concrete has a severe effect on permeability. Figure 7.33 shows the typical relationship between the water–cement ratio and the coefficient of permeability of mature cement paste (Powers, 1954). It can be seen from the figure that increasing the water–cement ratio from 0.3 to 0.7 increases the coefficient of permeability by a factor of 1000. For a concrete to be watertight, the water–cement ratio should not exceed 0.48 for exposure to fresh water and should not be more than 0.44 for exposure to seawater (American Concrete Institute, 1975).

Other factors that affect the permeability include age of concrete, fineness of cement particles, and air-entraining agents. Age reduces the permeability, since hydration products fill the spaces between cement grains. The finer the cement particles, the faster is the rate of hydration and the faster is the development of impermeable concrete. Air-entraining agents indirectly reduce the permeability, since they allow the use of a lower water–cement ratio.

7.4.4 ▨ Stress–Strain Relationship

Typical stress–strain behavior of 28-day-old concrete with different water–cement ratios are shown in Figure 7.34 (Hognestad et al., 1955). It can be seen that increasing the water–cement ratio decreases both strength and stiffness of the concrete. The figure also shows that the stress–strain behavior is close to linear at low stress levels, then becomes nonlinear as stress increases. With a water–cement ratio of 0.50 or less and a strain of up to 0.0015, the stress–strain behavior is almost linear. With higher water–cement ratios, the stress–strain behavior becomes nonlinear at smaller strains. The curves also show that high-strength concrete has sharp peaks and sudden failure characteristics when compared to low-strength concrete.

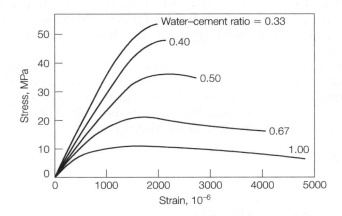

FIGURE 7.34   Typical stress–strain relationships for compressive tests on 0.15 × 0.30-m concrete cylinders at an age of 28 days.

As discussed in Chapter 1, the elastic limit can be defined as the largest stress that does not cause a measurable permanent strain. When the concrete is loaded slightly beyond the elastic range and then unloaded, a small amount of strain might remain initially, but it may recover eventually due to creep. Also, since concrete is not perfectly elastic, the rate of loading affects the stress–strain relationship to some extent. Therefore, a specific rate of loading is required for testing concrete. It is interesting to note that the shape of the stress–strain relationship of concrete is almost the same for both compression and tension, although the tensile strength is much smaller than the compressive strength. In fact, the tensile strength of concrete typically is ignored in the design of concrete structures.

The modulus of elasticity of concrete is commonly used in designing concrete structures. Since the stress–strain relationship is not exactly linear, the classic definition of modulus of elasticity (Young's modulus) is not applicable. The initial tangent modulus of concrete has little practical importance. The tangent modulus is valid only for a low stress level where the tangent is determined. Both secant and chord moduli represent "average" modulus values for certain stress ranges. A chord modulus in compression is more commonly used to represent the modulus of elasticity of concrete. It is determined according to ASTM C469. The method requires three or four loading and unloading cycles, after which the chord modulus is determined between a point corresponding to a very small strain value and a point corresponding to either 40% of the ultimate stress or a specific strain value. Normal-weight concrete has a modulus of elasticity of 14 to 40 GPa.

Poisson's ratio can also be determined using ASTM C469. Poisson's ratio is used in advanced structural analysis of shell roofs, flat-plate roofs, and mat foundations. Poisson's ratio of concrete varies between 0.11 and 0.21, depending on aggregate type, moisture content, concrete age, and compressive strength. A value of 0.15 to 0.20 is commonly used.

It is interesting to note that both aggregate and cement paste, when tested individually, exhibit linear stress–strain behavior. However, the stress–strain relationship of concrete is nonlinear. The reason for this behavior is attributed to the microcracking in concrete at the interface between aggregate particles and the cement paste (Shah and Winter, 1968).

The modulus of elasticity of concrete increases when the compressive strength increases, as demonstrated in Figure 7.34. There are several empirical relations between the modulus of elasticity of concrete and the compressive strength. For normal-weight concrete, the relationship used in the United States for designing concrete structures is defined by the ACI Building Code as:

$$E_c = 4{,}731\sqrt{f'_c} \qquad\qquad (7.3)$$

where

E_c = the modulus of elasticity, MPa
f'_c = the compressive strength, MPa.

Sample Problem 7.8

A normal-weight concrete has an average compressive strength of 30 MPa. What is the estimated modulus of elasticity?

Solution

$$E_c = 4731\sqrt{f'_c} = 4731(30)^{1/2} = 25913 \text{ MPa} = 25.9 \text{ GPa}$$

This relation is useful, since it relates the modulus of elasticity (needed for designing concrete structures) with the compressive strength, which can be measured easily in the laboratory.

7.5   Testing of Hardened Concrete

Many tests are used to evaluate the hardened concrete properties, either in the laboratory or in the field. Some of these tests are destructive, while others are nondestructive. Tests can be performed for different purposes; however, they are mostly conducted to control the quality of the concrete and to check specification compliance. Probably the most common test performed on hardened concrete is the compressive strength test, since it is relatively easy to perform and since there is a strong correlation between the compressive strength and many desirable properties (Neville, 1996; Mehta and Monteiro, 2013). Other tests include split tension, flexure strength, rebound hammer, penetration resistance, ultrasonic pulse velocity, and maturity tests.

7.5.1   Compressive Strength Test

The compressive strength test is the test most commonly performed on hardened concrete. Compressive strength is one of the main structural design requirements to ensure that the structure will be able to carry the intended load. As indicated earlier, compressive strength increases as the water–cement ratio decreases. Since the water–cement ratio is directly related to the concrete quality, compressive strength is also used as a measure of quality, such as durability and resistance to weathering. Thus, in many cases, designers specify a high compressive strength of the concrete to ensure high quality, even if this strength is not needed for structural support. The compressive strength f'_c of normal-weight concrete is between 20 and 40 MPa. In the United States, the test is performed on cylindrical specimens and is standardized by ASTM C39. The specimen is prepared, either in the lab or in the field, according to ASTM C192 or C31, respectively. Cores could also be drilled from the structure following ASTM C42. The standard specimen size is 0.15 m in diameter and 0.30 m high, although other sizes with a height–diameter ratio of two can also be used. The diameter of the specimen must be at least three times the nominal maximum size of the coarse aggregate in the concrete.

Specimens are prepared in three equal layers and are rodded 25 times per layer. After the surface is finished, specimens are kept in the mold for the first 24 ± 8 hours. Specimens are then removed from the mold and cured at 23 ± 1.7°C, either in saturated-lime water or in a moist cabinet having a relative humidity of 95% or higher, until the time of testing. Before testing, specimens are capped at the two bases to ensure parallel surfaces. High-strength gypsum plaster, sulfur mortar, or a special capping compound can be used for capping and is applied with a special alignment device (ASTM C617). The specimens are tested by applying axial compressive load with a specified rate of loading until failure (Figure 7.35). The compressive strength of the specimen is determined by dividing the maximum load carried by the specimen during the test by the average cross-sectional area. The number of specimens and the number of test batches depend on established practice and the nature of the test program. Usually three or more specimens are tested for each test age and test condition. Test ages often used are 7 days and 28 days for normal concrete.

Note that the test specimen must have a height–diameter ratio of two. The main reason for this requirement is to eliminate the end effect due to the friction between the loading heads and the specimen. Thus, producing a zone of uniaxial compression

FIGURE 7.35   Compressive strength test.

within the specimen. If the height–diameter ratio is less than two, a correction factor can be applied to the results as indicated in ASTM C39.

The compressive strength of the specimen is affected by the specimen size. Increasing the specimen size reduces the strength because there is a greater probability of weak elements where failure starts in large specimens than in small specimens. In general, large specimens have less variability and better representation of the actual strength of the concrete than small specimens. Therefore, the 0.15-m by 0.30-m size is the most suitable specimen size for determining the compressive strength. However, some agencies use 0.10-m diameter by 0.20-m high specimens. The advantages of using smaller specimens are the ease of handling, less possibility of accidental damage, less concrete needed, the ability to use a low-capacity testing machine, and less space needed for curing and storage. Because of the strength variability of small specimens, more specimens should be tested for smaller specimens than are tested for standard-sized specimens. In some cases, five 0.10-m by 0.20-m replicate specimens are used instead of the three replicates commonly used for the standard-sized specimens. Also, when small-sized specimens are used, the engineer should understand the limitations of the test and consider these limitations in interpreting the results. For example research by the Missouri Department of Transportation recommended a strength correction factor of 0.94 when the smaller samples are used. This recommendation was for a specific application of one type of concrete and it should not be applied to other situations (http://library.modot.mo.gov/RDT/reports/Ri03038/Brf03038.pdf).

The interface between the hardened cement paste and aggregate particles is typically the weakest location within the concrete material. When concrete is stressed beyond the elastic range, microcracks develop at the cement paste–aggregate interface and continuously grow until failure. Figure 7.36 shows a scanning electron microscope micrograph of the fractured surface of a hardened cement mortar cylinder at 500×. The figure shows that the cleavage fracture surfaces where sand particles were dislodged during loading. The figure also shows the microcracks around some sand particles developed during loading.

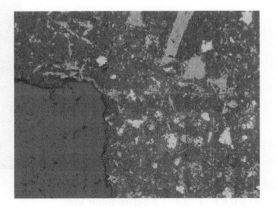

FIGURE 7.36   Scanning electron image showing the interface between a sand grain (lower left corner) and the paste.

7.5.2 ▨ Split-Tension Test

The split-tension test (ASTM C496) measures the tensile strength of concrete. In this test, a 0.15-m by 0.30-m concrete cylinder is subjected to a compressive load at a constant rate along the vertical diameter until failure, as shown in Figure 7.37. Failure of the specimen occurs along its vertical diameter, due to tension developed in the transverse direction. The split tensile (indirect tensile) strength is computed as

$$T = \frac{2P}{\pi L d} \qquad\qquad (7.4)$$

where

T = tensile strength, MPa,
P = load at failure, N,
L = length of specimen, mm, and
d = diameter of specimen, mm

Typical indirect tensile strength of concrete varies from 2.5 to 3.1 MPa (Neville, 1996). The tensile strength of concrete is about 10% of its compressive strength.

7.5.3 ▨ Flexure Strength Test

The flexure strength test (ASTM C78) is important for design and construction of road and airport concrete pavements. The specimen is prepared either in the lab or in the field in accordance with ASTM C192 or C31, respectively. Several specimen sizes can be used. However, the sample must have a square cross section and a span

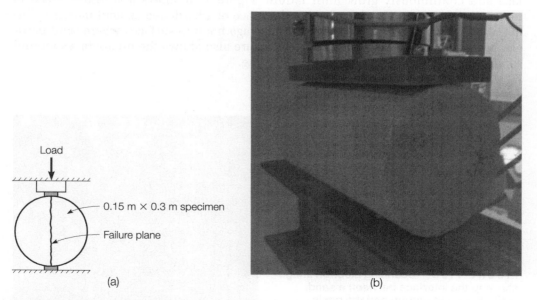

Load

0.15 m × 0.3 m specimen

Failure plane

(a)

(b)

FIGURE 7.37 Split-tension test.

of three times the specimen depth. Typical dimensions are 0.15-m by 0.15-m cross section and 0.30-m span. After molding, specimens are kept in the mold for the first 24 ± 8 hours, then removed from the mold and cured at 23 ± 1.7°C, either in saturated-lime water or in a moist cabinet with a relative humidity of 95% or higher until testing. The specimen is then turned on its side and centered in the third-point loading apparatus, as illustrated in Figure 7.38. The load is continuously applied at a specified rate until rupture. If fracture initiates in the tension surface within the middle third of the span length, the flexure strength (modulus of rupture) is calculated as

$$R = \frac{Mc}{I} \tag{7.5}$$

where

R = flexure strength, MPa,
M = maximum bending moment = $PL/6$, N·mm,
c = d/2, mm,
I = moment of inertia = $bd^3/12$, mm^4
P = maximum applied load, which is distributed evenly (1/2 to each) over the two loading points, N,
L = span length, mm,
b = average width of specimen, mm, and
d = average depth of specimen, mm

Note that third-point loading ensures a constant bending moment without any shear force applied in the middle third of the specimen. Thus, Equation 7.5 is valid as long as fracture occurs in the middle third of the specimen. If fracture occurs slightly outside the middle third, the results can still be used with some corrections. Otherwise the results are discarded.

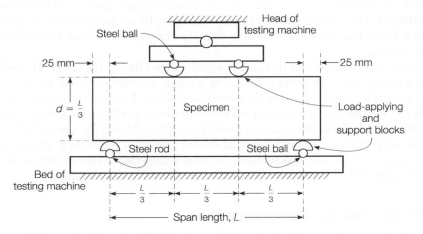

FIGURE 7.38   Apparatus for flexure test of concrete by third-point loading method. (reprinted, with permission, from ASTM C78, Figure 1, copyright ASTM International, 100 Barr Harbor Drive, West Conshohocken, PA 19428)

FIGURE 7.39 Rebound hammer for nondestructive evaluation of hardened concrete.

For normal-weight concrete, the flexure strength can be approximated as:

$$R = (0.62 \text{ to } 0.83)\sqrt{f'_c} \text{ MPa} \tag{7.6}$$

7.5.4 ▓ Rebound Hammer Test

The rebound hammer test, also known as the Schmidt hammer test, is a nondestructive test performed on hardened concrete to determine the hardness of the surface (Figure 7.39). The hardness of the surface can be correlated, to some extent, with the concrete strength. The rebound hammer is commonly used to get an indication of the concrete strength. The device is about 0.3 m long and encloses a mass and a spring. The spring-loaded mass is released to hit the surface of the concrete. The mass rebounds, and the amount of rebound is read on a scale attached to the device. The larger the rebound, the harder is the concrete surface and, therefore, the greater is the strength. The device usually comes with graphs prepared by the manufacturer to relate rebound to strength. The test can also be used to check uniformity of the concrete surface.

The test is very simple to run and is standardized by ASTM C805. To perform the test, the hammer must be perpendicular to a clean, smooth concrete surface. In some cases, it would be hard to satisfy this condition. Therefore, correlations, usually provided by the manufacturer, can be used to relate the strength to the amount of rebound at different angles. Rebound hammer results are also affected by several other factors, such as local vibrations, the existence of coarse aggregate particles at the surface, and the existence of voids near the surface. To reduce the effect of these factors, it is desirable to average 10 to 12 readings from different points in the test area.

7.5.5 ▓ Penetration Resistance Test

The penetration resistance test, also known as the Windsor Probe test, is standardized by ASTM C803. The instrument (Figure 7.40) is a gunlike device that shoots probes into the concrete surface in order to determine its strength. The amount

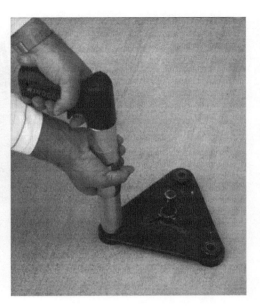

FIGURE 7.40   Windsor probe test device.

of penetration of the probe in the concrete is inversely related to the strength of concrete. The test is almost nondestructive since it creates small holes in the concrete surface.

The device is equipped with a special template with three holes, which is placed on the concrete surface. The test is performed in each of the holes. The average of the penetrations of the three probes through these holes is determined, using a scale and a special plate. Care should be exercised in handling the device to avoid injury. As a way of improving safety, the device cannot be operated without pushing hard on the concrete surface to prevent accidental shooting. The penetration resistance test is expected to provide better strength estimation than the rebound hammer, since the penetration resistance measurement is made not just at the surface but also in the depth of the sample.

7.5.6 ▦ Ultrasonic Pulse Velocity Test

The ultrasonic pulse velocity test (ASTM C597) measures the velocity of an ultrasonic wave passing through the concrete (Figure 7.41). In this test, the path length between transducers is divided by the travel time to determine the average velocity of wave propagation. Attempts have been made to correlate pulse velocity data with concrete strength parameters. No good correlations were found, since the relationship between pulse velocity and strength data is affected by a number of variables, such as age of concrete, aggregate–cement ratio, aggregate type, moisture condition, and location of reinforcement (Mehta and Moneiro, 2013). This test is used to detect cracks, discontinuities, or internal deterioration in the structure of concrete.

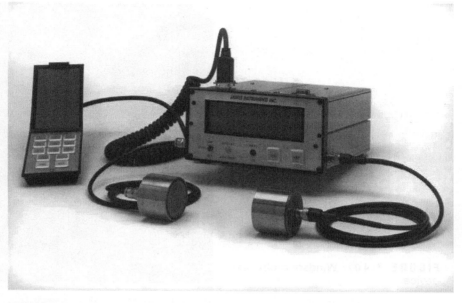

FIGURE 7.41 Ultrasonic pulse velocity apparatus. (Courtesy of James Instruments Inc.)

7.5.7 ■ Maturity Test

Maturity of a concrete mixture is defined as the degree of cement hydration, which varies as a function of both time and temperature. Therefore, it is assumed that, for a particular concrete mixture, strength is a function of maturity. Maturity meters (Figure 7.42) have been developed to provide an estimate of concrete strength by monitoring the temperature of concrete with time. This test (ASTM C1074) is performed on fresh concrete and continued for several days. The maturity meter must be calibrated for each concrete mix.

7.6 Alternatives to Conventional Concrete

Several alternatives increase the flexibility and applications of concrete. While a technical presentation of materials for each of these technologies is beyond the scope of this book, the engineer should be aware of some of the materials used to provide additional capabilities of concrete. Some of these alternatives include the following:

  self-consolidating concrete
  flowable fill
  shotcrete
  lightweight concrete
  heavyweight concrete

FIGURE 7.42   Checking the maturity of fresh concrete using the maturity meter.

high-strength concrete
shrinkage-compensating concrete
polymers and concrete
fiber-reinforced concrete
roller-compacted concrete
high-performance concrete
pervious concrete

7.6.1 ▨ Self-Consolidating Concrete

Self-consolidating concrete (SCC), also known as self-compacting concrete (Figure 7.43), is a highly flowable, nonsegregating concrete that can spread into place, fill the formwork, and encapsulate the reinforcement, without any mechanical consolidation (NRMCA). Some of the advantages of using SCC are the following:

1. It can be placed at a faster rate, with no mechanical vibration and less screeding, resulting in savings in placement costs.
2. It improves and makes more uniform the architectural surface finish with little or no remedial surface work.
3. It improves ease of filling restricted sections and hard-to-reach areas. This allows the designer to create structural and architectural shapes not achievable with conventional concrete.
4. It improves consolidation around reinforcement and bond with reinforcement.
5. It improves pumpability.

FIGURE 7.43 Placing self-consolidating concrete in a heavily reinforced slab. (Courtesy of FHWA)

Two important properties specific to SCC in its plastic state are its *flowability* and *stability*. The high flowability of SCC is achieved by using high-range water-reducing admixtures without adding extra mixing water. The stability, or resistance to segregation, is attained by increasing the amount of fines and/or by using admixtures that modify the viscosity of the mixture. Fines could be either cementitious materials or mineral fines.

The most common field test that has been used to measure the flowability and stability of SCC is a modified version of the slump test, discussed earlier. In this slump flow test, the slump cone is completely filled, without consolidation. The cone is lifted, and the spread of the concrete is measured, as shown in Figure 7.44. The flowability is measured by the spread, or slump flow. The spread typically ranges from 455 to 810 mm, depending on the requirements of the project. The resistance to segregation, or stability, is measured with the visual stability index (VSI). The VSI is established on the basis of whether bleed water is observed at the leading edge of the spreading concrete or aggregates pile at the center. VSI values range from 0 for "highly stable" to 3 for unacceptable stability.

During the slump flow test, the viscosity can be measured by the rate at which the concrete spreads. The time taken for the concrete to reach a spread diameter of 50 cm from the moment the sump cone is lifted up is measured. This is called the T_{50} measurement and typically varies between 2 and 10 seconds for SCC. A T_{50} value

FIGURE 7.44 Measuring the spread during the slump flow test.

indicates a more viscous mix, which is more appropriate for concrete in applications with congested reinforcement or in deep sections. A lower T_{50} value may be appropriate for concrete that has to travel long horizontal distances without much obstruction.

7.6.2 ▧ Flowable Fill

Flowable fill is a self-leveling and self-compacting, cementitious material with an unconfined compressive strength of 8.3 MPa or less. Flowable fill is primarily used as a backfill material in lieu of compacted granular fill (Figure 7.45). Flowable fill is also commonly referred to as controlled low-strength material (CLSM), controlled density fill (CDF), flowable compacting fill, lean fill, unshrinkable fill, flow mortar, fly ash flow, and liquid dirt (NRMCA).

The flowable fill mix consists of cement, sand, and water typically mixed with fly ash, ground granulated blast furnace (GGBF) slag, and/or air generating admixtures. The type and proportions of ingredients used to make the flowable fill can largely change its properties to match its intended use.

One of the unique properties of flowable fill that makes it advantageous compared with compacted granular fill is its flowability. High flowability and self-leveling characteristics allow flowable fill to eliminate voids and access spaces that prove to be difficult or impossible with compacted granular fill.

When the flowable fill is placed in the cavity, its initial in-place volume is slightly reduced in the order of 10 mm/m of depth. This settlement is caused by the displacement of water and the release of entrapped air as a result of consolidation. Once the flowable fill is hardened, the settlement stops and the volume does not change.

FIGURE 7.45 Filling a sink hole with flowable fill.

Flowable fill can be proportioned to have a wide range of compressive strength. However, the most commonly used flowable fill mixtures are proportioned with consideration of possible excavation in future years and range in compressive strengths between 0.35 and 1 MPa. Flowable fill with a compressive strength less than 1 MPa may be excavated manually, while at the same time having adequate bearing capacity to support external loads, such as the weight of a vehicle. Flowable fill exhibiting strengths between 1 and 2 MPa will require mechanical equipment for excavation. If the strength exceeds 2 MPa, the material is not considered excavatable.

Flowable fill has several advantages over compacted granular backfill. Flowable fill does not require compaction, which is a main concern with granular backfill. Flowable fill can also reach inaccessible locations, such as places around pipes, which are hard to reach with granular backfill. Flowable fill also has a greater bearing capacity than compacted granular fill and no noticeable settlement. It can even be placed in standing water. These advantages result in reduced in-place costs for labor and equipment, as well as time saving during construction.

Flowable fill is typically used as backfill for utility trenches, retaining walls, pipe bedding, and building excavation. It is also used as structural fill for poor soil conditions, mud jacking, floor and footing support, and bridge conversion. Other uses include pavement base, erosion control, and void filling. Flowable fill has been getting more common in recent years and is available at most ready-mix producers.

7.6.3 ▇ Shotcrete

Shotcrete is mortar or small-aggregate concrete that is sprayed at high velocity onto a surface (see Figures 7.46 and 7.47). Shotcrete, also known as "gunite" or "sprayed

FIGURE 7.46   Constructing a swimming pool using shotcrete. (Courtesy of the American Shotcrete Association)

FIGURE 7.47   Shotcrete used in lining a tunnel. (Courtesy of the American Shotcrete Association)

concrete," is a relatively dry mixture that is consolidated by the impact force and can be placed on vertical or horizontal surfaces without sagging.

Shotcrete is applied by either the dry or wet process. In the dry process, a premixed blend of cement and damp aggregate is propelled through a hose by compressed air to a nozzle, while the water is added at the nozzle. In the wet process, all ingredients are premixed and pumped through a hose to the nozzle and forced to the surface using compressed air. In either case, the nozzle should be held perpendicular to the surface to reduce the rebound of coarse aggregates off the surface. The nozzle is held about 0.5 to 1.5 m away from the surface.

Supplementary cementitious materials, such as fly ash and silica fume, can be used in shotcrete to improve workability, chemical resistance, and durability. Accelerating admixtures can also be used to reduce the time of initial set and to allow buildup of thicker layers of shotcrete in a single pass. Steel fibers may also be used to improve flexural strength, ductility, and toughness (Kosmatka et al., 2011).

7.6.4 ■ Lightweight Concrete

Students competing in the annual ASCE concrete canoe competition frequently produce concrete with a unit weight less than water. The *ACI Guide for the Structural Lightweight Aggregate Concrete* requires a 28-day compressive strength of 17 MPa and an air-dried unit weight of less than 1850 kg/m^3 for structural lightweight concrete. The use of lightweight concrete in a structure is usually predicated on the overall cost of the structure; the concrete may cost more, but the reduced dead weight can reduce structural and foundation costs.

The mix proportions for lightweight concrete must compensate for the absorptive nature of the aggregates. Generally, lightweight aggregates are highly absorptive and can continue to absorb water for an extended period of time (Figures 7.48 and 7.49). This makes the determination of a water–cement ratio problematical. In addition, the lightweight aggregates tend to segregate by floating to the surface. A minimum slump mix, with air entraining, is used to mitigate this effect.

Nonstructural applications of very lightweight concrete have also been developed. Concrete made with styrofoam "aggregates" has been used for insulation in some building construction.

7.6.5 ■ Heavyweight Concrete

Biological shielding used for nuclear power plants, medical units, and atomic research and test facilities requires massive walls to contain radiation. Concrete is an excellent shielding material. For biological shields, the mass of the concrete can be increased with the use of heavyweight aggregates. The aggregates can be either natural or manufactured. Natural heavyweight aggregates include barite, magnetite, hematite, geothite, illmenite, and ferrophosphorus. The specific gravity of these aggregates ranges from 3.4 to 6.5. Steel, with a specific gravity of 7.8, can be used as aggregate for heavyweight concrete (Figure 7.50). However, the specific gravity of the aggregates makes workability and segregation of heavyweight concrete problematical. Using a higher proportion of sand improves the workability. The workability

FIGURE 7.48   Aggregate used for lightweight concrete.

FIGURE 7.49   ASCE students at Arizona State University making a concrete canoe using lightweight concrete.

FIGURE 7.50 Steel used as aggregate for heavyweight concrete.

problem can be avoided by preplacing aggregate, then filling the voids between aggregate particles with grout of cement, sand, water, and admixtures. ASTM C637, *Specifications for Aggregates for Radiation Shielding Concrete* and ASTM C637, *Nomenclature of Constituents of Aggregates for Radiation Shielding Concrete* provide further information on heavyweight concrete practices.

7.6.6 ▓ High-Strength Concrete

Concrete made with normal-weight aggregate and having compressive strengths greater than 40 MPa is considered to be high-strength concrete. Producing a concrete with more than 40 MPa compressive strength requires care in the proportioning of the components and in quality control during construction. The microstructure of concrete with a compressive strength greater than 40 MPa is considerably different from that of conventional concrete. In particular, the porosity of the cement paste and the transition zone between the cement paste and the aggregate are the controlling factors for developing high strength. This porosity is controlled by the water–cement ratio. The development of superplasticizers has permitted the development of high-strength concrete that is workable and flowable at low water–cement ratios. In addition, high-strength concrete has excellent durability due to its tight pore structure. In the United States, high-strength concrete is used primarily for skyscrapers. The high-strength and corresponding high elastic modulus allow for reduced structural element size.

7.6.7 ▓ Shrinkage-Compensating Concrete

Normal concrete shrinks at early ages, especially if it is not properly cured, as discussed earlier in this chapter. The addition of alumina powders to the cement can cause the concrete to expand at early ages. Shrinkage-compensating cement is marketed as Type K cement. Expansive properties can be used to advantage by restraining the concrete, either by reinforcing or by other means, at early ages. As the restrained concrete tries to expand, compressive stresses are developed. These compressive stresses reduce the tensile stresses developed by drying shrinkage, and the chance of the concrete cracking due to drying shrinkage is reduced. Details on the design and use of shrinkage-compensating concrete are available in ACI Committee 223 report *Recommendations for the Use of Shrinkage-Compensating Cements.*

7.6.8 ▨ Polymers and Concrete

Polymers can be used in several ways in the production of concrete. The polymer can be used as the sole binding agent to produce *polymer concrete*. Polymers can be mixed with the plastic concrete to produce *polymer–portland cement concrete*. Polymers can be applied to hardened concrete to produce *polymer-impregnated concrete*.

Polymer concrete is a mixture of aggregates and a polymer binder. There are a wide variety of polymers that can be mixed with aggregates to make polymer concrete. Some of these can be used to make rapid-setting concrete that can be put in service in under an hour after placement. Others are formulated for high strength; 140 MPa is possible. Some have good resistance to chemical attack. A common characteristic is that most polymer concretes are expensive, limiting their application to situations in which their unique characteristics make polymer concrete a cost-effective alternative to conventional concrete.

Polymer–portland cement concrete incorporates a polymer into the production of the portland cement concrete. The polymer is generally an elastomeric emulsion, such as latex.

7.6.9 ▨ Fiber-Reinforced Concrete

The brittle nature of concrete is due to the rapid propagation of microcracking under applied stress. However, with fiber-reinforced concrete, failure takes place due to fiber pull-out or debonding. Unlike plain concrete, fiber-reinforced concrete can sustain load after initial cracking. This effectively improves the toughness of the material. In addition, the flexural strength of the concrete is improved by up to 30%. For further information on the design and applications of fiber-reinforced concrete, consult the *ACI Guide for Specifying, Mixing, Placing and Finishing Steel Fiber Reinforced Concrete.*

Fibers are available in a variety of sizes, shapes, and materials (Figure 7.51). The fibers can be made of steel, plastic, glass, and natural materials. Steel fibers are the

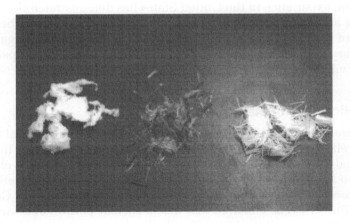

FIGURE 7.51   Fibers used for fiber-reinforced concrete.

most common. The shape of fibers is generally described by the aspect ratio, length/ diameter. Steel fibers generally have diameters from 0.25 to 0.9 mm with aspect ratios of 30 to 150. Glass fiber elements' diameters range from 0.013 to 1.3 mm.

The addition of fibers to concrete reduces the workability. The extent of reduction depends on the aspect ratio of the fibers and the volume concentration. Generally, due to construction problems, fibers are limited to a maximum of 2% by volume of the mix. Admixtures can be used to restore some of the workability to the mix.

Since the addition of fibers does not greatly increase the strength of concrete, its use in structural members is limited. In beams, columns, suspended floors, and so on, conventional reinforcing must be used to carry the total tensile load. Fiber-reinforced concrete has been successfully used for floor slabs, pavements, slope stabilization, and tunnel linings.

7.6.10 ■ Roller-Compacted Concrete

Based on the unique requirements for mass concrete used for dam construction, roller-compacted concrete (RCC) was developed. This material uses a relatively low cement factor, relaxed gradation requirements, and a water content selected for construction considerations rather than strength. RCC is a no-slump concrete that is transported, placed, and compacted with equipment used for earth and rockfill dam construction. The RCC is hauled by dump trucks, spread with bulldozers, and compacted with vibration compactors. Japanese experience using RCC in construction found several advantages:

1. The mix is economical, because of the low cement content.
2. Formwork is minimal, because of the layer construction method.
3. The low cement factor limits the heat of hydration, reducing the need for external cooling of the structure.
4. The placement costs are lower than those for conventional concrete methods, due to the use of high-capacity equipment and rapid placement rates.
5. The construction period is shorter than that for conventional concrete.

In addition, experience in the United States has demonstrated that RCC in-place material costs are about one-third those of conventional concrete. The two primary applications of RCC have been for the construction of dams and large paved areas, such as military tank parking aprons. Figure 7.52 shows roller compacted concrete paved with a vibratory screed.

7.6.11 ■ High-Performance Concrete

While the current specifications for concrete have provided a material that performs reasonably well, there is concern that the emphasis on strength in the mix design process has led to concrete that is inadequate in other performance characteristics. This has led to an interest in developing specifications and design methods for what has been termed *high-performance concrete* (HPC). The American Concrete Institute (ACI) defines HPC as concrete that meets special performance and uniformity requirements, which cannot always be obtained using conventional ingredients,

FIGURE 7.52   Placing roller compacted concrete with a vibratory screed. (Courtesy of A.G. Peltz Group, LLC)

normal mixing procedures, and typical curing practices. These requirements may include the following enhancements:

ease of placement and compaction
long-term mechanical properties
early-age strength
toughness
volume stability
extended life in severe environments

These enhanced characteristics may be accomplished by altering the aggregate gradation, including special admixtures, and improving mixing and placement practices. Currently, a compressive strength in the order of 70 to 175 MPa can be obtained. As the need for HPC is better understood and embraced by the engineering community, there will probably be a transition in concrete specification from the current prescriptive method to the performance-based or performance-related specifications. A Strategic Highway Research Program (SHRP) study (Zia et al., 1991) defined HPC as:

maximum water–cementitious materials ratio of 0.35;
minimum durability factor of 80%, as determined by ASTM C 666, Procedure A; and
minimum strength criteria of either:

a. 21 MPa within 4 hours after placement (Very Early Strength, VES),
b. 34 MPa within 24 hours (High Early Strength, HES), or
c. 69 MPa within 28 days (Very High Strength, VHS)

Thus, high-performance concrete is characterized by special performance, both short-term and long-term, and uniformity in behavior (Nawy, 2000). Such requirements cannot always be achieved by using only conventional materials or applying conventional practices. Since concrete is the most widely used construction material worldwide, new concrete construction has to utilize the currently available new technology in order to eliminate costly future rehabilitation.

Revolutionary new construction materials and modifications and improvements in the behavior of traditional materials have been developed in the past three decades. These developments have been considerably facilitated by increased knowledge of the molecular structure of materials, studies of long-term failures, development of more powerful instrumentation and monitoring techniques, and the need for stronger and better performing materials, suitable for larger structures, longer spans, and more ductility.

Beyond the current advances of the concrete technology and the development of high-performance concretes, it is expected that the concrete industry will continue to improve through the development of new components and admixtures, microstructural studies, blended cement compositions, better material selection and proportioning techniques, and more efficient placement techniques.

7.6.12 ■ Pervious Concrete

Concrete is generally designed to have limited amount of air voids in order to increase its strength and durability, which is required in most concrete structures. However, in response to sustainability considerations, pervious concrete has been developed that is specifically designed to allow water to pass through it (Figure 7.53) (NRMCA, 2011). This is advantageous in some structures that do not require high strength, particularly in urban areas. The pervious concrete may be used as the surface layer

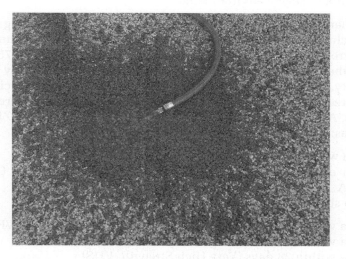

FIGURE 7.53 Water flow through pervious concrete.

for light-traffic pavement designed to capture and store rainfall and runoff, which is then allowed to percolate into the subgrade soil.

Pervious concrete materials are the same as conventional concrete with the exception that the amount of fines are limited to create the void space in the mix needed for permeability. Pervious concrete is typically designed with a void content of 15 to 30%. Due to the low amount of fine aggregates and the high air voids, a modified mix design procedure is needed.

The surface texture and appearance is determined by the maximum aggregate size. For parking lots, low-traffic pavements, and pedestrian walkways the smallest feasible aggregate is used for aesthetic reasons. Coarse aggregate size has been used extensively for parking lot and pedestrian applications, dating from 1990s in Florida (NRMCA. 2011).

Typical material properties of pervious concrete include (NRMCA, 2011):

- Permeability—120 to 770 L/m^2/min.
- Compressive strength—3.5 to 28 MPa typically about 17 MPa. The relatively low compressive strength is due to the need to have high air voids for permeability. Compressive strength should not be used as an acceptance criteria.
- Flexural strength—1 to 3.8 MPa; flexural strength is generally not used as a design parameter for pervious pavements.
- Drying shrinkage—0.002, approximately $\frac{1}{2}$ the shrinkage of conventional concrete. The lower shrinkage and the surface texture allow construction without control joints; the pavement is allowed to crack randomly.
- Freeze–thaw resistance—the rapid draining of the pavement limits the potential freeze–thaw resistance. Conventional testing methods that maintain the concrete in a saturated condition are overly harsh for pervious concrete. Freeze–thaw resistance is dramatically improved with air-entrainer admixtures.

7.7 Concrete Sustainability

7.7.1 ▓ LEED Considerations

The concrete industry has embraced the LEED concept and has developed a wealth of information for designers to consider. Ashley and Lemay (2009) compiled a reference guide for claiming LEED credits with the use of concrete. The following is a consolidation of the information in this document. The authors provide guidance on potential credits with respect to the LEED 2009 New Construction rating system.

Development Density and Community Connectivity—LEED credits can be claimed for high-density development, which essentially requires multistory buildings. Ashley and Lemay claim concrete is advantageous for multistory buildings construction. (Of course the steel industry can make a similar claim.)

Brownfield Redevelopment—Contaminated soils can be stabilized with cementitious materials and leaching concentrations can be controlled to below regulatory limits. The cementitious material is mixed with the contaminated soil and allowed to cure. Various methods can be used to perform the mixing.

Protect and Restore Habitat—Minimizing disturbance of natural areas or restoration of damaged areas to provide habitat and promote biodiversity can be accomplished with parking structures, either integrated into the building or as separate structures. The use of pervious concrete for parking areas can control runoff and reduce the need for retention basins.

Maximize Open Space—The potential LEED credits for this category are similar to the protection and restoration of habitat.

Stormwater Design: Quantity Control—Two potential uses of concrete were identified for this LEED credit, pervious pavements and reinforced concrete roof decks to support vegetated roofs. Pervious concrete systems generally remove over 80% of total suspended solids and are considered a best management practice for treating stormwater.

Heat Island Effect: Non-Roof—Strategies for minimizing the thermal differences between developed and undeveloped areas include providing shade for at least 50% of hardscaped surfaces or provide hardscaped surfaces with solar reflectance index, SRI, of at least 29. The requirement for SRI can be met by using concrete 50 % of sidewalks, parking lots, drives, and other impervious surfaces. Where paved surfaces are required, using materials with higher SRI can reduce the heat island effect resulting in saving energy savings in adjacent buildings by reducing the demand for air conditioning. Concrete generally has an SRI of greater than 29. Concretes made with white cements or slag can have SRI greater than 78. Covering a minimum of 50 % of parking spaces including underground, under deck, under roof, or under building significantly reduces the heat island effect.

Heat Island Effect: Roof—Concrete with high SRI can be used for roofing and concrete structural systems are ideal for supporting the heavy loads of vegetated roofs.

Water Efficient—There are several LEED credits available for efficient water use and reuse including landscaping, wastewater technologies, and water use reduction. In each case, water is captured in a cistern for reuse. The cisterns may be constructed of concrete. Capturing water in a previous pavement and transferring the water to a cistern for subsequent use is also an option.

Energy Performance—To achieve any level of LEED certification, proposed buildings must demonstrate a minimum of a 10% improvement for new buildings, 5% existing building renovation, compared to the baseline building performance per ASHRAE/IESNA Standard 90.1-2007. The use of thermal mass for building components is an option for achieving this performance. Buildings constructed of concrete and concrete masonry possess thermal mass which helps moderate indoor temperatures, reducing heating and cooling loads.

Further LEED credit can be earned by optimizing the energy performance of a proposed building by demonstrating through analysis that the energy cost savings exceed the baseline requirements. Insulated concrete systems provide thermal efficiencies that can help earn this credit.

Construction Waste Management—Concrete is a relatively heavy material that can be crushed and recycled into aggregate for road bases or construction fill. In addition, concrete diverted from landfills and recycling into new concrete, or an alternative product such as landscaping blocks, can be considered for this credit.

Recycled Content—This LEED requirement encourages use of recycled materials including post-consumer recycled content and pre-consumer content. Supplementary cementitious materials, such as fly ash, silica fume, and slag cement, are pre-consumer recycled content. Recycled concrete as aggregate qualifies as post-consumer recycled content. Generally, reinforcing bars, manufactured from recycled steel, are considered post-consumer recycled content.

Regional Materials—The requirements of these credits are to use building materials or products that have been extracted, harvested, or recovered, as well as manufactured, within 805 km (computed on a weighted average of the transportation mode) of the project. Ready-mixed concrete will almost always qualify since ready-mixed concrete plants are generally within 805 km of a job site, and most of the ingredient materials are harvested within 805 km.

Daylight & Views—The strategy is to design the building to maximize interior daylighting and views to the outdoors through building orientation, shallow floor plates, and increased building perimeter. Concrete floor systems can span large distances with shallow floor plates and column free spaces to help achieve these credits. Exposed concrete ceilings can reflect light deep into interior spaces.

Innovation in Design—If an innovative green design strategy is used that does not fit into the point structure of the five LEED categories, or if it goes significantly beyond a credit requirement in one of the existing credit categories then additional LEED credit can be earned. For example, if the extracted, processed and manufactured regionally content is 30% of the materials, then the project could receive for exceeding the minimum requirements. Concrete contributes significantly to this credit category. Using materials with over 30% recycled content exceeds Materials and Resources Credit minimums. Concrete made some percentage of recycled aggregate in combination with supplementary materials can contribute significantly to this credit. Concrete with high volumes of fly ash, slag, or silica reduces the embodied CO_2 and may qualify for this credit. Use of exposed concrete for walls, floors, and ceilings can reduce other wall, floor and ceiling coverings which reduces the amount of material used. Since these coverings are common sources of volatile organic compounds (VOCs) credit may be earned for improved indoor air quality.

7.7.2 ■ Other Sustainability Considerations

Concrete is depended on the constituent materials of aggregates and cement. The sections in the aggregate and cement chapters should be considered when considering the sustainability of concrete.

Summary

The design of durable portland cement concrete materials is the direct responsibility of civil engineers. Selection of the proper proportions of portland cement, water, aggregates, and admixtures, along with good construction practices, dictates the quality of concrete used in structural applications. Using the volumetric mix design method presented in this chapter will lead to concrete with the required strength and durability. However, the proper design of portland cement concrete is irrelevant unless proper construction procedures are followed, including the appropriate mixing, transporting, placing, and curing of the concrete. To ensure that these processes produce concrete with the desired properties, civil engineers perform a variety of quality control tests, including slump tests, air content tests, and strength-gain-with-time tests.

While the vast majority of concrete projects are constructed with conventional materials, there are a variety of important alternative concrete formulations available for specialty applications. These alternatives are introduced in this chapter; however, the technology associated with these alternatives is relatively complex, and further study is required in order to fully understand the behavior of these materials.

Questions and Problems

7.1   The design engineer specifies a concrete strength of 37.92 MPa. Determine the required average compressive strength for:
a. a new plant where s is unknown
b. a plant where $s = 3.45$ MPa for 22 test results
c. a plant with extensive history of producing concrete with $s = 2.76$ MPa
d. a plant with extensive history of producing concrete with $s = 4.14$ Mpa

7.2   A project specifies a concrete strength of 24.1 MPa. Materials engineers will design the mix for a strength higher than that to account for variables.
a. Calculate the required average compressive strength of the mix design if the mixing plant has a standard deviation of $s = 3.8$ MPa.
b. Using the ACI code equation, estimate what would be the modulus of elasticity of this concrete at the *required* compressive strength.

7.3   A project specifies a concrete strength of at least 20.7 MPa. Materials engineers will design the mix for a strength higher than that. Calculate the required average compressive strength of the mix design if the standard deviation is $s = 2.4$ MPa. Estimate the modulus of elasticity of the concrete at the required average compressive strength (the calculated strength, not the given strength).

7.4   What is your recommendation for the maximum size of coarse aggregate for the following situation? A continuously reinforced concrete pavement cross section contains a layer of 19-mm reinforcing bars at 150 mm. centers, such that the steel is just above mid-depth of a 250 mm. thick slab. Cover over the top of the steel is therefore about 100 mm.

7.5   A concrete mix with a 75 mm slump, w/c ratio of 0.50, and sand with a fineness modulus of 2.4 contains 1009 kg/m³ of coarse aggregate. Compute the required mass of coarse aggregate per cubic meter. To adjust the mix so as to increase the compressive strength, the water–cement ratio is reduced to 0.45. Will the quantity of coarse aggregate increase, decrease, or stay the same? Explain your answer.

7.6　Using the following data for an unreinforced PCC Pavement slab:
- Design strength f'_c = 28 MPa
- Slab thickness = 300 mm
- Standard deviation of f'_c obtained from 20 samples = 1.4 MPa
- Ignore any exposure requirement
- Use air-entrained concrete
- Fineness modulus of fine aggregate = 2.60
- Maximum aggregate size = 50 mm and nominal maximum aggregate size = 37.5 mm
- Bulk oven-dry specific gravity of coarse aggregate = 2.6
- Oven-dry rodded density of coarse aggregate = 2002 kg/m^3

Find the following:

a. Required compressive strength
b. w/c ratio
c. Coarse aggregate amount (kg/m^3)
d. If the w/c ratio is 10% reduced, will the quantity of coarse aggregate increase, decrease or remain the same? Explain your answer.

7.7　You are working on a concrete mix design that requires each cubic yard of concrete to have 0.45 water–cement ratio, 1165 kg/m^3 of dry gravel, 4% air content, and 335 kg/m^3 of cement. The available gravel has a specific gravity of G_{gravel} = 2.7, a moisture content of 1.6%, and absorption of 2.4%. The available sand has a specific gravity of G_{sand} = 2.5, a moisture content of 4.8%, and absorption of 1.5%. For each cubic meter of concrete needed on the job, calculate the mass of cement, moist gravel, moist sand, and water that should be added to the batch. Summarize and total the mix design when finished (don't include air entrainer in summary).

7.8　Design the concrete mix according to the following conditions:

> *Design Environment*
> Building frame
> Required design strength = 27.6 MPa
> Minimum dimension = 150 mm
> Minimum space between rebar = 40 mm
> Minimum cover over rebar = 40 mm
> Statistical data indicate a standard deviation of compressive strength of 2.1 MPa is expected (more than 30 samples).
> Only air entrainer is allowed.
> *Available Materials*
> Air entrainer: Manufacture specification 6.3 mL/1% air/100 kg cement.
> Coarse aggregate: 19 mm nominal maximum size, river gravel (rounded)
> 　　Bulk oven-dry specific gravity = 2.55, absorption = 3.6%
> 　　Oven-dry rodded density = 1761 kg/m^3
> 　　Moisture content = 2.5%
> Fine aggregate: Natural sand
> 　　Bulk oven-dry specific gravity = 2.659, absorption = 0.5%
> 　　Moisture content = 2%
> 　　Fineness modulus = 2.47

7.9   Design the concrete mix according to the following conditions:

      Design Environment
      Pavement slab, subjected to freezing
      Required design strength = 21 MPa
      Slab thickness = 300 mm
      Statistical data indicate a standard deviation of compressive strength of
           1.7 MPa is expected (more than 30 samples).
      Only air entrainer is allowed.
      Available Materials
      Air entrainer: Manufacture specification is 9.5 mL/1% air/100 kg
           cement.
      Coarse aggregate: 2 in. nominal maximum size, crushed stone
           Bulk oven-dry specific gravity = 2.573, absorption = 2.8%
           Oven-dry rodded density = 1922 kg/m^3
           Moisture content = 1%
      Fine aggregate: Natural sand
           Bulk oven-dry specific gravity = 2.540, absorption = 3.4%
           Moisture content = 4.5%
           Fineness modulus = 2.68

7.10   The design of a concrete mix requires 1173 kg/m^3 of gravel in dry condition, 582 kg/m^3 of sand in dry condition, and 157 kg/m^3 of free water. The gravel available at the job site has a moisture content of 0.8% and absorption of 1.5%, and the available sand has a moisture content of 1.1% and absorption of 1.3%. What are the masses of gravel, sand, and water per cubic meter that should be used at the job site?

7.11   Design a non-air-entrained concrete mix for a small job with a nominal maximum gravel size of 25 mm. Show the results as follows:
      a. masses of components to produce 2000 kg of concrete
      b. volumes of components to produce 1 m^3 of concrete

      Use one system of units only as specified by the instructor.

7.12   Design a non-air-entrained concrete mix for a small job with a nominal maximum gravel size of 19 mm. Show the results as follows:
      a. mass of components to produce 2263 kg of concrete
      b. volumes of components to produce 1 m^3 of concrete

7.13   Students in the materials lab mixed concrete with the following ingredients: 9.7 kg of cement, 18.1 kg of sand, 28.2 kg of gravel, and 6.5 kg of water. The sand has a moisture content of 3.1% and an absorption of 4.2%. The gravel has a moisture content of 3.4% and an absorption of 4.4%. Since water absorbed in the aggregate does not react with the cement or improve the workability of the plastic concrete, what is the water–cement ratio of this mix according to the American Concrete Institute's weight mix design method? If a water–cement ratio of 0.5 is required using the same materials and ingredients but different amount of mixing water, what is the mass of the mixing water to use?

7.14 Why is it necessary to measure the air content of concrete at the job site rather than at the batch plant? Name one of the methods used to measure the air content of concrete.

7.15 What do we mean by curing concrete? What will happen if concrete is not cured?

7.16 Discuss five different methods of concrete curing.

7.17 Draw a graph showing the typical relation between the compressive strength and age for continuously moist-cured concrete and concrete cured for 3 days only. Label all axes and curves.

7.18 Why is extra water harmful to fresh concrete, but good for concrete after it reaches its final set?

7.19 Discuss the change in volume of concrete at early ages.

7.20 Discuss the creep response of concrete structures. Provide examples of the effect of creep on concrete structures.

7.21 What is the effect of water–cement ratio on the permeability of hardened concrete? How does permeability affect the durability of the concrete structure? Using Figure 7.33, what is the approximate percent increase in the coefficient of permeability if the water–cement ratio increases from 0.5 to 0.6? Using the same figure, what is the approximate percent increase in the coefficient of permeability if the water–cement ratio increases from 0.5 to 0.7?

7.22 On one graph, draw a sketch showing the typical relationship between the stress and strain of concrete specimens with high and low water–cement ratios. Label all axes and curves. Comment on the effect of increasing the water–cement ratio on the stress–strain response.

7.23 Using Figure 7.34,
    a. Determine the ultimate stress at each water–cement ratio.
    b. Determine the secant modulus at 40% of the ultimate stress at each water–cement ratio.
    c. Plot the relationship between the secant moduli and the ultimate stresses.
    d. Plot the relationship between the moduli and the ultimate stresses on the same graph of part (c), using the relation of the ACI Building Code (Equation 7.3).
    e. Compare the two relations in questions c and d and comment on any discrepancies.
    f. Determine the toughness at each water–cement ratio and comment on the effect of increasing water–cement ratio on the toughness of concrete.

7.24 Three concrete mixes with the same ingredients, except the amount of mixing water, and their slump values were obtained. Three 100 mm × 200 mm concrete cylinders were prepared for each mix. The cylinders were cured for 7 days and then tested for compressive strength. The test results are as shown in Table P7.24. Assume that the aggregate was at the saturated surface-dry condition before adding mixing water.

It is required to do the following:

a. Plot the relationship between slump and amount of mixing water for all mixes. Comment on the effect of increasing the amount of water on workability.

b. Determine the compressive strength of each cylinder after 7 days.

c. Determine the average compressive strength of each mix after 7 days.

d. Using Figure 7.24, estimate the compressive strength after 28 days (f'_C) for each mix.

e. Determine the w/c ratio for each mix. Plot the average f'_C values versus w/c ratios for all mixes. Comment on the effect of increasing the w/c ratio on the compressive strength.

TABLE P7.24

Mix No.	Mass of Cement (kg)	Mass of Water* (kg)	Slump (mm)	Cylinder No.	Maximum Load (kN)	Compressive Strength after 7 Days (MPa)	Average Compressive Strength after 7 Days (MPa)	Estimated Compressive Strength after 28 Days (MPa)
1	10	5	40	1	138.2			
				2	175.9			
				3	136.9			
2	10	5.5	55	1	100.7			
				2	115.2			
				3	113.6			
3	10	6	75	1	73.3			
				2	78.4			
				3	76.0			

* Assume that the water listed here is above the saturated surface-dry (SSD) condition of the aggregate.

7.25   Three 100 mm × 200 mm concrete cylinders with water to cement ratios of 0.50, 0.55, and 0.60, respectively. After curing for 7 days, the specimens were subjected to increments of compressive loads. The load versus deformation results were as shown in Table P7.25.

Assuming that the gauge length is the whole specimen height, determine the following:

a. The compressive stresses and strains for each specimen at each load increment.

b. Plot stresses versus strains for each specimen on one graph.

c. If the ultimate stress is 30, 25 and 20 MPa at 0.50, 0.55 and 0.60, respectively, determine modulus of elasticity as the secant modulus at 40% of the ultimate stress at each water–cement ratio.

TABLE P7.25

Specimen No.	w/c Ratio	Load (kN)	Deformation (mm)	Stress (MPa)	Strain (m/m)	Secant Modulus (MPa)
1	0.50	0	0			
		25	0.0007			
		50	0.0014			
		75	0.0021			
		100	0.0029			
2	0.55	0	0			
		25	0.0008			
		50	0.0016			
		75	0.0025			
		100	0.0036			
3	0.60	0	0			
		25	0.001			
		50	0.0021			
		75	0.0034			
		100	0.0051			

    d. Comment on the effect of increasing the water–cement ratio on the modulus of elasticity.

7.26   Three 150 mm × 300 mm concrete cylinders with water to cement ratios of 0.4, 0.6, and 0.8, respectively. After curing for 28 days, the specimens were subjected to increments of compressive loads until failure. The load versus deformation results were as shown in Table P7.26.

TABLE P7.26

	Specimen No.		
	1	2	3
w/c Ratio	0.4	0.6	0.8
Deformation (mm)	Load (kN)		
0	0	0	0
0.3	514	348	244
0.6	853 (failure)	472	304
0.9		433 (failure)	263
1.2			235 (failure)

Assuming that the gauge length is the whole specimen height, it is required to do the following:

a. The compressive stresses and strains for each specimen at each load increment.
b. Plot stresses versus strains for all specimens on one graph.
c. The ultimate strength for each specimen.
d. The modulus of elasticity as the secant modulus at 40% of the ultimate stress for each specimen.
e. The strain at failure for each specimen.
f. The toughness for each specimen.
g. Comment on the effect of increasing the water–cement ratio on the following:
   i.   Ultimate strength
   ii.   Modulus of elasticity
   iii. Ductility
   iv. Toughness. Curves may be approximated with a series of straight lines.

7.27   A normal-weight concrete has an average compressive strength of 31 MPa. What is the estimated modulus of elasticity?

7.28   Discuss the significance of the compressive strength test on concrete. What is the typical range of the compressive strength of concrete? Draw a graph to show the relationship between compressive strength and water–cement ratio for different curing times (label all axes and curves).

7.29   What is the standard size of PCC specimens to be tested for compressive strength? If a smaller size is used, which size is expected to provide higher compressive strengths? Why? Which size provides strengths close to that of an actual concrete structure?

7.30   A short plain concrete column with dimensions of 300 mm × 300 mm × 900 mm is to be constructed. If the compressive strength of the concrete is 34.5 MPa, what is the maximum load that can be applied to this column using a factor of safety of 1.2?

7.31   What is the purpose of performing the flexure test on concrete? How are the results of this test related to the compressive strength test results?

7.32   What are the advantages of using a third-point loading flexure test over a center-point loading flexure test? Draw a shear force diagram and a bending moment diagram for each case to support your answer.

7.33   Consider a standard flexural strength specimen of length L, width a, and depth a. Assume third point loading where the load at failure from the test machine is P, which is distributed evenly (1/2 to each) over the two loading points. Derive the equation for calculating the modulus of rupture of the beam in terms of P, L, and a.

7.34   To evaluate the effect of a certain admixture on the flexure strength of concrete, two mixes were prepared, one without admixture and one with admixture.

Three beams were prepared of each mix. All the beams had a cross section of 0.15 m by 0.15 m and a span of 0.45 m. The third-point loading flexure strength test was performed on each beam after 7 days of curing. The loads at failure of the beams without admixture were 32.8, 34.5, and 31.7 kN, while the loads at failure of beams with admixture were 39.4, 35.6, and 35.0 kN. Determine:

a. The modulus of rupture of each beam in MPa.

b. The average moduli of rupture of the beams without and with admixture.

c. The percent of increase of the average modulus of rupture due to adding the admixture.

7.35 To evaluate the effect of a certain admixture on the flexure strength of concrete, two mixes were prepared, one without admixture and one with admixture. Three beams were prepared of each mix. All the beams had a cross section of 100 mm × 100 mm and a span of 300 mm. A center-point loading flexure strength test was performed on each beam after 7 days of curing. The loads at failure of the beams without admixture were 26.89, 22.56, and 26.40 kN, while the loads at failure of beams with admixture were 32.47, 32.49, and 31.06. Determine:

a. The modulus of rupture of each beam in GPa.

b. The average moduli of rupture of the beams without and with admixture.

c. The percent of increase of the average modulus of rupture due to adding the admixture.

7.36 A center-point loading flexural test was performed on a concrete beam with a cross section of 100 mm × 100 mm and a span of 200 mm. If the failure load was 26.9 kN, find the flexure strength.

7.37 A normal-weight concrete has an average compressive strength of 20 MPa. What is the estimated flexure strength?

7.38 Three batches of concrete were prepared using the same materials and ingredients, except that they have water–cement ratios of 0.50, 0.55, and 0.60, respectively. The following tests were performed on specimens made of the three batches:

- Compressive strength test on 100 mm × 200 mm cylinders
- Center-point flexure test on 100 mm × 100 mm × 300 mm beams
- Split tension test on 150 mm × 300 mm cylinders

Three replicates were tested for each test. Table P7.38 shows the average failure loads for the three replicates of each case.

It is required to do the following:

a. Complete Table P7.38

b. Using an Excel sheet, plot the relationships between water–cement ratio and compressive strength, modulus of rupture, and tensile strength on the same graph. Label all axes and curves.

c. Comment on the effect of water–cement ratio on the compressive strength, modulus of rupture, and tensile strength.

TABLE P7.38

w/c Ratio	Compressive Strength Test		Flexure Test		Split Tension Test	
	Failure Load (kN)	Compres- sive Strength (MPa)	Failure Load (kN)	Modulus of Rupture (MPa)	Failure Load (kN)	Tensile Strength (MPa)
0.50	204.2		18.6		237.2	
0.55	141.7		12.2		216.1	
0.60	110.5		10.0		191.1	

7.39   150 mm × 300 mm concrete cylinders were made of the same batch and separated at random to six equal groups. The six groups of cylinders were submerged in water for different times before testing for compressive strength
and the results are shown in Table P7.39.

It is required to do the following:
a. Determine the compressive strength at different curing times.
b. Using an Excel sheet, plot the relationship between curing time and
compressive strength. Label all axes and curves.
c. Comment on the effect of curing time on the compressive strength.
d. What is the approximate ratio between the compressive strengths at 7
days and 28 days?
e. What is the approximate ratio between the compressive strengths at 28
days and 180 days?

TABLE P7.39

Group no.	1	2	3	4	5	6
Water submergence (days)	3	7	14	28	90	180
Average maximum load before failure (kN)	189.6	379.1	543.5	632.0	733.0	789.9

7.40   Discuss two nondestructive tests to be performed on hardened concrete. Show
the basic principles behind the tests and how they are performed.

7.41   Discuss the concept of concrete maturity meters.

7.42   Discuss four alternatives that increase the use and application of conventional
concrete.

7.43   What is self-consolidating concrete? How are its properties achieved? How
are these properties measured?

7.44 What is flowable fill? Discuss its ingredients and its advantages.

7.45 Two 150 mm × 300 mm concrete cylinders with randomly oriented steel fiber contents of 0 and 2% by weight, respectively. After curing for 28 days, the specimens were subjected to increments of compressive loads until failure. The load versus deformation results were as shown in Table P7.45.

TABLE P7.45

Specimen no.	1	2
Fiber content (%)	0	2
Deformation (mm)	Load (kN)	
0	0	0
0.0305	507	507
0.0610	645 (failure)	681
0.0915		578
0.0915		396
1.525		338 (failure)

Assuming that the gauge length is the whole specimen height, determine the following:
a. The compressive stresses and strains for each specimen at each load increment.
b. Plot stresses versus strains for the two specimens on one graph.
c. The initial modulus of elasticity for each specimen.
d. The ultimate strength for each specimen.
e. The strain at failure for each specimen.
f. Toughness. Curves may be approximated with a series of straight lines.
g. Comment on the effect of adding fiber on the following:
   i. Modulus of elasticity
   ii. Ultimate strength
   iii. Ductility
   iv. Toughness

7.46 Discuss the concept of high-performance concrete. Discuss some of its properties that make it preferred over conventional concrete.

7.47 Comparing PCC with mild steel, answer the following questions:
a. Which one is stronger?
b. Which one has a higher modulus or stiffness?
c. Which one is more brittle?
d. What is the range of compressive strength for a typical PCC?
e. What is the compressive strength for a high-strength concrete?
f. What would be a reasonable range for PCC modulus?

7.48   In a ready-mix plant, cylindrical samples are prepared and tested periodically to detect any mix problem and to ensure that the compressive strength is higher than the lower specification limit. The minimum target value was set at 27.6 MPa. The compressive strength data shown in Table P7.48 were collected. Use one system of units only as specified by the instructor.

TABLE P7.48

Sample No.	Compressive Strength, MPa	Sample No.	Compressive Strength, MPa
1	33.9	14	39.8
2	32.6	15	29.4
3	39.1	16	35.1
4	29.7	17	32.2
5	42.1	18	35.7
6	29.8	19	37.5
7	36.1	20	25.5
8	34.1	21	31.1
9	36.1	22	25.4
10	28.9	23	28.3
11	39.8	24	25.4
12	31.2	25	27.0
13	28.0		

a. Calculate the mean, standard deviation, and the coefficient of variation of the data.

b. Using a spreadsheet program, create a control chart for these data showing the target value and the lower specification limit. Is the plant production meeting the specification requirement? If not, comment on possible reasons. Comment on the data scatter.

7.49   Using Internet resources, prepare a 1-page typed, single-space summary discussing how the principles of sustainable development are/can be incorporated in concrete structures.

7.50   Using Internet resources, write a paragraph on the use of porous concrete pavements for stormwater management.

7.8 References

American Concrete Institute. Specifications for Structural Concrete. ACI Committee 301 Report, ACI 301–99. Farmington Hills, MI: American Concrete Institute, 1999.

American Concrete Institute. *Guide for Concrete Floor and Slab Construction.* ACI Committee 302 Report, ACI 302.1R-96. Farmington Hills, MI: American Concrete Institute, 1996.

American Concrete Institute. *Hot-Weather Concreting.* ACI Committee 305 Report, ACI 305R-99. Farmington Hills, MI: American Concrete Institute, 1999.

American Concrete Institute. *Cold-Weather Concreting.* ACI Committee 306 Report, ACI 306R-88. Farmington Hills, MI: American Concrete Institute, 1997.

American Concrete Institute. *Standard Practice for Selecting Proportions for Normal, Heavyweight and Mass Concrete.* ACI Committee 211 Report, ACI 211.1–91. Farmington Hills, MI: American Concrete Institute, 1991.

American Concrete Institute. *Standard Practice for Curing Concrete.* ACI Committee 308 Report, ACI 308–92. Farmington Hills, MI: American Concrete Institute, 1997.

American Concrete Institute. *Building Code Requirements for Reinforced Concrete.* ACI Committee 318 Report, ACI 318–08. Farmington Hills, MI: American Concrete Institute, 2008.

Ashley, E. and L. Lemay, *Concrete's Contribution to LEED 2009.* Ready Mined Concrete Industry LEED Reference Guide Update, Silver Spring MD, 2009.

Hognestad, E., N. W. Hanson, and D. McHenry. *Concrete Stress Distribution in Ultimate Strength Design.* Development Department Bulletin DX006. Skokie, IL: Portland Cement Association, 1955.

Kosmatka, S. H., and M. L. Wilson. *Design and Control of Concrete Mixtures.* 15th ed. (Revised). Skokie, IL: Portland Cement Association, 2011.

Mehta, P. K. and P. J. M. Monteiro. *Concrete: Microstructure, Properties, and Materials.* 4th ed. New York: McGraw-Hill Professional, 2013.

National Ready Mixed Concrete Association (NRMCA). Concrete in Practice, CIP 37, "Self Consolidating Concrete." Silver Springs, MD: NRMCA, www.nrmca.org

National Ready Mixed Concrete Association (NRMCA). "Flowable Fill Materials." Silver Spring, MD: NRMCA, www.nrmca.org

National Ready Mixed Concrete Association (NRMCA). "Pervious Concrete Pavements." Silver Spring, MD: NRMCA, www.nrmca.org, 2011.

Nawy, E. G. *Fundamentals of High Strength High Performance Concrete.* Concrete Design and Construction Series. Harlow, UK: Longman Group Limited, 2000.

Neville, A. M. *Properties of Concrete.* 4th ed. New York: John Wiley and Sons, 1996.

Portland Cement Association. *Concrete for Small Jobs.* IS174 T, 1988. Skokie, IL: Portland Cement Association, http://www.portcement.org/pdf_files/IS174.

Powers, T. C., L. E. Copeland, J. C. Hayes and H. M. Mann. "Permeability of portland cement paste. *Journal of American Concrete Institute,*" 51 (11) 285–298, Nov 1954.

Shah, S. P. and G. Winter. Inelastic Behavior and Fracture of Concrete. *Symposium on Causes, Mechanism, and Control of Cracking in Concrete.* American Concrete Institute Special Publication no. 20. Farmington Hills, MI: American Concrete Institute, 1968.

Zia, P., M. L. Leming, and S. H. Ahmad. *High Performance Concretes: A State-of-the-Art Report.* SHRP-C/FR-91-103, Washington, DC: Strategic Highway Research Program, National Research Council, 1991. http://www.tfhrc.gov/structur/hpc/hpc2/exec.htm.

CHAPTER

MASONRY

A masonry structure is formed by combining masonry units, such as stone, blocks, or brick, with mortar. Masonry is one of the oldest construction materials. Examples of ancient masonry structures include the pyramids of Egypt, the Great Wall of China, and Greek and Roman ruins. Bricks of nearly uniform size became commonly used in Europe during the beginning of the 13th century. The first extensive use of bricks in the United States was around 1600. In the last two centuries, bricks have been used in constructing sewers, bridge piers, tunnel linings, and multistory buildings. Masonry units (Figure 8.1) are a popular construction material throughout the world and competes favorably with other materials, such as wood, steel, and concrete for certain applications (Adams, 1979).

Masonry units can be classified as

- concrete masonry units
- clay bricks
- structural clay tiles
- glass blocks
- stone

Concrete masonry units can be either solid or hollow, but clay bricks, glass blocks, and stone are typically solid. Structural clay tiles are hollow units that are larger than clay bricks and are used for structural and nonstructural masonry applications, such as partition walls and filler panels. They can be used with their webs in either a horizontal or a vertical direction. Figure 8.2 shows examples of concrete masonry units, clay bricks, and structural clay tiles. Concrete masonry units and clay bricks are commonly used in the United States.

FIGURE 8.1 Masonry units used in construction.

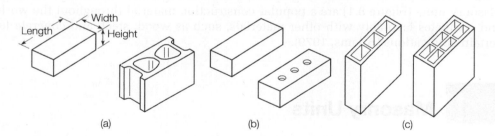

FIGURE 8.2 Examples of masonry units: (a) concrete masonry units, (b) clay bricks, and (c) structural clay tiles.

8.1.1 ■ Concrete Masonry Units

Solid concrete units are commonly called concrete bricks, while hollow units are known as concrete blocks, hollow blocks, or cinder blocks. Hollow units have a net cross-sectional area in every plane parallel to the bearing surface less than 75% of the gross cross-sectional area in the same plane. If this ratio is 75% or more, the unit is categorized as solid (Portland Cement Association, 1991).

Concrete masonry units are manufactured in three classes, based on their density: *lightweight units, medium-weight units, and normal-weight units,* with dry unit weights as shown in Table 8.1. Well-graded sand, gravel, and crushed stone are used to manufacture normal-weight units. Lightweight aggregates such as pumice, scoria,

TABLE 8.1   Mass Classifications and Allowable Maximum Water Absorption of Concrete Masonry Units (Reprinted, with permission, from ASTM C90, Table 3 and C129, Table 2, copyright ASTM International, 100 Barr Harbor Drive, West Conshohocken, PA 19428)

Weight Classification	Unit Mass Mg/m^3	Maximum Water Absorption kg/m^3 (Average of 3 units)
Lightweight	Less than 1.68	288
Medium weight	1.68–2.00	240
Normal weight	2.00 or more	208

cinders, expanded clay, and expanded shale are used to manufacture lightweight units. Lightweight units have higher thermal and fire resistance properties and lower sound resistance than normal weight units.

Concrete masonry units are manufactured using a relatively dry (zero-slump) concrete mixture consisting of portland cement, aggregates, water, and admixtures. Type I cement is usually used to manufacture concrete masonry units; however, Type III is sometimes used to reduce the curing time. Air-entrained concrete is sometimes used to increase the resistance of the masonry structure to freeze and thaw effects and improve workability, compaction, and molding characteristics of the units during manufacturing. The units are molded under pressure, then cured, usually using low-pressure steam curing. After manufacturing, the units are stored under controlled conditions so that the concrete continues curing.

Concrete masonry units can be classified as load bearing (ASTM C90) and non–load bearing (ASTM C129). Load-bearing units must satisfy a higher minimum compressive strength requirement than non–load-bearing units, as shown in Table 8.2. The compressive strength of individual concrete masonry units is determined by capping the unit and applying load in the direction of the height of the unit

TABLE 8.2   Strength Requirements of Load Bearing and Non–Load-Bearing Concrete Masonry Units (Reprinted, with permission, from ASTM C90, Table 3 and C129, Table 2, copyright ASTM International, 100 Barr Harbor Drive, West Conshohocken, PA 19428)

Type	Minimum Compressive Strength Based on Net Area MPa	
	Average of Three Units	Individual Units
Load bearing	13.1	11.7
Non–load-bearing	4.1	3.5

until failure (ASTM C140). A full-size unit is recommended for testing, although a portion of a unit can be used if the capacity of the testing machine is not large enough. The *gross area compressive strength* is calculated by dividing the load at failure by the gross cross-sectional area of the unit. The *net area compressive strength* is calculated by dividing the load at failure by the net cross-sectional area. The net cross-sectional area is calculated by dividing the net volume of the unit by its average height. The net volume is determined using the water displacement method according to ASTM C140.

Sample Problem 8.1

A hollow concrete masonry unit has actual gross dimensions of 190 mm $\times$ 190 mm $\times$ 390 mm (width $\times$ height $\times$ length). The unit is tested in a compression machine with the following results:

Failure Load = 1110 kN

Net volume of 6×10^6 mm^3

   a. Calculate the gross area compressive strength.
   b. Calculate the net area compressive strength.

Solution

   a. Gross area = 190 $\times$ 390 = 74100 mm^2

       Gross area compressive strength = 1110000/74100 = 15 MPa
   b. Net area = 6×10^6/190 = 31579 mm^2

       Net area compressive strength = 1110000/31579 = 35 MPa

---

The amount of water absorption of concrete masonry units is controlled by ASTM standards to reduce the effect of weathering and limit the amount of shrinkage due to moisture loss after construction (ASTM C90). The absorption of concrete masonry units is determined by immersing the unit in water for 24 hours (ASTM C140). The absorption and moisture content are calculated as follows

$$\text{Absorption (kg/m}^3) = \frac{W_s - W_d}{W_s - W_i} \times 1000 \tag{8.1}$$

$$\text{Absorption (\%)} = \frac{W_s - W_d}{W_d} \times 100 \tag{8.2}$$

$$\text{Moisture content as a percent of total absorption} = \frac{W_r - W_d}{W_s - W_d} \times 100 \tag{8.3}$$

where

W_s = saturated mass of specimen, kg

W_d = oven-dry mass of unit, kg

W_i = immersed mass of specimen, kg, and

W_r = mass of specimen as received

Table 8.1 shows the allowable maximum water absorption for load-bearing concrete masonry units.

Sample Problem 8.2

A concrete masonry unit was tested according to ASTM C140 procedure and produced the following results:

mass of unit as received = 10354 g

saturated mass of unit = 11089 g

oven-dry mass of unit = 9893 g

immersed mass of unit = 5136 g

Calculate the absorption in kg/m³, percent absorption, and moisture content of the unit as a percent of total absorption.

Solution: Absorption $= \dfrac{11089 - 9893}{11089 - 5136} \times 1000 = 200.1 \text{ kg/m}^3$

Percent absorption $= \dfrac{11089 - 9893}{9893} \times 100 = 12.1\%$

Moisture content as a percent of total absorption

$= \dfrac{10354 - 9893}{11089 - 9893} \times 100 = 38.5\%$

Concrete masonry units are available in different sizes, colors, shapes, and textures. Concrete masonry units are specified by their nominal dimensions. The *nominal dimension* is greater than its *specified* (or *modular*) dimension by the thickness of the mortar joint, usually 10 mm. For example, a 200 mm × 200 mm × 400 mm block has an actual width of 190 mm, height of 190 mm, and length of 390 mm, as illustrated in Figure 8.3. Load-bearing concrete masonry units are available in nominal widths of 100, 150, 200, 250, and 300 mm, heights of 100 and 200 mm, and lengths of 300, 400, and 600 mm. Also, depending on the position within the masonry wall, they are manufactured as stretcher, single-corner, and double-corner units, as depicted in Figure 8.4.

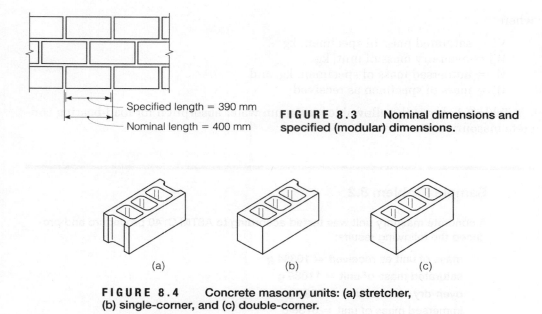

Specified length = 390 mm

Nominal length = 400 mm

FIGURE 8.3   Nominal dimensions and specified (modular) dimensions.

FIGURE 8.4   Concrete masonry units: (a) stretcher, (b) single-corner, and (c) double-corner.

Solid concrete masonry units (concrete bricks) are manufactured in two types based on their exposure properties: concrete building bricks (ASTM C55) and concrete facing bricks (ASTM C1634). The concrete building bricks are manufactured for general use in nonfacing, utilitarian applications, while the concrete facing bricks are typically used in applications where one or more faces of the unit is intended to be exposed. The concrete facing bricks have stricter requirements than the concrete building bricks. The maximum allowable water absorption of the concrete facing bricks is less than that of the concrete building bricks. Also, the minimum net area compressive strength of the concrete facing bricks is higher than that of the concrete building bricks as shown in Table 8.3.

TABLE 8.3   Strength Requirements of Building and Facing Concrete Bricks (Reprinted, with permission, from ASTM C55 and C1634, copyright ASTM International, 100 Barr Harbor Drive, West Conshohocken, PA 19428.)

Type	Minimum Compressive Strength Based on Net Area MPa	
	Average of Three Units	Individual Units
Building bricks	17.3	13.8
Facing bricks	24.1	20.7

8.1.2 ▒ Clay Bricks

Clay bricks are small, rectangular blocks made of fired clay. The clays used for brick making vary widely in composition from one place to another. Clays are composed mainly of silica (grains of sand), alumina, lime, iron, manganese, sulfur, and phosphates, with different proportions. Bricks are manufactured by grinding or crushing the clay in mills and mixing it with water to make it plastic. The plastic clay is then molded, textured, dried, and finally fired. Bricks are manufactured in different colors, such as dark red, purple, brown, gray, pink, or dull brown, depending on the firing temperature of the clay during manufacturing. The firing temperature for brick manufacturing varies from 900°C to 1200°C. Clay bricks have an average density of 2 Mg/m³.

Clay bricks are used for different purposes, including building, facing and aesthetics, floor making, and paving. *Building bricks* (*common bricks*) are used as a structural material and are typically strong and durable. *Facing bricks* are used for facing and aesthetic purposes and are available in different sizes, colors, and textures. *Floor bricks* are used on finished floor surfaces and are generally smooth and dense, with high resistance to abrasion. Finally, *paving bricks* are used as a paving material for roads, sidewalks, patios, driveways, and interior floors. Paving bricks are available in different colors, such as red, gray, or brown, are typically abrasion resistant and could usually be *vitrified* (glazed to render it impervious to water and highly resistant to corrosion).

Absorption is one of the important properties that determine the durability of bricks. Highly absorptive bricks can cause efflorescence and other problems in the masonry. According to ASTM C67, absorption by 24-hr submersion, absorption by 5-hr boiling, and saturation coefficient are calculated as:

$$\text{Absorption by 24-hr submersion (\%)} = \frac{(W_{s24} - W_d)}{W_d} \times 100 \qquad (8.4)$$

$$\text{Absorption by 5-hr boiling (\%)} \;\; = \frac{(W_{b5} - W_d)}{W_d} \times 100 \qquad (8.5)$$

$$\text{Saturation coefficient} = \frac{(W_{s24} - W_d)}{(W_{b5} - W_d)} \qquad (8.6)$$

where

W_d = dry weight of specimen,
W_{s24} = saturated weight after 24-hr submersion in cold water, and
W_{b5} = saturated weight after 5-hr submersion in boiling water.

Clay bricks are very durable and fire resistant and require very little maintenance. They have moderate insulating properties, which make brick houses cooler in summer and warmer in winter, compared with houses built with other construction materials. Clay bricks are also noncombustible and poor conductors.

The compressive strength of clay bricks is an important mechanical property that controls their load-carrying capacity and durability. The compressive strength of clay bricks is dependent on the composition of the clay, method of brick manufacturing, and the degree of firing. The compressive strength is determined by capping and testing a half unit "flatwise" (load applied in the direction of the height of the unit) and is calculated by dividing the load at failure by the cross-sectional area (ASTM C67). In determining the compressive strength, either the net or gross cross-sectional area is used. Net cross-sectional area is used only if the net cross section is less than 75% of the gross cross section. A quarter of a brick can be tested if the capacity of the testing machine is not large enough to test a half brick. Other mechanical properties of bricks include modulus of rupture, tensile strength, and modulus of elasticity. Most clay bricks have modulus of rupture between 3.5 and 26.2 MPa. The tensile strength is typically between 30% to 49% of the modulus of rupture. The modulus of elasticity ranges between 10.3 and 34.5 GPa.

Clay building bricks are graded according to properties related to durability and resistance to weathering, such as compressive strength, water absorption, and saturation coefficient (ASTM C62). Table 8.4 shows the three available grades and their requirements: SW, MW, and NW, standing for severe weathering, moderate weathering, and negligible weathering, respectively. Grade SW bricks are intended for use in areas subjected to frost action, especially at or below ground level. Grade NW bricks are recommended for use in areas with no frost action and in dry locations, even where subfreezing temperatures are expected. Grade NW bricks can be used in interior construction, where no freezing occurs. Note that higher compressive strengths, lower water absorptions, and lower saturation coefficients are required as the weathering condition gets more severe, so as to reduce the effect of freezing and thawing and wetting and drying.

TABLE 8.4   Physical Requirements for Clay Building Bricks (ASTM C62) (Copyright ASTM, Reprinted with Permission)

Grade	Min. Compressive Strength, Gross Area, MPa		Max. Water Absorption by 5-hr Boiling, %		Max. Saturation Coefficient	
	Average of Five Bricks	Individual	Average of Five Bricks	Individual	Average of Five Bricks	Individual
SW[1]	20.7	17.2	17.0	20.0	0.78	0.80
MW[2]	17.2	15.2	22.0	25.0	0.88	0.90
NW[3]	10.3	8.6	No limit	No limit	No limit	No limit

[1] Severe weathering
[2] Moderate weathering
[3] Negligible weathering

Sample Problem 8.3

The 5-hr boiling test was performed on a medium weathering clay brick according to ASTM C67 and produced the following masses:

dry mass of specimen = 1.788 kg

saturated mass after 5-hr submersion in boiling water = 2.262 kg

Calculate percent absorption by 5-hr boiling and check whether the brick satisfies the ASTM requirements.

Solution

$$\text{Absorption by 5-hr boiling} = \frac{2.262 - 1.788}{1.788} \times 100 = 26.5\%$$

From Table 8.4, the maximum allowable absorption by 5-hr boiling = 25.0%. Therefore, the brick does not satisfy the ASTM requirements.

Facing clay bricks (ASTM C216) are manufactured in two durability grades for severe weathering (SW) and moderate weathering (MW). Each durability grade is manufactured in three appearance types: FBS, FBX, and FBA. These three types stand for *face brick standard, face brick extra,* and *face brick architecture.* Type FBS bricks are used for general exposed masonry construction. Type FBX bricks are used for general exterior and interior masonry construction, where a high degree of precision and a low permissible variation in size are required. The FBA type bricks are manufactured to produce characteristic architectural effects resulting from nonuniformity in size and texture of the individual units.

Similar to concrete masonry units, clay bricks are designated by their nominal dimensions. The *nominal dimension* of the brick is greater than its *specified* (or *modular*) dimension by the thickness of the mortar joint, which is about 10 mm and could go up to 12.5 mm. The *actual size* of the brick depends on the nominal size and the amount of shrinking that occurs during the firing process, which ranges from 4% to 15%.

Clay bricks are specified by their nominal width times nominal height times nominal length. For example, a 100 × 70 × 200 brick has nominal width of 100 mm, height of 70 mm, and length of 200 mm. Clay bricks are available in nominal widths ranging from 75 to 300 mm, heights from 50 to 200 mm, and lengths up to 400 mm. Bricks can be classified as either modular or nonmodular, where modular bricks have widths and lengths of multiples of 100 mm.

8.2   Mortar

Mortar is a mixture of cementitious material, aggregate, and water. Mortar can be classified as cement-lime mortar, cement mortar, or masonry cement mortar. Mortar is used for the following functions:

- bonding masonry units together, either non-reinforced or reinforced
- serving as a seating material for the units
- leveling and seating the units
- providing aesthetic quality of the structure

Mortar is manufactured in four types: M, S, N, and O. Mortar also needs to satisfy either proportion specifications or property specifications (ASTM C270). The proportion specifications specify the ingredient quantities, while the property specifications specify the compressive strength, water retention, air content, and the aggregate ratio.

Mortar can be evaluated either in the laboratory or in the field. In the laboratory evaluation, the compressive strength of mortar is tested using 50-mm cubes according to ASTM C109. The minimum average compressive strengths of types M, S, N, and O at 28 days are 17.2, 12.4, 5.2, and 2.4 MPa (ASTM C270). The field evaluation involves the preparation of one or more trial batches before construction. These trial batches are sampled and used in establishing the plastic and hardened properties of the mixtures (ASTM C780).

Mortar starts to bind masonry units when it sets. During construction, bricks and blocks should be rubbed and pressed down in order to force the mortar into the pores of the masonry units to produce maximum adhesion. It should be noted, however, that mortar is the weakest part of the masonry wall. Therefore, thin mortar layers generally produce stronger walls than do thick layers.

Unlike concrete, the compressive strength is not the most important property of mortar. Since mortar is used as an adhesive and sealant, it is very important that it forms a complete, strong, and durable bond with the masonry units and with the rebars that might be used to reinforce masonry walls. The ability to bond individual units is measured by the *tensile bond strength* of mortar (ASTM C952), which is related to the force required to separate the units. The tensile bond strength affects the shear and flexural strength of masonry. The tensile bond strength is usually between 0.14 and 0.55 MPa and is affected by the amount of lime in the mix.

Other properties that affect the performance of mortar are workability, tensile strength, compressive strength, resistance to freeze and thaw, and water retention. ASTM C91 defines water retention as a measure of the rate at which water is lost to the masonry units.

8.3   Grout

Grout is a high-slump concrete consisting of portland cement, sand, fine gravel, water, and sometimes lime. Grout is used to fill the cores or voids in hollow masonry units for the purpose of (1) bonding the masonry units, (2) bonding the reinforcing steel to the masonry, (3) increasing the bearing area, (4) increasing fire resistance, and (5) improving the overturning resistance by increasing the weight. The minimum compressive strength of grout is 14 MPa at 28 days, according to ASTM C476.

8.4 Plaster

Plaster is a fluid mixture of portland cement, lime, sand, and water, which is used for finishing either masonry walls or framed (wood) walls. Plaster is used for either exterior or interior walls. Stucco is plaster used to cover exterior walls. The average compressive strength of plaster is about 13.8 MPa at 28 days.

8.5 Masonry Sustainability

8.5.1 LEED Considerations

Subasic (2013) developed a summary table of the sustainable design with concrete masonry units (Table 8.5).

8.5.2 Other Sustainability Considerations

As a traditional material, it is difficult to make the case that masonry should be credited as a sustainable material. Subasic (2013) points out that LEED uses surrogate measures of environmental impact, such as recycled content, and measures of on-site construction practices. He argues for a more comprehensive evaluation of environmentally preferable products which uses the following strategies:

- Abundance of raw materials
- Efficient use of raw materials
- Use of recycled materials
- Sustainable measures in acquisition or manufacture
- Use of regionally available materials (near to the building project site)
- Regional manufacture or fabrication (near to the building project site)
- Recyclable
- Salvageable
- Durable
- Non-toxic
- Avoidance of construction waste.

Subasic makes the case that when these factors are considered concrete masonry, pavers, and segmented retaining walls are sustainable materials and provide sustainable structures.

TABLE 8.5   LEED 2009 Credits Related to Concrete Masonry Products

Sustainable sites	Concrete Masonry Strategy
SS c2: Development Density and Community Connectivity	Concrete masonry provides benefits of fire safety, impact resistance, acoustic isolation, flexible design, and minimal access requirements in urban locations.
SS cc6: Stormwater Design	Concrete masonry pavers can be used as part of permeable paving system for pedestrian sidewalks, courtyard and other hardscape areas as well as for roads and parking lots.
SS c7.1 Heat Island Effect—non-roof	Concrete masonry pavers with a solar reflectance index of 29 or higher meet the requirement for this credit.

Energy and Atmosphere	
EA p2: Minimum Energy Performance	Concrete masonry walls provide thermal mass to help meet the minimum energy efficiency requirements.
EA c1: Optimize Energy Performance	Concrete masonry walls provide thermal mass to help exceed the minimum energy efficiency requirements.

Materials and Resources	
MR c1: Building Reuse	Masonry buildings are often reused due to their longevity and beauty.
MR c2: Construction Waste Management	Scrap or waste concrete masonry products can easily be salvaged or recycled, as can their packaging material.
MR c3: Material Reuse	Sand-set concrete masonry pavers and segmented retaining wall units can easily be salvaged and reused.
MR c4: Recycled Content	Recycled aggregate can be found in many concrete masonry products, and fly ash and slag cement are commonly used in grout. Steel reinforcing bars contain high recycled content.
MR c5: Regional Materials	Concrete masonry producers are regionally located and often source their raw materials from less than 320 km of the manufacturing facility.

Indoor Environmental Quality	
EQ c4: Low-Emitting Materials	Concrete masonry walls and pavers used on the interior of a building are inherently low emitting if no surface applied coatings are used.

Summary

Masonry is one of the oldest building technologies, dating back to use of sun-dried adobe blocks in ancient times. Modern masonry units are produced to high standards in the manufacturing process. While the strength of the masonry units is important for quality control, the strength of masonry construction is generally limited by the ability to bond the units together with mortar. The ability of masonry units to resist environmental degradation is an important quality consideration. This ability is closely related to the absorption of the masonry units.

Questions and Problems

8.1   Define solid and hollow masonry units according to ASTM C90.

8.2   What are the advantages of masonry walls over framed (wood) walls?

8.3   A concrete masonry unit is tested for compressive strength and produces the following results:

Failure load = 726 kN
Gross area = 0.081 m^2
Gross volume = 0.015 m^3
Net volume = 0.007 m^3

Is the unit categorized as solid or hollow? Why? What is the compressive strength? Does the compressive strength satisfy the ASTM requirements for load bearing units shown in Table 8.2?

8.4   A half-block concrete masonry unit is tested for compressive strength. The outside dimensions of the specimen are 190 mm × 190 mm × 190 mm. The cross section is a hollow square with a wall thickness of 25 mm. The load is applied perpendicular to the hollow cross section and the maximum load is 225 kN.
a. Determine the gross area compressive strength.
b. Determine the net area compressive strength.

8.5   A concrete masonry unit has actual gross dimensions of 190 mm × 190 mm × 190 mm. The unit is tested in a compression machine with the following results:

Failure load = 436 kN
Net volume of 224516 mm^3

a. Is the unit categorized as solid or hollow?
b. Calculate the gross area compressive strength.
c. Calculate the net area compressive strength.

8.6   A half-block concrete masonry unit is subjected to compression until failure. The outside dimensions of the specimen are 190 mm × 190 mm × 190 mm. The cross section is a hollow square with a wall thickness of 38 mm. The load is applied perpendicular to the hollow cross section and the maximum load is 296 kN.

    a. Determine the gross area compressive strength.

    b. Determine the net area compressive strength.

8.7   A symmetrical hollow concrete masonry unit, with a cross-sectional area as shown in Figure P8.7, is tested for compressive strength. The compressive load at failure is 329 kN. What is the compressive strength in MPa?

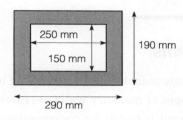

FIGURE P8.7

8.8   A concrete masonry unit has actual gross dimensions of 190 mm × 190 mm × 190 mm. The unit is tested in a compression machine with the following results:

    Maximum failure load = 370 kN
    Net volume = 5125 cm³

    a. Is the unit categorized as solid or hollow? Why?

    b. Calculate the gross area compressive strength.

    c. Calculate the net area compressive strength.

8.9   a Why is it important that the concrete masonry units meet certain absorption requirements?

    b. A portion of a normal-weight concrete masonry unit was tested for absorption and moisture content and produced the following masses:

    mass of unit as received = 1.95 kg
    saturated mass of unit = 2.13 kg
    oven-dry mass of unit = 1.91 kg
    immersed mass of unit = 0.95 kg

    Calculate the absorption in kg/m³ and the moisture content of the unit as a percent of total absorption. Does the absorption meet the ASTM C90 requirement for absorption?

8.10   A portion of a medium-weight concrete masonry unit was tested for absorption and moisture content and produced the following results:

mass of unit as received = 5435 g
saturated mass of unit = 5776 g
oven-dry mass of unit = 5091 g
immersed mass of unit = 2973 g

Calculate the absorption in kg/m^3 and the moisture content of the unit as a percent of total absorption. Does the absorption meet the ASTM C90 requirement for absorption?

8.11 A portion of a concrete masonry unit was tested for absorption and moisture content according to ASTM C140 procedure and produced the following results:

mass of unit as received = 7024 g
saturated mass of unit = 7401 g
oven-dry mass of unit = 6916 g
immersed mass of unit = 3624 g

Determine the following:
a. percent absorption
b. moisture content of the unit as a percent of total absorption
c. dry density
d. weight classification according to ASTM C90 (lightweight, medium weight, or normal weight)

8.12 A concrete masonry unit was tested according to ASTM C140 procedure and produced the following results:

mass of unit as received = 8271 g
saturated mass of unit = 8652 g
oven-dry mass of unit = 7781 g

For this unit, calculate (a) absorption in percent and (b) moisture content as a percent of absorption.

8.13 Define the nominal, specified (modular), and actual dimensions of clay bricks.

8.14 State and define the three grades of clay bricks.

8.15 A severe weathering clay brick was tested for absorption and saturation coefficient according to ASTM C67 procedure and produced the following data:

dry mass of specimen = 2.186 kg
saturated mass after 24-hr submersion in cold water = 2.453 kg
saturated mass after 5-hr submersion in boiling water = 2.472 kg

Calculate absorption by 24-hr submersion, absorption by 5-hr boiling, and saturation coefficient. Does the brick satisfy the ASTM requirements?

8.16 What is mortar made of? What are the functions of mortar?

8.17 What is grout? What is grout used for?

8.18 What are the ingredients of plaster? What is it used for?

8.6   References

Adams, J. T. *The Complete Concrete, Masonry and Brick Handbook.* New York: Arco, 1979.

Portland Cement Association, Masonry, http://www.cement.org/for-concrete-books-learning/materials-applications/masonry.

Portland Cement Association. *Mortars for Masonry Walls.* Skokie, IL: Portland Cement Association, 1987.

Portland Cement Association. *Masonry Information.* Skokie, IL: Portland Cement Association, 1991.

Somayaji, S. *Civil Engineering Materials.* 2nd ed., Upper Saddle River, NJ: Prentice Hall, 2001.

Subsacic, C.A., *A White Paper Report on Green Building Sustainable Design and Concrete Masonry as Manufactured and Constructed in California and Nevada,* Prepared for Concrete Masonry Association of California and Nevada, http://www.cmacn.org/PDF/White_Papter_Report_Green_Building_Sustainable_Design_Concrete_Masonfy.pdf, 2013.

C H A P T E R

9

ASPHALT BINDERS AND
ASPHALT MIXTURES

Asphalt is one of the oldest materials used in construction. Asphalt binders were used in 3000 B.C., preceding the use of the wheel by 1000 years. Before the mid-1850s, asphalt came from natural pools found in various locations throughout the world, such as the Trinidad Lake asphalt, which is still mined. However, with the discovery and refining of petroleum in Pennsylvania, use of asphalt cement became widespread. By 1907, more asphalt cement came from refineries than came from natural deposits. Today, practically all asphalt cement is from refined petroleum.

Bituminous materials are classified as asphalts and tars, as shown in Figure 9.1. Several asphalt products are used; asphalt is used mostly in pavement construction but is also used as sealing and waterproofing agents. Tars are produced by the destructive distillation of bituminous coal or by cracking petroleum vapors. In the United States, tar is used primarily for waterproofing membranes, such as roofs. Tar may also be used for pavement treatments, particularly where fuel spills may dissolve asphalt cement, such as on parking lots and airport aprons.

The fractional distillation process of crude petroleum is illustrated in Figure 9.2. Different products are separated at different temperatures. Figure 9.2 shows the main products, such as gasoline, kerosene, diesel oil, and asphalt residue (asphalt cement). Since asphalt is a lower valued product than other components of crude oil, refineries are set up to produce the more valuable fuels at the expense of asphalt production. The quantity and quality of the asphalt depends on the crude petroleum source and the refining method. Some crude sources, such as the Nigerian oils, produce little asphalt, while others, such as many of the Middle Eastern oils, have a high asphalt content.

This chapter reviews the types, uses, and chemical and physical properties of asphalt. The asphalt concrete used in road and airport pavements, which is a mixture of asphalt cement and aggregates (Figure 9.3), is also presented. This chapter discusses the recently developed Performance Grade asphalt binder specifications and Super-pave™ (a registered trademark of the National Academy of Sciences) mix design. Recycling of pavement materials and additives used to modify the asphalt properties are also included.

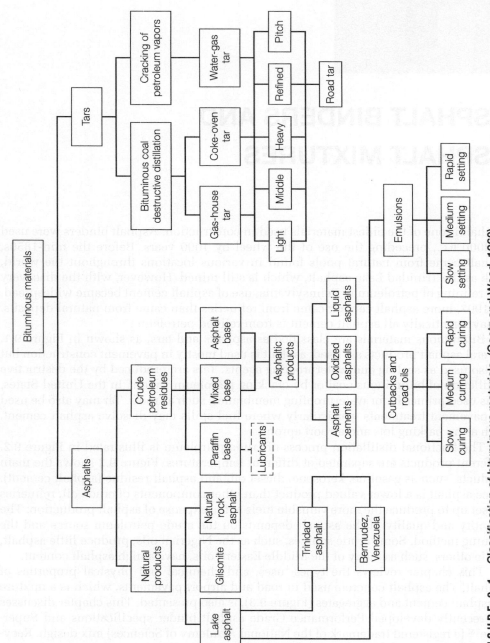

FIGURE 9.1 Classification of bituminous materials. (Goetz and Wood, 1960)

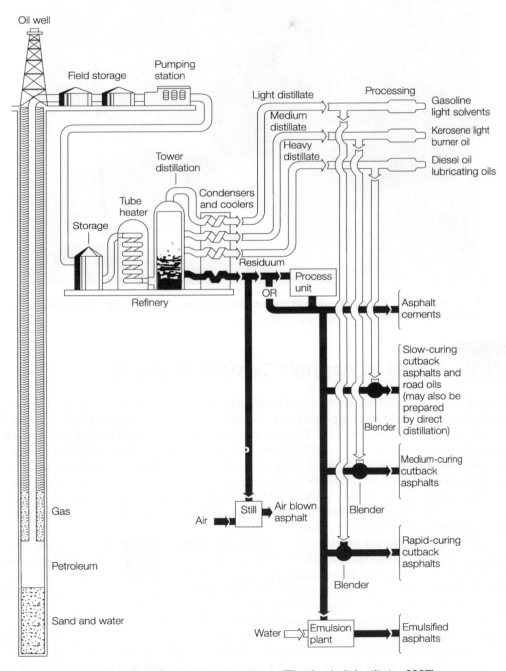

FIGURE 9.2   Distillation of crude petroleum. (The Asphalt Institute, 2007)

FIGURE 9.3   Photo showing asphalt cement, aggregate, and asphalt concrete. (Courtesy of FHWA)

9.1   Types of Asphalt Cement Products

Asphalt cement used in pavements is produced in three forms: *asphalt cement, asphalt cutback,* and *asphalt emulsion.* Asphalt cement is a blend of hydrocarbons of different molecular weights. The characteristics of the asphalt depend on the chemical composition and the distribution of the molecular weight hydrocarbons. As the distribution shifts toward heavier molecular weights, the asphalt becomes harder and more viscous. At room temperatures, asphalt cement is a semisolid material that cannot be applied readily as a binder without being heated. Liquid asphalt products, cutbacks and emulsions, have been developed and can be used without heating (The Asphalt Institute, 2007).

Although the liquid asphalts are convenient, they cannot produce a quality of asphalt concrete comparable to what can be produced by heating neat asphalt cement and mixing it with carefully selected aggregates. Asphalt cement has excellent adhesive characteristics, which make it a superior binder for pavement applications. In fact, it is the most common binder material used in pavements.

A cutback is produced by dissolving asphalt cement in a lighter molecular weight hydrocarbon solvent. When the cutback is sprayed on a pavement or mixed with aggregates, the solvent evaporates, leaving the asphalt residue as the binder. In the past, cutbacks were widely used for highway construction. They were effective and could be applied easily in the field. However, three disadvantages have severely limited the use of cutbacks. First, as petroleum costs have escalated, the use of these expensive solvents as a carrying agent for the asphalt cement is no longer cost effective. Second, cutbacks are hazardous materials due to the volatility of the solvents. Finally, application of the cutback releases environmentally unacceptable hydrocarbons into the atmosphere. In fact, many regions with air pollution problems have outlawed the use of any cutback material.

An alternative to dissolving the asphalt in a solvent is dispersing the asphalt in water as emulsion as shown in Figure 9.4. In this process, the asphalt cement is physically broken down into micron-sized globules that are mixed into water containing an emulsifying agent. Emulsified asphalts typically consist of about 60% to 70% asphalt cement, 30% to 40% water, and a fraction of a percent of emulsifying agent. There are many types of emulsifying agents; basically they are a soap material. The emulsifying molecule has two distinct components, the head portion, which has an electrostatic charge, and the tail portion, which has a high affinity for asphalt. The charge can be either positive to produce a *cationic* emulsion or negative to produce an *anionic* emulsion. When asphalt is introduced into the water with the emulsifying agent, the tail portion of the emulsifier attaches itself to the asphalt, leaving the head exposed. The electric charge of the emulsifier causes a repulsive force between the asphalt globules, which maintains their separation in the water. Since the specific gravity of asphalt is very near that of water, the globules have a neutral buoyancy and, therefore, do not tend to float or sink. When the emulsion is mixed with aggregates or used on a pavement, the water evaporates, allowing the asphalt globs to come together, forming the binder. The phenomenon of separation between the asphalt residue and water is referred to as *breaking* or *setting*. The rate of emulsion setting can be controlled by varying the type and amount of the emulsifying agent.

Since most aggregates bear either positive surface charges (such as limestone) or negative surface charges (such as siliceous aggregates), they tend to be compatible with

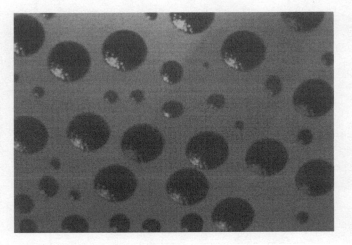

FIGURE 9.4 Photo of magnified asphalt emulsion showing minute droplets of asphalt cement dispersed in a water medium. (Courtesy of FHWA)

anionic or cationic emulsions, respectively. However, some emulsion manufacturers can produce emulsions that bond well to aggregate-specific types, regardless of the surface charges.

Although emulsions and cutbacks can be used for the same applications, the use of emulsions is increasing because they do not include hazardous and costly solvents.

9.2   Uses of Asphalt

The main use of asphalt is in pavement construction and maintenance. In addition, asphalt is used in sealing and waterproofing various structural components, such as roofs and underground foundations.

The selection of the type and grade of asphalt depends on the type of construction and the climate of the area. Asphalt cements, also called asphalt binders, are used typically to make hot-mix asphalt and warm mix concrete, HMA and WMA respectively, for the surface and base layers of asphalt pavements (see Figures 9.5 and 9.6). Asphalt concrete is also used in patching and repairing both asphalt and portland cement concrete pavements. Liquid asphalts (emulsions and cutbacks) are used for pavement maintenance and preservation applications, such as fog seals, chip seals, slurry seals, and microsurfacing (see Figures 9.7–9.10)

FIGURE 9.5   Placing asphalt concrete used for the surface layer of asphalt pavement.

F I G U R E   9 . 6   Compaction of asphalt concrete.

F I G U R E   9 . 7   Applying fog seal (diluted emulsion) for preserving existing
pavement. (Courtesy of FHWA)

FIGURE 9.8 Spraying tack coat (emulsion) on existing asphalt pavement before placing an asphalt overlay. (Courtesy of FHWA)

FIGURE 9.9 Applying chip seal (emulsion followed by aggregates) for preserving existing pavement. (Courtesy of FHWA)

FIGURE 9.10   Applying microsurfacing for preserving existing pavement. (Courtesy of FHWA)

(The Asphalt Institute, 2007; Mamlouk and Zaniewski, 1998). Liquid asphalts may also be used to seal the cracks in pavements. Liquid asphalts are mixed with aggregates to produce cold mixes, as well. Cold mixtures are normally used for patching (when asphalt concrete is not available), base and subbase stabilization, and surfacing of low-volume roads. Table 9.1 shows common paving applications for asphalts.

9.3   Temperature Susceptibility of Asphalt

The consistency of asphalt is greatly affected by temperature. Asphalt gets hard and brittle at low temperatures and soft at high temperatures. Figure 9.11 shows a conceptual relation between temperature and logarithm of viscosity. The viscosity of the asphalt decreases when the temperature increases. Asphalt's temperature susceptibility can be represented by the slope of the line shown in Figure 9.11; the steeper the slope the higher the temperature susceptibility of the asphalt. However, additives can be used to reduce this susceptibility.

When asphalt is mixed with aggregates, the mixture will perform properly only if the asphalt viscosity is within an optimum range. If the viscosity of asphalt is higher than the optimum range, the mixture will be too brittle and susceptible to

TABLE 9.1 Paving Applications of Asphalt

Term	Description	Application
Hot-mix and warm-mix asphalt	Carefully designed mixture of asphalt cement and aggregates	Pavement surface and base, patching
Cold mix	Mixture of aggregates and liquid asphalt	Patching, low volume road surface, asphalt stabilized base
Fog seal	Spray of diluted asphalt emulsion on existing pavement surface	Seal existing pavement surface
Prime coat	Spray coat asphalt emulsion to bond aggregate base and asphalt concrete surface	Construction of flexible pavement
Tack coat	Spray coat asphalt emulsion between lifts of asphalt concrete	Construction of new pavements or between an existing pavement and an overlay
Chip seal	Spray coat of asphalt emulsion (or asphalt cement or cutback) followed with aggregate layer	Maintenance of existing pavement or low volume road surfaces
Slurry seal	Mixture of emulsion, well-graded fine aggregate and water	Resurface low volume roads
Microsurfacing	Mixture of polymer modified emulsion, well-graded crushed fine aggregate, mineral filler, water, and additives	Texturing, sealing, crack filling, rut filling, and minor leveling

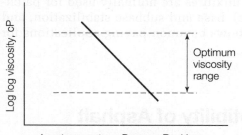

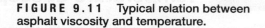

FIGURE 9.11 Typical relation between asphalt viscosity and temperature.

low-temperature (thermal) cracking (Figure 9.12). On the other hand, if the viscosity is below the optimum range, the mixture will flow readily, resulting in permanent deformation (rutting) as shown in Figure 9.13.

Due to temperature susceptibility, the grade of the asphalt cement should be selected according to the climate of the area. The viscosity of the asphalt should be mostly within the optimum range for the area's annual temperature range; soft-grade asphalts are used for cold climates and hard-grade asphalts for hot climates (see Figure 9.14).

FIGURE 9.12   Thermal cracking resulting from the use of too-stiff asphalt in a cold climate area. (Courtesy of FHWA)

FIGURE 9.13   Rutting that could result from the use of too-soft asphalt.

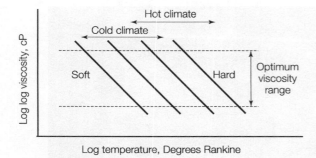

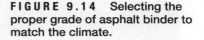

FIGURE 9.14 Selecting the proper grade of asphalt binder to match the climate.

9.4 Chemical Properties of Asphalt

Asphalt is a mixture of a wide variety of hydrocarbons primarily consisting of hydrogen and carbon atoms, with minor components such as sulfur, nitrogen, and oxygen (heteroatoms), and trace metals. The percentages of the chemical components, as well as the molecular structure of asphalt, vary depending on the crude oil source (Peterson, 1984).

The molecular structure of asphalt affects the physical and aging properties of asphalt, as well as how the asphalt molecules interact with each other and with aggregate. Asphalt molecules have three arrangements, depending on the carbon atom links: (1) *aliphatic* or paraffinic, which form straight or branched chains, (2) *saturated rings,* which have the highest hydrogen to carbon ratio, and (3) *unsaturated rings* or aromatic. Heteroatoms attached to carbon alter the molecular configuration. Since the number of molecular structures of asphalt is extremely large, research on asphalt chemistry has focused on separating asphalt into major fractions that are less complex or more homogeneous. Each of these fractions has a complex chemical structure.

Asphalt cement consists of asphaltenes and maltenes (petrolenes). The maltenes consist of resins and oils. The asphaltenes are dark brown, friable solids that are chemically complex, with the highest polarity among the components. The asphaltenes are responsible for the viscosity and the adhesive property of the asphalt. If the asphaltene content is less than 10%, the asphalt concrete will be difficult to compact to the proper construction density. Resins are dark and semisolid or solid, with a viscosity that is largely affected by temperature. The resins act as agents to disperse asphaltenes in the oils; the oils are clear or white liquids. When the resins are oxidized, they yield asphaltene-type molecules. Various components of asphalt interact with each other to form a balanced or compatible system. This balance of components makes the asphalt suitable as a binder.

Three fractionation schemes are used to separate asphalt components, as illustrated in Figure 9.15. The first scheme [Figure 9.15(a)] is partitioning with partial solvents in which *n*-butanol is added to separate (precipitate) the asphaltics. The butranol is then evaporated, and the remaining component is dissolved

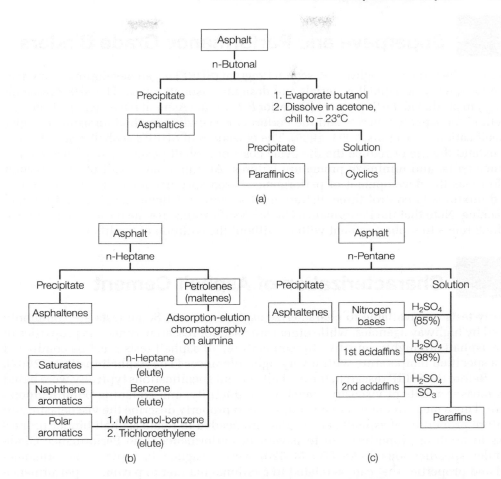

FIGURE 9.15   Schematic diagrams of three asphalt fractionation schemes: (a) partitioning with partial solvents, (b) selective-adsorption-description, and (c) chemical precipitation (J. Claine Petersen. Chemical Composition of Asphalt as Related to Asphalt Durability: State of the Art. In *Transportation Research Record 999*, Transportation Research Board, National Research Council, Washington, DC, 1984).

in acetone and chilled to −23°C to precipitate the paraffinics and leave the cyclics in solution. The second scheme [Figure 9.15(b)] is selective adsorption–desorption, in which *n*-heptane is added to separate asphaltene. The remaining maltene fraction is introduced to a chromatographic column and desorbed using solvents with increasing polarity to separate other fractions. The third scheme [Figure 9.15(c)] is chemical precipitation in which *n*-pentane is added to separate the asphaltenes. Sulfuric acid (H_2SO_4) is added in increasing strengths to precipitate other fractions.

In addition, asphalt can be separated based on the molecular size with the use of high-pressure liquid chromatography (gel-permeation chromatography).

9.5   Superpave and Performance Grade Binders

In 1987, the Strategic Highway Research Program (SHRP) began developing a new system for specifying asphalt materials and designing asphalt mixes. The SHRP research program produced the Superpave (*Superior Performing Asphalt Pavements*) mix design method for asphalt concrete and the Performance Grading method for asphalt binder specification (McGennis, 1994; 1995). The objectives of SHRP's asphalt research were to extend the life or reduce the life-cycle costs of asphalt pavements, reduce maintenance costs, and minimize premature failures. An important result of this research effort was the development of performance-based specifications for asphalt binders and mixtures to control three distress modes: rutting, fatigue cracking, and thermal cracking. Note that the Performance Grade specifications use the term *asphalt binder,* which refers to asphalt cement with or without the addition of modifiers.

9.6   Characterization of Asphalt Cement

Many tests are available to characterize asphalt cement. Some tests are commonly used by highway agencies, while others are used for research. Since the properties of the asphalt are highly sensitive to temperature, all asphalt tests must be conducted at a specified temperature within very tight tolerances (The Asphalt Institute, 2007).

Before the SHRP research, the asphalt cement specifications typically were based on measurements of viscosity, penetration, ductility, and softening point temperature. These measurements are not sufficient to properly describe the viscoelastic and failure properties of asphalt cement that are needed to relate asphalt binder properties to mixture properties and to pavement performance. The Performance Grade binder specifications (AASHTO M 320) were designed to provide performance-related properties that can be related in a rational manner to pavement performance (McGennis, 1994). Concerns with the ability of AASTO M 320 to properly characterize rutting behavior, especially for modified binders, lead to the development of standards and test procedures for using the Multiple Stress Creep Recovery method for grading binders, AASHTO MP 19; this is a provisional standard and the FHWA is encouraging states to implement the new method.

9.6.1   Performance Grade Characterization Approach

Both AASHTO M 320 and MP 19 Performance Grade tests used to characterize the asphalt binder are performed at pavement temperatures to represent the upper, middle, and lower ranges of service temperatures. The measurements are obtained at temperatures in keeping with the distress mechanisms. Therefore, unlike previous specifications that require performing the test at a fixed temperature and varying the requirements for different grades of asphalt, the Performance Grade specifications require performing the test at the critical pavement temperatures and fixing the criteria for all asphalt grades. Thus, the Performance Grade philosophy ensures that the asphalt properties meet the specification criteria at the critical pavement temperatures.

Three pavement design temperatures are required for the binder specifications: a maximum, an intermediate, and a minimum temperature. The maximum and minimum pavement temperatures for a given geographical location in the United States can be generated using algorithms contained within the SHRP software, based on weather information from 7500 weather stations (http://www.fhwa.dot.gov/ research/tfhrc/programs/infrastructure/pavements/ltpp/dwnload.cfm). The maximum pavement design temperature is selected as the highest successive seven-day average maximum pavement temperature. The minimum pavement design temperature is the minimum pavement temperature expected over the life of the pavement. The minimum and maximum pavement temperatures, along with a reliability level as described in the following, are used to select the binder grade temperatures. The intermediate pavement design temperature is the average of the upper and lower binder grade temperatures plus 4°C.

Laboratory tests that evaluate rutting potential use the upper binder grade temperature, whereas tests that evaluate fatigue potential use the intermediate binder grade temperature. Thermal-cracking tests use the minimum binder grade temperature plus 10°C. The minimum binder grade temperature is increased by 10°C to reduce the testing time. Through the use of the time–temperature superposition principle, test results obtained at a higher temperature and shorter load duration are equivalent to tests performed at a lower temperature and longer load duration (McGennis, 1994).

9.6.2 ■ Performance Grade Binder Characterization

Several tests are used in the Performance Grade method to characterize the asphalt binder. Some of these tests have been used before for asphalt testing, while others are new. The discussion that follows summarizes the main steps and the significance of SHRP tests. With the exception of the rotational (Brookfield) viscometer, solubility, and flash point tests, the test temperatures are selected based on the temperature at the design location. The binder specification indicates the specific test temperatures used for various binders for each test (McGennis, 1994). Four tests are performed on the neat or tank asphalt: flash point, solubility, rotational viscosity, and dynamic shear rheometer (DSR). To simulate the effect of aging on the properties of the binder, the rolling thin-film oven (RTFO) and pressure-aging vessel (PAV) condition the binder for short-term and long-term effects. Samples are conditioned with both the RTFO and PAV before determining their characteristics with respect to fatigue and low-temperature cracking.

Rolling Thin-Film Oven   The RTFO procedure is used to simulate the short-term aging that occurs in the asphalt during production of asphalt concrete. In the RTFO method (ASTM D2872), the asphalt binder is poured into special bottles. The bottles are placed in a rack in a forced-draft oven, as shown in Figure 9.16, at a temperature of 163°C for 75 minutes. The rack rotates vertically, continuously exposing fresh asphalt. The binder in the rotating bottles is also subjected to an air jet to speed up the aging process. The Performance Grade specifications limit the amount of mass loss during RTFO conditioning. RTFO conditioning is used to prepare samples for evaluation for rutting potential with the DSR and prior to conditioning with the PAV. Under the penetration and viscosity grading methods, the aged binder is usually tested for penetration or viscosity and the results are compared with those of un-conditioned asphalt.

FIGURE 9.16   Rolling thin-film oven test apparatus.

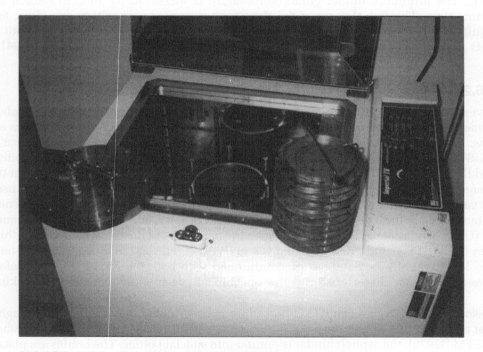

FIGURE 9.17   Pressure-aging vessel apparatus.

Pressure-Aging Vessel   The PAV consists of a temperature-controlled chamber, and pressure- and temperature-controlling and measuring devices, as illustrated in Figure 9.17 (ASTM D6521). The asphalt binder is first aged, using the RTFO (ASTM D2872). A specified thickness of residue from the RTFO is placed in the PAV pans. The asphalt is then aged at the specified aging temperature for 20 hours in a vessel under 2.10 MPa of air pressure. Aging temperature, which ranges between 90°C and 110°C, is selected according to the grade of the asphalt binder. Since the procedure forces oxygen into the sample, it is necessary to use a vacuum oven to remove any air bubbles from the sample prior to testing.

The PAV is designed to simulate the oxidative aging that occurs in asphalt binders during pavement service. Residue from this process may be used to estimate the physical or chemical properties of an asphalt binder after 5 to 10 years in the field.

Flash Point   At high temperatures, asphalt can flash or ignite in the presence of open flame or spark. The flash point test is a safety test that measures the temperature at which the asphalt flashes; asphalt cement may be heated to a temperature below this without becoming a fire hazard. The Cleveland open cup method (ASTM D92) requires partially filling a standard brass cup with asphalt cement. The asphalt is then heated at a specified rate and a small flame is periodically passed over the surface of the cup, as shown in Figure 9.18. The flash point is the temperature of the asphalt when the volatile fumes coming off the sample will sustain a flame for a short period of time. The minimum temperature at which the volatile fumes are sufficient to sustain a flame for an extended period of time is the fire point.

Rotational Viscometer Test   The rotational (Brookfield) viscometer test (ASTM D4402) consists of a rotational coaxial cylinder viscometer and a temperature control unit, as shown in Figure 9.19. The test is performed on unaged binders. The asphalt binder

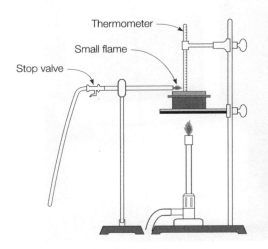

Thermometer
Small flame
Stop valve

FIGURE 9.18   Cleveland open cup flash point test apparatus.

FIGURE 9.19   Rotational viscometer. (Courtesy of FHWA)

sample is placed in the sample chamber at 135°C; then both are placed in the thermocell. A spindle is placed in the asphalt sample and rotated at a specified speed. The viscosity is determined by the amount of torque required to rotate the spindle at the specified speed. The spindle size used is determined based on the viscosity being measured. The Performance Grade specification limit is stated in Pascal seconds (Pa · s), which is equal to cP multiplied by 1000. The viscosity is recorded as the average of three readings, at one-minute intervals, to the nearest 0.1 Pa · s. In addition to testing at the temperature required by the specification, additional tests are performed at higher temperatures to establish the temperature susceptibility relationship used to determine the compaction and mixing temperatures required for the mix design process.

Dynamic Shear Rheometer Test   The DSR, test system, Figure 9.20, consists of two parallel metal plates, an environmental chamber, a loading device, and a control and data acquisition system (AASHTO T315). The DSR is used to measure three specification requirements in the AASHTO M 320 Performance Grading system. For testing the neat binder and for the rutting potential test, the test temperature is equal to the upper temperature for the grade of the asphalt binder (e.g., a PG 64-22 is tested at 64°C). For these tests, the sample size is 25 mm in diameter and 1 mm thick. Prior to testing for rutting potential, the sample is conditioned in the RTFO. For evaluating fatigue potential, the intermediate temperature is used; 25°C for PG 64-22. The sample size is 8 mm in diameter by 2 mm thick. Prior to testing, the sample is conditioned in the RTFO, followed by the PAV.

FIGURE 9.20   Dynamic shear rheometer apparatus. (Courtesy of FHWA)

During testing, one of the parallel plates is oscillated with respect to the other at preselected frequencies and rotational deformation amplitudes (or torque amplitudes). The required amplitude depends on the value of the complex shear modulus of the asphalt binder being tested. Specification testing is performed at an angular frequency of 10 radians/seconds. The complex shear modulus (G^*) and phase angle (δ) are calculated automatically by the rheometer's computer software. The complex shear modulus and the phase angle define the asphalt binder's resistance to shear deformation in the linear viscoelastic region.

AASHTO MP 19, Performance-Graded Asphalt Binder Using Multiple Stress Creep Recovery, augments the requiements of M 320 using the DSR to determine a nonrecoverable creep compliance factor to evaluate rutting potential. Per AASHTO TP 70, the test is performed at the high rated temperature of the binder using the 25-mm loading head. The DSR is configured to apply cycles of stress consisting of a constant stress for 1.0 second followed by a 9.0-second rest period; the strain of the sample is continously recorded. The analysis method for determining the percent nonrecoverable creep compliance are covered in AASHTO TP 70 and are beyond the scope of an introductory materials book.

Bending Beam Rheometer Test The bending beam rheometer measures the midpoint deflection of a simply supported prismatic beam of asphalt binder subjected to a constant load applied to its midpoint (ASTM D6648). The bending beam rheometer test system consists of a loading frame, a controlled temperature bath, and a computer-controlled automated data acquisition unit, as shown in Figure 9.21. The test temperature is 10°C higher than the lower temperature rating of the binder (e.g., a PG 64–22 is tested at −12°C). The sample is conditioned in both the RTFO and PAV prior to testing. An asphalt binder beam is placed in the bath and loaded with a constant force of 980 ± 50 mN for 240 seconds. As the beam creeps, the midpoint deflection is monitored after 8, 15, 30, 60, 120, and 240 seconds. The constant maximum stress in the beam is calculated from the load magnitude and the dimensions of the beam. The maximum strain is calculated from the deflection and the dimensions of the beam. The flexural creep stiffness of the beam is then calculated by dividing the maximum stress by the maximum strain for each of the specified loading times.

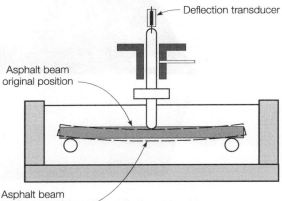

Deflection transducer

Asphalt beam original position

Asphalt beam deflected position

FIGURE 9.21 Schematic of the bending beam rheometer.

The low-temperature thermal-cracking performance of paving mixtures is related to the creep stiffness and the slope of the logarithm of the creep stiffness versus the logarithm of the time curve of the asphalt binder contained in the mix.

Direct Tension Test   The direct tension test system consists of a displacement-controlled tensile loading machine with gripping system, a temperature-controlled chamber, measuring devices, and a data acquisition system, as shown in Figure 9.22 (ASTM D6723). In this test, an asphalt binder specimen, conditioned in the RTFO and PAV, is pulled at a constant rate of deformation of 1 mm/min. The test temperature equals the low-temperature rating plus 10°C. A noncontact extensometer measures the elongation of the specimen. The maximum load developed during the test is monitored. The tensile strain and stress in the specimen when the load reaches a maximum is reported as the failure strain and failure stress, respectively.

The strain at failure is a measure of the amount of elongation that the asphalt binder can sustain without cracking. Strain at failure is used as a criterion for specifying the low-temperature properties of the binder.

9.6.3 ■ Traditional Asphalt Characterization Tests

Traditional tests that were used to characterize asphalt before the development of the Performance Grade system include the penetration and absolute and kinematic viscosities. These tests are currently used to evaluate the asphalt cement in emulsions.

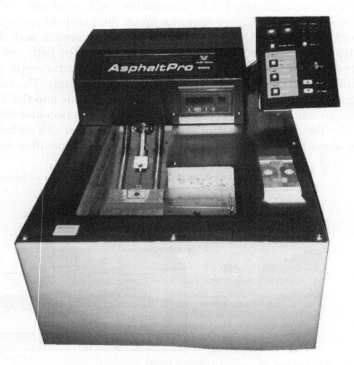

FIGURE 9.22   Direct tension test apparatus.

Penetration   The penetration test (ASTM D5) measures asphalt cement consistency. An asphalt sample is prepared and brought to 25°C. A standard needle with a total mass of 100 g is placed on the asphalt surface. The needle is released and allowed to penetrate the asphalt for 5 seconds, as shown in Figure 9.23. The depth of penetration, in units of 0.1 mm, is recorded and reported as the penetration value. A large penetration value indicates soft asphalt.

Absolute and Kinematic Viscosity Tests   Similar to the penetration test, the viscosity test is used to measure asphalt consistency. Two types of viscosity are commonly measured: absolute and kinematic. The absolute viscosity procedure (ASTM D2171) requires heating the asphalt cement and pouring it into a viscometer placed in a water or oil bath at a temperature of 60°C (Figure 9.24). The viscometer is a U-tube, with a reservoir where the asphalt is introduced and a section with a calibrated diameter and timing marks. For absolute viscosity tests, vacuum is applied at one end. The time during which the asphalt flows between two timing marks on the viscometer is measured using a stopwatch. The flow time, measured in seconds, is multiplied by the viscometer calibration factor to obtain the absolute viscosity in units of Pa·s. Different-sized viscometers are used for different asphalt grades to meet minimum and maximum flow time requirements of the test procedure.

The kinematic viscosity test procedure (ASTM D2170) is similar to that of the absolute viscosity test, except that the test temperature is 135°C. Since the viscosity

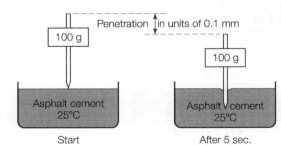

FIGURE 9.23   Penetration test.

FIGURE 9.24   Absolute viscosity test apparatus.

of the asphalt at 135°C is fairly low, vacuum is not used. The time it takes the asphalt to flow between the two timing marks is multiplied by the calibration factor to obtain the kinematic viscosity in units of centistokes (cSt)

9.7   Classification of Asphalt

Several methods are used to characterize asphalt binders (asphalt cements), asphalt cutbacks, and asphalt emulsions.

9.7.1 ■ Asphalt Binders

Asphalt binder is produced in several grades or classes. There are four methods for classifying asphalt binders:

1. performance grading
2. penetration grading
3. viscosity grading
4. viscosity of aged residue grading

Performance Grade Specifications and Selection   Several grades of binder are available, based on their performance in the field. Names of grades start with PG (Performance Graded) followed by two numbers representing the maximum and minimum pavement design temperatures in Celsius. For example, an asphalt binder PG 52–28 would meet the specification for a design high pavement temperature up to 52°C and a design low-temperature warmer than −28°C. The temperatures are calculated 20 mm below the pavement surface. The high and low pavement temperatures are related to the air temperature as well as other factors. Table 9.2 shows the binder grades in the Performance Grade specifications.

The performance-graded asphalt binder specifications are shown in Table 9.3 (AASHTO M 320, ASTM D6373). AASHTO M 320, updated in 2010, includes a

TABLE 9.2   Binder Grades in the Performance Grade Specifications

High Temperature Grades (°C)	Low Temperature Grades (°C)
PG 46	−34, −40, −46
PG 52	−10, −16, −22, −28, −34, −40, −46
PG 58	−16, −22, −28, −34, −40
PG 64	−10, −16, −22, −28, −34, −40
PG 70	−10, −16, −22, −28, −34, −40
PG 76	−10, −16, −22, −28, −34
PG 82	−10, −16, −22, −28, −34

TABLE 9.3 Performance Graded Asphalt Binder Specifications (Table 1 of AASHTO M 320)

Performance Grade	PG 46-			PG 52-							PG 58-					PG 64-					
	34	40	46	10	16	22	28	34	40	46	16	22	28	34	40	10	16	22	28	34	40
Average seven-day maximum pavement design temperature, °C		<46					<52						<58						<64		
Minimum pavement design temperature, °C	>−34	>−40	>−46	>−10	>−16	>−22	>−28	>−34	>−40	>−46	>−16	>−22	>−28	>−34	>−40	>−10	>−16	>−22	>−28	>−34	>−40
								Original Binder													
Flash point temperature: minimum, °C								230													
Viscosity, ASTM D4402: maximum, 3 Pa·s, test temperature, °C								135													
Dynamic Shear: G*/sin δ, minimum, 1.00 kPa test temperature @ 10 radians/seconds, °C		46					52						58						64		
							Rolling Thin-Film Oven Residue														
Mass loss, maximum, percent								1.00													
Dynamic Shear: G*/sin δ, minimum, 2.20 kPa test temperature @ 10 radians/seconds, °C		46					52						58						64		
							Pressure-Aging Vessel (PAV) Residue														
PAV aging temperature, °C		90					90						100						100		
Dynamic Shear: G*·sin δ, maximum, 5000 kPa test temperature @ 10 radians/seconds, °C	10	7	4	25	22	19	16	13	10	7	25	22	19	16	13	31	28	25	22	19	16
Creep stiffness: S, maximum, 300 MPa, m-value, minimum, 0.300 test temperature @ 60s, °C	−24	−30	−36	0	−6	−12	−18	−24	−30	−36	−6	−12	−18	−24	−30	0	−6	−12	−18	−24	−30
Direct tension: failure strain, minimum, 1.0% test temperature @ 1.0 mm/minutes, °C	−24	−30	−36	0	−6	−12	−18	−24	−30	−36	−6	−12	−18	−24	−30	0	−6	−12	−18	−24	−30

TABLE 9.3 (*Continued*)

Performance Grade	PG 70-						PG 76-					PG 82-				
	10	16	22	28	34	40	10	16	22	28	34	10	16	22	28	34
Average seven day maximum pavement design temperature, °C	<70						<76					<82				
Minimum pavement design temperature, °C	>−10	>−16	>−22	>−28	>−34	>−40	>−10	>−16	>−22	>−28	>−34	>−10	>−16	>−22	>−28	>−34
Original Binder																
Flash point temperature: minimum °C	230															
Viscosity, ASTM D4402: maximum, 3 Pa·s, test temperature, °C	135															
Dynamic Shear: G*/sin δ, minimum, 1.00 kPa test temperature @ 10 radians/seconds, °C	70						76					82				
Rolling Thin-Film Oven Residue																
Mass loss, maximum, percent	1.00															
Dynamic Shear: G*/sin δ, minimum, 2.20 kPa test temperature@ 10 radians/seconds, °C	70						76					82				
Pressure-Aging Vessel (PAV) Residue																
PAV aging temperature, °C	100 (110)						100 (110)					100 (110)				
Dynamic Shear, G*·sin δ, maximum, 5000 kPa test temperature @ 10 radians/seconds, °C	34	31	28	25	22	19	37	34	31	28	25	40	37	34	31	28
Creep stiffness: S, maximum, 300 MPa, m-value, minimum, 0.300 test temperature @ 60 s, °C	0	−6	−12	−18	−24	−30	0	−6	−12	−18	−24	0	−6	−12	−18	−24
Direct tension: failure strain, minimum, 1.0% test temperature @ 1.0 mm/minutes, °C	0	−6	−12	−18	−24	−30	0	−6	−12	−18	−24	0	−6	−12	−18	−24

second specification table that replaces the creep stiffness and direct tension parameters with a critical temperature parameter. Table 9.3 shows the design criteria of various test parameters at the specified test temperatures. One important difference between the Performance Grade specifications and the traditional specifications is in the way the specifications work. As shown in Table 9.3, the physical properties (criteria) remain constant for all grades, but the temperatures at which these properties must be achieved vary, depending on the climate at which the binder is expected to be used. The temperature ranges shown in Table 9.3 encompass all pavement temperature regimes that exist in the United States and Canada.

The binder is selected to satisfy the maximum and minimum design pavement temperature requirements. The average seven-day maximum pavement temperature is used to determine the design maximum, whereas the design minimum pavement temperature is the lowest pavement temperature. Since the maximum and minimum pavement temperatures vary from one year to another, a reliability level is considered. As used in the Performance Grade, reliability is the percent probability in a single year that the actual pavement temperature will not exceed the design high pavement temperature or be lower than the design low pavement temperature.

It is assumed that the design high and design low pavement temperatures throughout the years follow normal distributions as illustrated in Figure 9.25(a). In this example, the average seven-day maximum pavement temperature is 56°C and the standard deviation is 2°C. Similarly, the average one-day minimum pavement temperature is −23°C and the standard deviation is 4°C. Since the area under the normal distribution curve represents the probability as illustrated in Figure 1.21, the range of temperature that satisfies the assumed probability can be calculated. For example, the range between −23°C and 56°C results in a 50% reliability for both high and low temperatures. By subtracting two standard deviations from the minimum

---

Sample Problem 9.1

What standard PG asphalt binder grade should be selected under the following conditions:

   The seven-day maximum pavement temperature has a mean of 57°C and a standard deviation of 2°C.

   The minimum pavement temperature has a mean of −6°C and a standard deviation of 3°C.

   Reliability is 98%.

Solution

High-temperature grade $\geq 57 + (2 \times 2) \geq 61°C$

Low-temperature grade $\leq -6 - (2 \times 3) \leq -12°C$

The closest standard PG asphalt binder grade that satisfies the two temperature grades is PG 64–16.

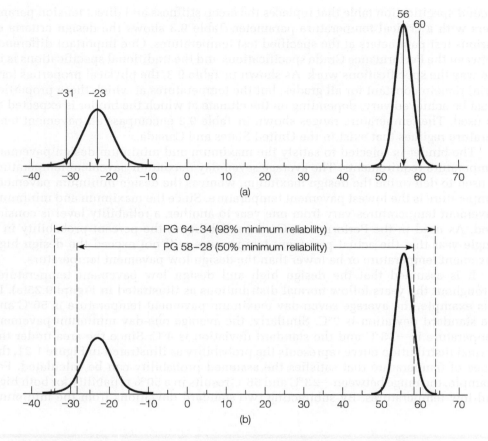

FIGURE 9.25 Example of the distribution of design pavement temperatures and selection of binder grades: (a) distribution of high and low design pavement temperatures and (b) binder grade selection. (Courtesy of FHWA)

pavement temperature and adding two standard deviations to the maximum pavement temperature, the range between −31°C and 60°C results in 98% reliability. In selecting the appropriate grade, the designer should select the standard PG grade that most closely satisfies the required reliability level. This "rounding" typically results in a higher reliability level than is intended, as shown in Figure 9.25(b). Note that the reliability levels at the high- and low-temperature grades do not need to be the same, depending on the specific pavement conditions.

It is logical that in addition to the pavement temperature, the amount of traffic loads and traffic speed should affect the selection of the binder. Therefore, an adjustment to the binder selection needs to be made according to the Superpave mix design specification, AASHTO M 323, using a binder "bump." Table 9.4 shows the number of increases in the high-temperature grade for different combinations of traffic loads, represented by the 82 kN equivalent single axle load (ESALs) and speeds. (ESAL is a pavement design factor used in the design of pavement that

TABLE 9.4 Number of increases in the high-temperature grade for different combinations of traffic loads and speeds

Design ESALs (millions)	Traffic Speed km/h		
	<20	20-70	>70
<0.3	—	—	—
0.3 – 30	2	1	—
>30	2	1	1

considers both traffic volume and loads (Huang, 2004)). For example, if a PG 64-22 was selected based on temperatures, and the traffic is 10 million ESALs and the speed is 50 km/h, a PG 70-22 should be selected for design. Also, if a PG 64-22 was selected based on temperatures, and the traffic is 40 million ESALs and the speed is 15 km/h, a PG 76-22 should be selected for design. There is no adjustment for the low-temperature grade.

The grade bumping process has not been entirely satisfactory. AASHTO has introduced a provisional specification, MP, that will use a different loading protocol for the dynamic shear rheometer that measures and specifies material properties that are directly related to the binder's properties to handle different traffic loads and speeds.

Other Asphalt Binder Grading Methods Table 9.5 shows various asphalt cement grades based on penetration and on their properties (ASTM D946). The grades correspond to the allowable penetration range; that is, the penetration of a 40–50 grade must be in the range of 40 to 50. Various AC grades based on viscosity and their properties are shown in Table 9.6 (ASTM D3381). The AC grade numbers are 1/10 of the middle of the allowable viscosity range; that is, an AC-5 has an absolute viscosity of 50 ± 10 Pa·s. Therefore, high viscosity asphalt cements have a high designation number. Aged-residue grades are based on the absolute viscosity of the asphalt after it has been conditioned (RTFO method) to simulate the effects of the aging that occur when the asphalt cement is heated to make asphalt concrete. The aged-residue grade

TABLE 9.5 Penetration Grading System of Asphalt Cement

Grade	Penetration	
	Minimum	Maximum
40–50	40	50
60–70	60	70
85–100	85	100
120–150	120	150
200–300	200	300

TABLE 9.6   Viscosity Grading System of Asphalt Cement

Grade	Absolute Viscosity (Pa·s)
AC-2.5	25 ± 5
AC-5	50 ± 10
AC-10	100 ± 20
AC-20	200 ± 40
AC-30	300 ± 60
AC-40	400 ± 80

numbers are 1/10 of the middle of the allowable viscosity range after conditioning, as shown in Table 9.7 (ASTM D3381). For example, an AR-1000 has an absolute viscosity of 100 ± 25 Pa·s.

9.7.2 ▨ Asphalt Cutbacks

Three types of cutbacks are produced, depending on the hardness of the residue and the type of solvent used. *Rapid-curing cutbacks* are produced by dissolving hard residue in a highly volatile solvent, such as gasoline. *Medium-curing cutbacks* use medium hardness residue and a less volatile solvent, such as kerosene. *Slow-curing cutbacks* are produced by either diluting soft residue in nonvolatile or low-volatility fuel oil or by simply stopping the refining process before all of the fuel oil is removed from the stock.

Curing the cutback refers to the evaporation of the solvent from the asphalt residue. Rapid-curing (RC) cutbacks cure in about 5 to 10 minutes, while medium-curing (MC) cutbacks cure in a few days. Slow-curing (SC) cutbacks cure in a few months. In addition to the three types, cutbacks have several grades defined by the kinematic viscosity at 60°C. Grades of 30, 70, 250, 800, and 3000 are manufactured, with higher grades indicating higher viscosities. Thus, cutback asphalts are designated by letters (RC, MC, or SC), representing the type, followed by a number that represents the grade. For example, MC-800 is a medium-curing cutback with a grade of 800. The different grades of cutback are produced by varying the amounts and types of solvent and base asphalt. The specifications of cutbacks are standardized by ASTM D2026, D2027, and D2028.

TABLE 9.7   Aged-Residue Grading System of Asphalt Cement

Grades	Absolute Viscosity (Pa·s)
AR-1000	100 ± 25
AR-2000	200 ± 50
AR-4000	400 ± 100
AR-8000	800 ± 200
AR-16000	1600 ± 400

9.7.3 ■ Asphalt Emulsions

Asphalt emulsions are produced in a variety of combinations of the electric charge of the emulsifying agent, the rate the emulsion sets (breaks), the viscosity of the emulsion, and the hardness of the asphalt cement. Both anionic and cationic emulsions are produced which have negative and positive charges, respectively. Depending on the emulsion concentration, the set or break time is varied from rapid to medium to slow set. Rapid-setting emulsion sets in about 5 to 10 minutes, medium-setting emulsion sets in several hours, and slow-setting emulsion sets in a few months. The viscosity of the emulsion is rated as normal flow or slow flow, based on the Saybolt–Furol viscosity (ASTM D244). The consistency of the asphalt in the emulsion is evaluated with the penetration test. Asphalt residues with a penetration of 100–200 are typically used. However, soft asphalts, with a penetration of greater than 200, or hard asphalts, with a penetration of 60–100, may be used.

Asphalt emulsion types are designated based on the setting rate as RS, MS, or SS for rapid, medium, and slow set, respectively. If a cationic emulsifying agent was used, the set designation is preceded by a C (e.g., a cationic rapid set emulsion is designated as CRS). If the first letter of the emulsion type is not a C, then it is an anionic emulsion. The flow rate of the emulsion is designated next with a 1 or 2 for a normal or slow flow respectively. Finally, the consistency of the asphalt is identified. Not all possible combinations of charge type, set rate, viscosity, and asphalt hardness are produced. Table 9.8 is a summary of the emulsion grades and types (Jansich and Gaillard, 1998).

Other emulsion types are also produced, such as the high float residue emulsion and the quick-set emulsion. The specifications of various asphalt emulsions are standardized by ASTM D977.

TABLE 9.8 Asphalt Emulsion Grades

Charge	Grade	Setting Speed	Viscosity of Emulsion at 25°C (ASTM D244)	Penetration of Residue
Anionic	RS-1	Rapid	20–100	100–200
	RS-2	Rapid	75–400*	100–200
	MS-1	Medium	20–100	100–200
	MS-2	Medium	≥100	100–200
	MS-2h	Medium	≥100	60–100
	SS-1	Slow	20–100	100–200
	SS-1h	Slow	20–100	60–100
Cationic	CRS-1	Rapid	20–100	100–250
	CRS-2	Rapid	100–400	100–250
	CMS-2	Medium	50–450	100–250
	CMS-2h	Medium	50–450	60–100
	CSS-1	Slow	20–100	100–250
	CSS-1h	Slow	20–100	60–100

* Test at 50°C

9.8 Asphalt Concrete

Asphalt concrete consists of asphalt binder and aggregates mixed together at a high temperature and placed and compacted on the road while still hot. Historically, asphalt concrete was called hot mix asphalt, HMA, as it was mixed at temperatures of 135 to 160°C. With energy costs and environmental concerns, technology has been developed to produce warm mix asphalt, WMA, at temperatures of 110 to 130°C. Asphalt (flexible) pavements cover approximately 93% of the 3.5 million miles of paved roads in the United States. The performance of asphalt pavements is largely a function of the asphalt concrete surface material.

The objective of the asphalt concrete mix design process is to provide the following properties (Roberts et al., 2009):

1. stability or resistance to permanent deformation under the action of traffic loads, especially at high temperatures
2. fatigue resistance to prevent fatigue cracking under repeated loadings
3. resistance to thermal cracking that might occur due to contraction at low temperatures
4. resistance to hardening or aging during production in the mixing plant and in service
5. resistance to moisture-induced damage that might result in stripping of asphalt from aggregate particles
6. skid resistance, by providing enough texture at the pavement surface
7. workability, to reduce the effort needed during mixing, placing, and compaction

Regardless of the set of criteria used to state the objectives of the mix design process, the design of asphalt concrete mixes requires compromises. For example, extremely high stability often is obtained at the expense of lower durability, and vice versa. Thus, in evaluating and adjusting a mix design for a particular use, the aggregate gradation and asphalt content must strike a favorable balance between the stability and durability requirements. Moreover, the produced mix must be practical and economical.

9.9 Asphalt Concrete Mix Design

The purpose of asphalt concrete mix design is to determine the design asphalt content using the available asphalt and aggregates. The design asphalt content varies for different material types, material properties, loading levels, and environmental conditions. To produce good-quality asphalt concrete, it is necessary to accurately control the design asphalt content. If the appropriate design asphalt content is not used, the pavement will lack durability or stability, resulting in premature pavement failure. For example, if not enough asphalt binder is used, not all the aggregate particles will be coated with asphalt, which will result in a less stable and less durable material. Also, if too much binder is used, aggregate particles may have too much "lubrication" and may move relative to each other upon application of the load, resulting in a less stable material. Typical design asphalt contents range from 4% to 7% by weight of total mix.

Before the Superpave mix design method was developed, there were two common asphalt concrete design methods: Marshall (ASTM D1559) and Hveem (ASTM D1560). The Marshall method was more commonly used than the Hveem method, due to its relative simplicity and ability to be used for field control. Both methods are empirical in nature; that is, they are based on previous observations. Both methods have been used satisfactorily for several decades and have produced long-lasting pavement sections. However, due to their empirical nature, they are not readily adaptable to new conditions, such as modified binders, large-sized aggregates, and heavier traffic loads.

The Superpave design system is performance based and is more rational than the Marshall and Hveem methods. Many highway agencies have implemented the Superpave system. The discussion below is limited to the Superpave and the Marshall methods.

9.9.1 ■ Specimen Preparation in the Laboratory

Asphalt concrete specimens are prepared in the laboratory for mix-design and quality-control tests. To prepare specimens in the laboratory, aggregates are batched according to a specified gradation and heated. Asphalt cement is also heated separately and added to the aggregate at a specified amount. Aggregates and asphalt are mixed until the aggregate particles are completely coated with asphalt. The samples may then be placed in an oven at a specified temperature to allow asphalt to be absorbed into the voids at the surface of the aggregate. Failure to oven-age the samples will result in incorrect estimates of the required binder content. Two types of samples are used in the mix design and quality control process, compacted samples for determining the bulk specific gravity and for other tests, and uncompacted or loose samples for determining the maximum theoretical specific gravity of the asphalt concrete.

Compacted Samples   Two compaction machines are commonly used:

1. Superpave gyratory compactor
2. Marshall hammer

Regardless of the compaction method, the procedure for preparing specimens basically follows the same four steps:

1. Heat and mix the aggregate and asphalt cement
2. Place the material into a heated mold
3. Apply compaction force
4. Allow the specimen to cool and extrude from the mold

The specific techniques for placing the material into the mold vary between the compaction methods, and the standards for the method must be followed.

The greatest difference among the compaction procedures is the manner in which the compaction effort is applied. For the gyratory compaction, the mixture in the mold is placed in the compaction machine at an angle to the applied force. As the force is applied the mold is gyrated, creating a shearing action in the mixture. Gyratory compaction devices have been available for a long time, but their use was limited due to the lack of a mix-design procedure based on this type of compaction. However, the Superpave mix design method (FHWA, 1995) uses a gyratory compactor; thus, this compaction method is now common. Figure 9.26 shows the Superpave gyratory compactor.

In the Marshall procedure, a slide hammer (Figure 9.27) weighing 4.45 kg is dropped from a height of 0.46 m to create an impact compaction force (ASTM D1559).

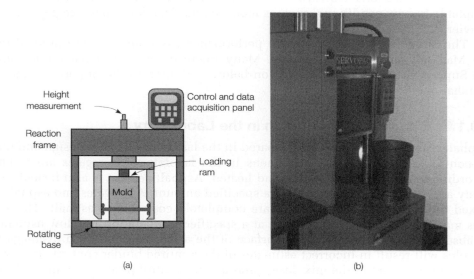

FIGURE 9.26 Superpave gyratory compactor. (Courtesy of Pavement Scientific International, Inc.)

FIGURE 9.27 Marshall compactor.

The head of the Marshall hammer has a diameter equal to the specimen size, and the hammer is held flush with the specimen at all times.

The Superpave gyratory compactor produces specimens 150 mm in diameter and 95 mm to 115 mm high, allowing the use of aggregates with a nominal maximum size of 37.5 mm. Specimens prepared with the Marshall compactor are typically 101.6 mm in diameter and 63.5 mm high, which limits the nominal maximum aggregate size to 25 mm. A special 152 mm Marshall mold is required for designing mixes with a 37.5 mm nominal maximum aggregate size.

Regardless of the compaction method, the samples are first used to determine the bulk specific gravity of the mix, G_{mb}. Traditionally, the saturated-surface dry method, ASTM D2726, was used to determine G_{mb}. However, this method is not reliable for samples with open surface voids, for example, an absorption of more than 2%. In response to this, a method has been developed for determining G_{mb} using a vacuum sealing method, AASHTO T 331. The following describes the traditional ASTM D2726 method. The dry weight of the sample is measured. The sample is suspended from a scale into a 25°C water bath for 3 to 5 minutes and the submerged weight is measured. The sample is removed from the water bath and the surface is "dried with a moist towel" to remove water at the surface of the sample, without removing water from the surface voids of the sample. This is the saturated surface dry weight. The bulk specific gravity of the mix is computed as:

$$G_{mb} = \frac{A}{B - C}$$
(9.1)

where

A = dry weight of sample
B = saturated surface dry weight
C = submerged weight

For mix design, compacted samples are prepared at multiple asphalt contents in order to determine the "optimum" asphalt content.

Once the bulk specific gravity of a sample is measured, the sample can be used for other tests depending on the mix design method.

Uncompacted Samples   ASTM D2041 is used to determine the theoretical maximum specific gravity of the mix. This test was developed by Jim Rice of The Asphalt Institute, so it is frequently called the Rice test and the results are called the Rice gravity. A loose mix is prepared using the same mixing and conditioning procedures used for preparing the compacted samples. Following conditioning the sample is spread out on a table to cool to the test temperature, 25°C. As the sample cools the asphalt coated aggregates are separated. The weight of the loose mixture specimen in air is measured. The sample is placed in a vacuum bowl and covered with water. A vacuum is applied to remove all air from the sample. While the sample is still covered with water, the submerged weight of the vacuum bowl with the sample is measured. In order to determine the submerged weight of the sample, the submerged weight of the empty vacuum bowl must be known. This is usually done periodically and is called the calibration weight of the bowl; it is not measured each time the test is performed. The theoretical maximum specific gravity, G_{mm}, is

$$G_{mm} = \frac{A}{A - (E - D)} = \frac{A}{A + D - E} \tag{9.2}$$

where

A = dry weight of the loose sample
D = calibration weight of the submerged vacuum bowl
E = submerged weight of the vacuum bowl with the sample

The theoretical maximum specific gravity of the sample at one asphalt content can be used to estimate G_{mm} at other asphalt contents by first computing the effective specific gravity, G_{se} of the aggregates. By definition, the theoretical maximum specific gravity of asphalt concrete is

$$G_{mm} = \frac{100}{\left(\dfrac{P_s}{G_{se}} + \dfrac{P_b}{G_b} \right)} \tag{9.3}$$

Solving this equation for G_{se} produces

$$G_{se} = \frac{P_s}{\left(\dfrac{100}{G_{mm}} - \dfrac{P_b}{G_b} \right)} \tag{9.4}$$

where

G_{mm} = theoretical maximum specific gravity of the asphalt concrete
P_s = percent weight of the aggregate
P_b = percent weight of the asphalt cement
G_{se} = effective specific gravity of aggregate coated with asphalt
G_b = specific gravity of the asphalt binder

Although G_{se} is determined for only one asphalt content, it is constant for all asphalt contents. Thus, once G_{se} is determined based on the results of the theoretical maximum specific gravity test, it can be used in Equation 9.3 to calculate G_{mm} for the different asphalt contents. The Marshall method uses this method for determining G_{mm} at each of the asphalt contents. The Superpave method requires directly measuring G_{mm} at each asphalt content evaluated during the mix design process.

9.9.2 ■ Density and Voids Analysis

It is important to understand the density and voids analysis of compacted asphalt mixtures for both mix design and construction control. Regardless of the method used, the mix design is a process to determine the volume of asphalt binder and aggregates required to produce a mixture with the desired properties. However, since volumes are difficult and not practical to measure, weights are used instead; the specific gravity is used to convert from weight to volume. Figure 9.28 shows

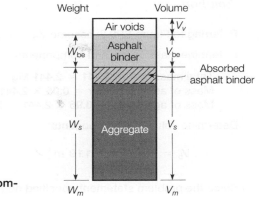

FIGURE 9.28 Components of compacted asphalt mixture.

that the asphalt mixture consists of aggregates, asphalt binder, and air voids. Note that a portion of the asphalt is absorbed by aggregate particles. Three important parameters commonly used are percent of air voids (voids in total mix) (VTM), voids in the mineral aggregate (VMA), and voids filled with asphalt (VFA). These are defined as

$$\text{VTM} = \frac{V_v}{V_m}100 \tag{9.5}$$

$$\text{VMA} = \frac{V_v + V_{be}}{V_m}100 \tag{9.6}$$

$$\text{VFA} = \frac{V_{be}}{V_{be} + V_V}100 \tag{9.7}$$

where

V_v = volume of air voids
V_{be} = volume of effective asphalt binder
V_m = total volume of the mixture

The effective asphalt is the total asphalt minus the absorbed asphalt.

Sample Problem 9.2

A compacted asphalt concrete specimen contains 5% asphalt binder (Sp. Gr. 1.023) by weight of total mix and aggregate with a specific gravity of 2.755. The bulk density of the specimen is 2.441 Mg/m³. Ignoring absorption, compute VTM, VMA, and VFA.

Solution

Referring to Figure SP9.2, assume $V_m = 1 \text{ m}^3$

Determine mass of mix and components:

Total mass $= 1 \times 2.441 = 2.441$ Mg
Mass of asphalt binder $= 0.05 \times 2.441 = 0.122$ Mg
Mass of aggregate $= 0.95 \times 2.441 = 2.319$ Mg

Determine volume of components:

$$V_b = \frac{0.122}{1.023} = 0.119 \text{ m}^3$$

Since the problem statement specified no absorption, $V_{be} = V_b$

$$V_s = \frac{2.319}{2.755} = 0.842 \text{ m}^3$$

$V_s = $ Volume of stone (aggregate)

Determine volume of voids:

$$V_v = V_m - V_b - V_s = 1 - 0.119 - 0.842 = 0.039 \text{ m}^3$$

Volumetric calculations:

$$\text{VTM} = \frac{V_v}{V_m}100 = \frac{0.039}{1.00}100 = 3.9\%$$

$$\text{VMA} = \frac{V_v + V_{be}}{V_m}100 = \frac{0.039 + 0.119}{1.00}100 = 15.8\%$$

$$\text{VFA} = \frac{V_{be}}{V_{be} + V_v}100 = \frac{0.119}{0.119 + 0.039}100 = 75\%$$

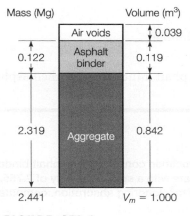

FIGURE SP9.2

The unit weight of each specimen is computed by multiplying the bulk specific gravity by the unit weight of water, 1 Mg/m^3.

Figure 9.28 and Equations 9.5 to 9.7 are used to demonstrate the volumetric relationships used for asphalt mix design analysis. However, the equations actually used for mix design have been derived to allow direct use of the measured parameters.

$$\text{VTM} = 100\left(1 - \frac{G_{mb}}{G_{mm}}\right) \tag{9.8}$$

$$\text{VMA} = \left(100 - G_{mb}\frac{P_s}{G_{sb}}\right) \tag{9.9}$$

$$\text{VFA} = 100\left(\frac{\text{VMA} - \text{VTM}}{\text{VMA}}\right) \tag{9.10}$$

Sample Problem 9.3

An asphalt concrete specimen has the following properties:

    asphalt content = 5.9% by total weight of mix
    bulk specific gravity of the mix = 2.457
    theoretical maximum specific gravity = 2.598
    bulk specific gravity of aggregate = 2.692

Calculate the percents VTM, VMA, and VFA.

Solution

$$\text{VTM} = 100\left(1 - \frac{G_{mb}}{G_{mm}}\right) = 100\left(1 - \frac{2.457}{2.598}\right) = 5.4\%$$

$$\text{VMA} = \left(100 - G_{mb}\frac{P_s}{G_{sb}}\right) = \left(100 - 2.457\frac{100 - 5.9}{2.692}\right) = 14.1\%$$

$$\text{VFA} = 100\left(\frac{\text{VMA} - \text{VTM}}{\text{VMA}}\right) = 100\left(\frac{14.1 - 5.4}{14.1}\right) = 61.7\%$$

9.9.3 ■ Superpave Mix Design

The Superpave mix-design process consists of (AASHTO R35-09)

- Selection of aggregates
- Selection of binder
- Determination of the design aggregate structure
- Determination of the design binder content
- Evaluation of moisture susceptibility

TABLE 9.9 Superpave Consensus Aggregate Properties

Design Level	Course Aggregate Angularity (% minimum)	Fine Aggregate Angularity (% minimum)	Flat and Elongated (% maximum)	Sand Equivalency (% minimum)
Light traffic	55/-	—	—	40
Medium traffic	75/-	40	10	40
Heavy traffic	85/80*	45	10	45

*85/80 denotes minimum percentages of one fractured face / two or more fractured faces

Aggregate Selection Aggregate for Superpave mixes must pass both source and consensus requirements. The source requirements are defined by the owner/purchasing agency and are specified for each stockpile. *Source* properties may include Los Angeles abrasion (see apparatus in Figure 5.8), soundness, and deleterious materials. Consensus requirements are part of the national Superpave specification and are on the blend of aggregates. The following *consensus* aggregate properties are required:

- coarse aggregate angularity measured by the percentage of fractured faces
- fine aggregate angularity (AASHTO TP33) (see apparatus in Figure 5.7)
- flat and elongated particles (ASTM D4791) (see apparatus in Figure 5.6)
- sand equivalency (ASTM D2419) (see apparatus in Figure 5.18)

Specification limits for these properties depend on the traffic level and how deep under the pavement surface the materials will be used, as shown in Table 9.9.

Aggregate used in asphalt concrete must be well graded. The 0.45 power chart discussed in Chapter 5 is recommended for Superpave mix designs. The gradation curve must go between control points. Figure 9.29 shows the gradation requirements for the 12.5 mm nominal-maximum sized Superpave mix.

To control segregation, aggregates are sorted into stockpiles based on size. The designer of an asphalt concrete mix must select a blend of stockpiles that meets the source, consensus, and gradation requirements.

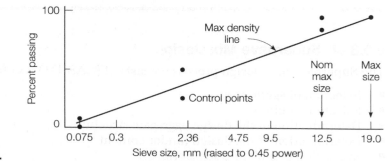

FIGURE 9.29
Superpave gradation limits for 12.5 mm nominal maximum size.

Binder Selection   The binder is selected based on the maximum and minimum pavement temperatures, amount of truck traffic, and traffic speed, as discussed earlier. In addition to the specification tests, the specific gravity and the viscosity versus temperature relationship for the selected asphalt binder must be measured. The specific gravity is needed for the volumetric analysis.

The viscosity–temperature relationship is needed to determine the required mixing and compaction temperatures. The Superpave method requires mixing the asphalt and aggregates at a temperature at which the viscosity of the asphalt binder is 0.170 ± 0.020 Pa · s and the compacting temperature corresponds to a viscosity of 0.280 ± 0.030 Pa · s. The rotational viscometer is used to determine the temperature-viscosity properties of the binder so the mixing and compaction temperatures can be determined. This information is generally provided by the binder supplier so the mix designer does not need to perform this test.

Design Aggregate Structure   After selecting the appropriate aggregate, binder, and modifiers (if any), trial specimens are prepared with three different aggregate gradations and asphalt contents. There are equations for estimating the asphalt content for use in the specimens prepared for determining the design aggregate structure. However, these equations are empirical and the designer is given latitude to estimate the asphalt content. Specimens are compacted using the Superpave gyratory compactor (Figure 9.26) with a gyration angle of 1.16 degrees (internal mold measurement) and a constant vertical pressure of 600 kPa. The number of gyrations used for compaction is determined based on the traffic level, as shown in Table 9.10.

TABLE 9.10   Number of Gyrations at Specific Design Traffic Levels (AASHTO R35-09)

	Traffic Level (10^6 ESAL)			
	< 0.3	0.3 to 3	3 to 30	> 30
N_{ini}	6	7	8	9
N_{des}	50	75	100	125
N_{max}	75	115	160	205
Typical Application	Very light traffic volumes where truck traffic is prohibited or very limited. Limited to local traffic or special purpose roads, for example, access to recreational sites.	Collector roads and access streets and county with medium traffic.	Many two-lane, multilane, divided, and partially or completely controlled access roadways, may include medium to highly trafficked city streets, many state routes, US highways and some rural interstates.	Majority of Interstate system and special applications such as truck weight stations or truck climbing lanes on two lane roads.

As shown in Table 9.10, the Superpave method recognizes three critical stages of compaction, initial, design, and maximum. The design compaction level N_{des} corresponds to the compaction that is anticipated at the completion of the construction process. The maximum compaction N_{max} corresponds to the ultimate density level that will occur after the pavement has been subjected to traffic for a number of years. The initial compaction level N_{ini} was implemented to assist with identifying "tender" mixes. A tender mix lacks stability during construction and hence will displace under the rollers rather than densifying.

For the initial stage of determining the design aggregate structure, samples are compacted with N_{des} gyrations. The volumetric properties are determined by measuring the bulk specific gravity G_{mb} of the compacted mix and the maximum theoretical specific gravity G_{mm} of a loose mix with the same asphalt content and aggregate composition. The volumetric parameters, VTM, VMA, and VFA, are determined and checked against the criteria in Table 9.11. Two additional parameters are evaluated

TABLE 9.11 Superpave Mix Design Criteria

Design Air Voids					4%	
Dust-to-Effective Asphalt[1]					0.6–1.2	
Tensile strength ratio					80% min	

	Nominal Maximum Size (mm)					
	37.5	25	19	12.5	9.5	4.75
Minimum VMA (%)	11	12	13	14	15	16

G_{mm} and VFA Requirements

Design ESAL in millions	Percent Maximum Theoretical Specific Gravity			Percent Voids Filled with Asphalt[2,3,4]
	N_{init}	N_{des}	N_{max}	
<0.3	≤ 91.5	96	≤ 98.0	70–80
0.3–3	≤ 90.5	96	≤ 98.0	65–78
3–10	≤ 89.0	96	≤ 98.0	65–75
10–30	≤ 89.0	96	≤ 98.0	65–75
≥30	≤ 89.0	96	≤ 98.0	65–75

Notes

[1.] Dust-to-binder ratio range is 0.9 to 2.0 for 4.75 mm mixes.

[2.] For 9.5 mm nominal maximum aggregate size mixes and design ESAL ≥3 million, VFA range is 73 to 76% and for 4.75 mm mixes, the range is 75 to 78%.

[3.] For 25 mm nominal maximum aggregate size mixes, the lower limit of the VFA range shall be 67% for design traffic levels <0.3 million ESALs.

[4.] For 37.5 mm nominal maximum aggregate size mixes, the lower limit of the VFA range shall be 64% for all design traffic levels.

in the Superpave method: percent of G_{mm} at N_{ini}, and the dust-to-effective binder content ratio. The percent of G_{mm} at N_{ini} is determined as

$$\text{Percent } G_{mm,Nini} = \text{Percent } G_{mm,Ndes} \frac{h_{des}}{h_{ini}} \tag{9.11}$$

where h_{ini} and h_{des} are the heights of the specimen at the initial and design number of gyrations, respectively. Note that the percent $G_{mm,Ndes}$ is equal to (100-VTM).

The dust-to-effective binder ratio is the percent of aggregate passing the 0.075 mm (#200) sieve divided by the effective asphalt content, computed as

$$P_{ba} = 100\left(\frac{G_{se} - G_{sb}}{G_{sb}G_{se}}\right)G_b \tag{9.12}$$

$$P_{be} = P_b - \left(\frac{P_{ba}}{100}\right)P_s \tag{9.13}$$

$$D/B = \frac{P_D}{P_{be}} \tag{9.14}$$

where

$\quad D/B$ = dust to binder ratio
$\quad P_{ba}$ = percent absorbed binder based on the mass of aggregates
$\quad G_{sb}$ = bulk specific gravity of aggregate
$\quad P_D$ = percent dust, or % of aggregate passing the 0.075 mm sieve
$\quad P_{be}$ = percent effective binder content

The Superpave method requires determining the volumetric properties at 4% VTM. The samples prepared for the evaluation of the design aggregate content are not necessarily at the binder content required for achieving this level of air voids. Therefore, the results of the volumetric evaluation are "corrected" to four percent air voids, as

$$P_{b,est} = P_{bt} - 0.4(4 - VTM_t) \tag{9.15}$$

$$VMA_{est} = VMA_t + C(4 - VTM_t)$$

$$C = -0.1 \text{ for } VTM_t < 4.0\%$$

$$C = 0.2 \text{ for } VTM_t \geq 4.0\% \tag{9.16}$$

$$VFA_{est} = 100\frac{VMA_{est} - 4.0}{VMA_{est}} \tag{9.17}$$

$$\% G_{mmNini.est} = \% G_{mmNini} - (4.0 - VTM_t) \tag{9.18}$$

$$P_{be,est} = P_{b,est} - \frac{P_s G_b(G_{se} - G_{sb})}{G_{se}G_{sb}} \tag{9.19}$$

$$D/B_{est} = \frac{P_D}{P_{be,est}} \tag{9.20}$$

where

$\quad P_{b,est}$ = *adjusted* estimated binder content
$\quad VMA_{est}$ = adjusted VMA
$\quad VMA_t$ = VMA determined from volumetric analysis
$\quad VFA_{est}$ = adjusted VFA

VTM_t = VTM determined from the volumetric analysis
$P_{be,est}$ = adjusted percent effective binder
D/B_{est} = adjusted dust-to–binder ratio
P_D = percent aggregate passing 0.075 mm sieve

In Equation 9.17, the air voids are assumed to be 4.0%, the target air void content for a mix design. The adjusted results of the design aggregate structure evaluation are compared to the Superpave mix-design criteria, as shown in Table 9.11.

The design aggregate blend is the one whose adjusted volumetric parameters meet all of the criteria. In the event that more than one of the blends meets all of the criteria, the designer can use discretion for the selection of the blend. A pair of samples is compacted at the adjusted binder content for the selected aggregate blend. The average percent of maximum theoretical specific gravity is determined and compared to the design criteria. If successful, the procedure continues with the determination of the design binder content. If unsuccessful, the design process is started over with the selection of another blend of aggregates.

Sample Problem 9.4

For a 12.5 million ESAL, select a 19-mm Superpave design aggregate structure based on the following data:

	Blend		
Data	1	2	3
G_{mb}	2.457	2.441	2.477
G_{mm}	2.598	2.558	2.664
G_b	1.025	1.025	1.025
$P_b(\%)$	5.9	5.7	6.2
$P_s(\%)$	94.1	94.3	93.8
$P_d(\%)$	4.5	4.5	4.5
G_{sb}	2.692	2.688	2.665
h_{ini} (mm)	125	131	125
h_{des} (mm)	115	118	115

Solution

In this problem, the lab data were entered into an Excel spreadsheet and the appropriate equations were entered, producing the results shown in the accompanying table. The steps in the calculation are shown for the first blend only; the reader can verify the calculations for the other blends.

Although both mixes 1 and 2 meet the criteria, blend 2 is preferred, since it has a higher VMA value.

Volumetric Analysis

Computed	Equation	Using Data for Blend 1	Blend 1	Blend 2	Blend 3	Criteria
G_{se}	9.4 $\dfrac{P_s}{\left(\dfrac{100}{G_{mm}} - \dfrac{P_b}{G_b}\right)}$	$\dfrac{94.1}{\left(\dfrac{100}{2.598} - \dfrac{5.9}{1.025}\right)}$	2.875	2.812	2.979	
VTM_t (%)	9.8 $100\left(1 - \dfrac{G_{mb}}{G_{mm}}\right)$	$100\left(1 - \dfrac{2.457}{2.598}\right)$	5.4	4.6	7.0	
VMA_t (%)	9.9 $\left(100 - G_{mb}\dfrac{P_s}{G_{sb}}\right)$	$\left(100 - 2.457\dfrac{94.1}{2.692}\right)$	14.1	14.4	12.8	
VFA_t (%)	9.10 $100\left(\dfrac{VMA - VTM}{VMA}\right)$	$100\left(\dfrac{14.1 - 5.4}{14.1}\right)$	61.7	68.1	45.3	
% $G_{mm,Nini}$	9.11 Percent $G_{mm,Ndes}\dfrac{h_{des}}{h_{ini}}$	$(100 - 5.4)\dfrac{115}{125}$	87.0	86.0	85.5	≤ 89.0
P_{ba} (%)	9.12 $100\left(\dfrac{G_{se} - G_{sb}}{G_{sb}G_{se}}\right)G_b$	$100\left(\dfrac{2.875 - 2.692}{2.692 * 2.875}\right)1.025$	2.42	1.68	4.05	
P_{be} (%)	9.13 $P_b - \left(\dfrac{P_{ba}}{100}\right)P_s$	$5.9 - \left(\dfrac{2.42}{100}\right)94.1$	3.62	4.12	2.4	
D/B	9.14 $\dfrac{P_D}{P_{be}}$	$\dfrac{4.5}{3.62}$	1.2	1.1	1.9	

Adjusted Values

Computed	Equation	Using Data for Blend 1	Blend 1	Blend 2	Blend 3	Criteria
$P_{b,est}$ (%)	9.15 $P_{bt} - 0.4(4 - VTM_t)$	$5.9 - 0.4(4 - 5.4)$	6.5	5.9	7.4	
VMA_{est} (%)	9.16 $VMA_t + C(4 - VTM_t)$	$14.1 + 0.2(4 - 5.4)$	13.8	14.3	12.2	≥ 13
VFA_{est} (%)	9.17 $100\dfrac{VMA_{est} - 4.0}{VMA_{est}}$	$100\dfrac{13.8 - 4.0}{13.8}$	71.0	72.0	67.2	65–75
% $G_{mm,Nini,est}$	9.18 $\% G_{mm,Nini} - (4.0 - VTM_t)$	$87 - (4.0 - 5.4)$	88.4	86.6		
$P_{be,est}$ (%)	9.19 $P_{b,est} - \dfrac{P_s G_b(G_{se} - G_{sb})}{G_{se}G_{sb}}$	$6.5 - \dfrac{94.1 * 1.025(2.875 - 2.692)}{2.875 * 2.692}$	4.2	4.3	3.6	
D/B_{est}	9.20 $\dfrac{P_D}{P_{be,est}}$	$\dfrac{4.5}{4.2}$	1.1	1.0	1.3	0.6–1.2

Design Binder Content   The design binder content is obtained by preparing eight specimens—two replicates at each of four binder contents: estimated design binder content (result of equation 9.15 for the selected design aggregate structure), 0.5% less than the design binder content, 0.5% more than the design binder content, and 1% more than the design binder content. Specimens are compacted using N_{des} gyrations. The volumetric properties are computed and plots are prepared of each volumetric parameter versus the binder content. The binder content that corresponds to 4% VTM is determined. Then the plots are used to determine the volumetric parameters at the selected binder content. If the properties meet the criteria in Table 9.11, two specimens are prepared and compacted using gyrations. If these specimens meet the criteria, then the moisture sensitivity of the mix is determined.

Sample Problem 9.5

Based on the numbers used in Sample Problem 9.4, determine the recommended asphalt content according to Superpave mix design for an equivalent single axle load (ESAL) of 20 million.

Solution

Samples were prepared at four asphalt contents. An Excel spreadsheet was prepared to present and analyze the data. The following results are obtained:
The results are plotted against binder content, as shown in Figure SP9.5.

Design Binder Content

Data	Asphalt Content Trial			
	1	2	3	4
P_b (%)	5.4	5.9	6.4	6.9
G_{mb}	2.351	2.441	2.455	2.469
G_{mm}	2.570	2.558	2.530	2.510
G_b	1.025	1.025	1.025	1.025
P_s (%)	94.6	94.1	93.6	93.1
P_d (%)	4.5	4.5	4.5	4.5
G_{sb}	2.688	2.688	2.688	2.688
h_{ini} (mm)	125	131	126	130
h_{des} (mm)	115	118	114	**112**

Design Binder Content

Volumetric Analysis

Computed	Equation	Asphalt Content Trial			
		1	2	3	4
G_{se}	9.4	2.812	2.812	2.812	2.812
VTM (%)	9.8	8.5	4.6	3.0	1.6
VMA (%)	9.9	17.3	14.5	14.5	14.5
VFA (%)	9.10	50.9	68.3	79.3	89.0
$\% G_{mm}$, N_{ini}	9.11	84.2	86.0	87.8	84.7
P_{ba} (%)	9.12	1.68	1.68	1.68	1.68
P_{be} (%)	9.13	3.81	4.32	4.83	5.34
D/B	9.14	1.2	1.0	0.9	0.8

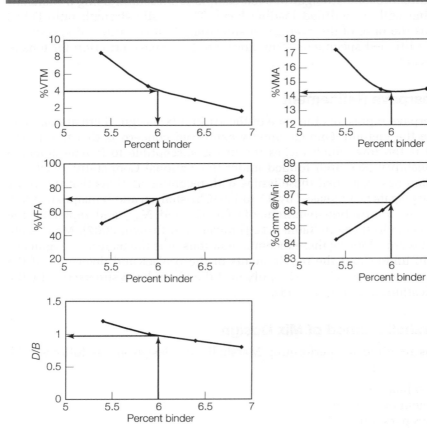

FIGURE SP9.5

The figure shows that, at 4% VTM, the binder content is 6%. The design values at 6% binder content are

- VMA = 14.5%
- VFA = 71%
- G_{mm} @ N_{ini} = 86.5%
- D/B ratio = 0.9

These values satisfy the design criteria shown in Table 9.11. Therefore, the design binder content is 6.0%.

Moisture Sensitivity Evaluation The moisture sensitivity of the design mixture is determined using the AASHTO T283 procedure on six specimens prepared at the design binder content and 7% air voids. Three specimens are conditioned by vacuum saturation, then freezing and thawing; three other specimens are not conditioned. The indirect tensile strength of each sample is measured using a Marshall stability machine with a modified loading head. The tensile strength ratio (TSR) is determined as the ratio of the average tensile strength of conditioned specimens to that of unconditioned specimens. The minimum Superpave criterion for tensile strength ratio is 80%.

9.9.4 ■ Superpave Refinement

Although Superpave represents the state of the art in mix design, there are ongoing efforts to refine the method. One commonly expressed concern is that the design binder content is too low which makes the mixes susceptible to fatigue cracking and high permeability. Research funded under the National Cooperative Highway Research Program has examined these issues with recommendations that the voids in the mineral aggregate be increased by 0.5 to 1% (Christensen and Bonaquist, 2006), elimination of the compaction requirements for N_{ini} and N_{max} and replacing the N_{design} requirements to those in Table 9.12 (Prowell and Brown, 2007). Many state highway agencies are adopting these recommendations. It is the materials engineer's responsibility to understand the requirements of the project and determine if the national standards, such as AASHTO, apply or if they have been superseded by the owner's specifications and test methods.

9.9.5 ■ Marshall Method of Mix Design

The basic steps required for performing Marshall mix design are as follows (The Asphalt Institute, 2001):

1. aggregate evaluation
2. asphalt cement evaluation
3. specimen preparation
4. Marshall stability and flow measurement
5. density and voids analysis
6. design asphalt content determination

TABLE 9.12   Recommendations for N_{Design} Levels

20-Year Design Traffic, 10^6 ESAL	2-Year Design Traffic, 10^5 ESAL	N_{design} for binders $<$ PG76-XX	N_{design} for binders $\geq$ PG 76-XX or mixes placed $>$100 mm from Surface
<0.3	<0.3	50	NA
0.3 to 3	0.3 to 2.3	65	50
3 to 10	2.3 to 9.25	80	65
10 to 30	9.25 to 25	80	65
>30	>25	100	80

1. **Aggregate Evaluation** The aggregate characteristics that must be evaluated before it can be used for an asphalt concrete mix include the durability, soundness, presence of deleterious substances, polishing, shape, and texture. Agency specifications define the allowable ranges for aggregate gradation. The standard Marshall method is applicable to densely graded aggregates with a maximum size of not more than 25 mm (AASHTO T245). A modified Marshall method is available using larger molds and a different compaction hammer for mixes with a nominal maximum aggregate size up to 50 mm (ASTM D5581). Only the standard method is described in the following.
2. **Asphalt Cement Evaluation** The grade of asphalt cement is selected based on the expected temperature range and traffic conditions. Most highway agencies have specifications that prescribe the grade of asphalt for the design conditions.
3. **Specimen Preparation** The full Marshall mix-design procedure requires 18 specimens 101.6 mm in diameter and 63.5 mm high. The stability and flow are measured for 15 specimens. In addition, three specimens are used to determine the theoretical maximum specific gravity G_{mm}. This value is needed for the void and density analysis. The specimens for the theoretical maximum specific gravity determination are prepared at the estimated design asphalt content. Samples are also required for each of five different asphalt contents; the expected design asphalt content, $\pm 0.5\%$ and $\pm 1.0\%$. Engineers use experience and judgment to estimate the design asphalt content.

   Specimen preparation for the Marshall method uses the Marshall compactor discussed earlier (Figure 9.27). The Marshall method requires mixing of the asphalt and aggregates at a temperature where the kinematic viscosity of the asphalt cement is 170 ± 20 cSt and compacting temperature corresponds to a viscosity of 280 ± 30 cSt.

   The Asphalt Institute permits three different levels of energy to be used for the preparation of the specimens: 35, 50, and 75 blows on each side of the sample. Most mix designs for heavy-duty pavements use 75 blows, since this better simulates the required density for pavement construction.

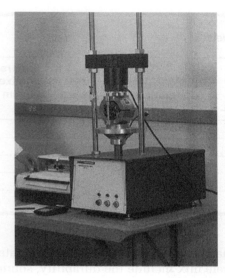

FIGURE 9.30 Marshall stability machine.

4. **Marshall Stability and Flow Measurement**—The Marshall stability of the asphalt concrete is the maximum load the material can carry when tested in the Marshall apparatus, Figure 9.30. The test is performed at a deformation rate of 51 mm/minute and a temperature of 60°C. The Marshall flow is the deformation of the specimen when the load starts to decrease. Stability is reported in newtons and flow is reported in units of 0.25 mm of deformation. The stability of specimens that are not 63.5 mm thick is adjusted by multiplying by the factors shown in Table 9.13. All specimens are tested and the average stability and flow are determined for each asphalt content.

5. **Density and Voids Analysis** The values of VTM, VMA, and VFA are determined using Equations 9.8, 9.9, and 9.10, respectively.

TABLE 9.13 Marshall Stability Adjustment Factors

Approximate Thickness of Specimen, mm	Adjustment Factor	Approximate Thickness of Specimen, mm	Adjustment Factor
50.8	1.47	65.1	0.96
52.4	1.39	66.7	0.93
54.0	1.32	68.3	0.89
55.6	1.25	69.8	0.86
57.2	1.19	71.4	0.83
58.7	1.14	73.0	0.81
60.3	1.09	74.6	0.78
61.9	1.04	76.2	0.76
63.5	1.00		

6. **Design Asphalt Content Determination** Traditionally, test results and calculations are tabulated and graphed to help determine the factors that must be
used in choosing the optimum asphalt content. Table 9.14 presents examples of
mix design measurements and calculations. Figure 9.31 shows plots of results
obtained from Table 9.14, which include asphalt content versus air voids, VMA,
VFA, unit weight, Marshall stability, and Marshall flow.

The design asphalt content is usually the most economical one that will satisfactorily meet all of the established criteria. Different criteria are used by different
agencies. Tables 9.15 and 9.16 depict the mix design criteria recommended by The
Asphalt Institute. Figure 9.32 shows an example of the narrow range of acceptable
asphalt contents. The asphalt content selection can be adjusted within this narrow
range to achieve a mix that satisfies the requirements of a specific project. Other
agencies, such as the National Asphalt Paving Association, use the asphalt cement

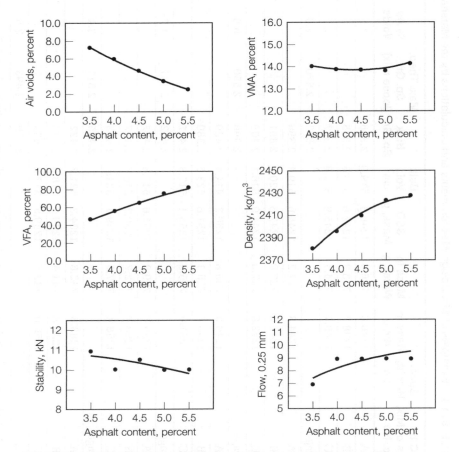

FIGURE 9.31   Graphs used for Marshall mix design analysis (The
Asphalt Institute, 2001). (See Table 9.14.)

TABLE 9.14 Examples of Mix Design Measurements and Calculations by the Marshall Method (The Asphalt Institute, 2001)

% AC by Mass of Mix, Spec. No.	Spec. Height, mm	Mass In Air, g	Mass In Water, g	SSD Mass, g	Bulk Vol., cm³	Bulk Sp. Gr.	Max. Theo. Sp. Gr. (loose mix)	% Air Voids	% VMA	% VFA	Measured Stability, kN	Adjusted Stability, kN	Flow, 0.25 mm
3.5–A		1240.6	726.4	1246.3	519.9	2.386					10.9	10.9	8
3.5–B		1238.7	723.3	1242.6	519.3	2.385					10.8	10.8	7
3.5–C		1240.1	724.1	1245.9	521.8	2.377					11.2	11.2	7
Average						2.383	2.570	7.3	14.0	48.0		10.9	7
4.0–A		1244.3	727.2	1246.6	519.4	2.396					9.7	9.7	9
4.0–B		1244.6	727.0	1247.6	520.6	2.391					10.1	10.1	9
4.0–C		1242.6	727.9	1244.0	516.1	2.408					10.3	10.3	8
Average						2.398	2.550	6.0	13.9	57.1		10.0	9
4.5–A		1249.3	735.8	1250.2	514.4	2.429					10.8	10.8	9
4.5–B		1250.8	728.1	1251.6	523.5	2.389					10.7	10.3	9
4.5–C		1251.6	735.3	1253.1	517.8	2.417					10.4	10.4	9
Average						2.412	2.531	4.7	13.9	66.1		10.5	9
5.0–A		1256.7	739.8	1257.6	517.8	2.427					10.2	10.2	9
5.0–B		1258.7	742.7	1259.3	516.6	2.437					9.7	9.7	8
5.0–C		1258.4	737.5	1259.1	521.6	2.413					10.0	10.0	9
Average						2.425	2.511	3.4	13.8	75.2		10.0	9
5.5–A		1263.8	742.6	1264.3	521.7	2.422					9.8	9.8	9
5.5–B		1258.8	741.4	1259.4	518.0	2.430					10.2	10.2	10
5.5–C		1259.0	742.5	1259.5	517.0	2.435					9.8	10.0	9
Average						2.429	2.493	2.5	14.1	82.1		10.0	9

Notes: AC-20 binder, G_b = 1.030, G_{sb} = 2.674, absorbed AC of aggregate: 0.6%, G_{se} = 2.717, compaction: 75 blows

TABLE 9.15   Asphalt Institute Criteria for Marshall Mix Design (The Asphalt Institute, 2001)

	Traffic Level					
	Light		Medium		Heavy	
Compaction (Blows)	35		50		75	
	Minimum	Maximum	Minimum	Maximum	Minimum	Maximum
Stability, kN	3.34	—	5.34	—	8.01	—
Flow, 0.25 mm	8	18	8	16	8	14
Air Voids, %	3	5	3	5	3	5
VMA, %	Use the criteria in Table 9.16					
VFA, %	70	80	65	78	65	75

TABLE 9.16   Minimum Percent Voids in Mineral Aggregate (VMA) (The Asphalt Institute, 2001)

	Minimum VMA, Percent		
Nominal Maximum Particle Size[1]	Design Air Voids[2]		
	3.0	4.0	5.0
2.36 mm	19.0	20.0	21.0
4.75 mm	16.0	17.0	18.0
9.5 mm	14.0	15.0	16.0
12.5 mm	13.0	14.0	15.0
19.0 mm	12.0	13.0	14.0
25.0 mm	11.0	12.0	13.0

[1]The nominal maximum particle size is one size larger than the first sieve to retain more than 10%.
[2]Interpolate minimum VMA for design air void values between those listed.

content at 4% air voids as the design value, and then check that the other factors meet the criteria. If the Marshall stability, Marshall flow, VMA, or VFA fall outside the allowable range, the mix must be redesigned using an adjusted aggregate gradation or new material sources.

The laboratory-developed mixture design forms the basis for the initial job mix formula (JMF). The initial JMF should be adjusted to reflect the slight differences between the laboratory-supplied aggregates and those used in the field.

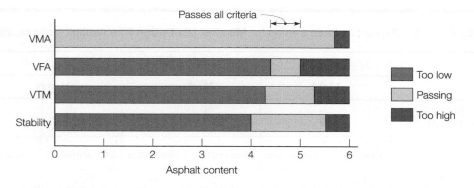

FIGURE 9.32   An example of the narrow range of acceptable asphalt contents. (The Asphalt Institute, 2001)

Sample Problem 9.6

The Marshall method was used to design an asphalt concrete mixture. A PG 64-22 asphalt cement with a specific gravity (G_b) of 1.031 was used. The mixture contains a 9.5 mm nominal maximum particle size aggregate with a bulk specific gravity (G_{sb}) of 2.696. The theoretical maximum specific gravity of the mix (G_{mm}) at asphalt content of 5.0% is 2.470. Trial mixes were made with average results as shown in the following table:

Asphalt Content (P_b) (% by Weight of Mix)	Bulk Specific Gravity (G_{mb})	Corrected Stability (kN)	Flow (0.25 mm)
4.0	2.360	6.3	9
4.5	2.378	6.7	10
5.0	2.395	5.4	12
5.5	2.405	5.1	15
6.0	2.415	4.7	22

Determine the design asphalt content using the Asphalt Institute design criteria for medium traffic (Table 9.15). Assume a design air void content of 4% when using Table 9.16.

Solution
Analysis steps:

1. Determine the effective specific gravity of the aggregates G_{se}, using Equation 9.4:

$$G_{se} \frac{P_s}{\left(\dfrac{100}{G_{mm}} - \dfrac{P_b}{G_b}\right)} = \frac{100 - 5.0}{\left(\dfrac{100}{2.470} - \dfrac{5.0}{1.031}\right)} = 2.666$$

The calculations in steps 2–5 are for 4.0% asphalt content as an example. Repeat for other asphalt contents.

2. Use G_{se} to determine G_{mm} for the other asphalt contents, using Equation 9.3:

$$G_{mm} = \frac{100}{\left(\dfrac{P_s}{G_{se}} + \dfrac{P_b}{G_b}\right)} = \frac{100}{\left(\dfrac{100 - 4.0}{2.666} + \dfrac{4.0}{1.031}\right)} = 2.507$$

3. Compute voids in the total mix for each asphalt content, using Equation 9.8:

$$VTM = 100(1 - {}^{G_3}/_{G_{mm}}) = 100(1 - {}^{2.360}/_{2.507}) = 5.9$$

4. Compute voids in mineral aggregate, using Equation 9.9:

$$VMA = 100 - ({}^{G_{mb}P_s}/_{G_{sb}}) = 100 - {}^{(2.360 \times (100-4.0))}/_{2.696} = 16.0$$

5. Compute voids filled with asphalt, using Equation 9.10:

$$VFA = 100\frac{(VMA - VTM)}{VMA} = 100\frac{(16.0 - 5.9)}{16.0} = 63.3$$

6. A summary of all calculations is given in the following table:

P_b (%)	G_{mb}	Corrected Stability (kN)	Flow, (0.25 mm)	G_{mm}	G_{se}	VTM (%)	VMA (%)	VFA (%)
4.0	2.360	6.3	9	2.507		5.9	16.0	63.3
4.5	2.378	6.7	10	2.488		4.4	15.8	71.9
5.0	2.395	5.4	12	2.470	2.666	3.0	15.6	80.5
5.5	2.405	5.1	15	2.452		1.9	15.7	87.8
6.0	2.415	4.7	22	2.434		0.8	15.8	95.0

7. Plot stability, flow, and volumetric parameters versus P_b. (See Figure SP9.6.)

8. Determine the asphalt content at VTM = 4% and the corresponding parameters. Compare with criteria:

	From Graphs	Criteria
P_b @ 4%	4.6	
Stability (kN)	6.6	5.34 (min)
Flow (0.25 mm)	10.5	8 to 16
G_{mb}	2.383	NA
VMA (%)	15.7	15.0 (min)
VFA (%)	75	65 to 78

Therefore, the design asphalt content is 4.6%.

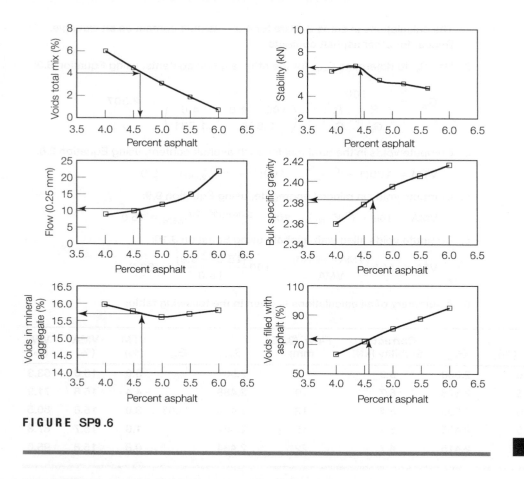

FIGURE SP9.6

9.9.6 ▩ Evaluation of Moisture Susceptibility

Since loss of bond between asphalt and aggregates (*stripping*) is a significant form of asphalt pavement distress, several methods have been developed for evaluating the susceptibility of a mix to water damage. However, the "modified Lottman" procedure, AASHTO T 283, *Resistance of Compacted Asphalt Mixtures to Moisture-Induced Damage,* is the procedure specified for Superpave mix design. The T 283 protocol is complex and requires 5 to 6 days to complete. Samples are compacted with either the Marshall hammer or the Superpave gyratory compactor, depending on the mix design method being used. Six samples are prepared at the specified air void content and dimensions. Three of the samples are reserved in an unconditioned state. The other three samples are conditioned with vacuum saturation and freezing and thawing. The samples are then conditioned in a water bath to achieve the test temperature. The indirect tensile strength of each sample is measured using the Marshall stabilometer. The tensile strength ratio, TSR, is determined as the average tensile strength of the conditioned samples divided by the average tensile strength of the unconditioned samples. The AASHTO Superpave mix design requires a minimum TSR of 0.8.

If the TSR indicates a potential problem, there are several ways to alter asphalt concrete's susceptibility to water damage. Methods identified by the Asphalt Institute include the following:

1. increasing asphalt content
2. using a higher viscosity asphalt cement
3. cleaning aggregate of any dust and clay
4. adding antistripping additives (ASA)
   a. Liquid ASA are surfacant chemicals that displace water from aggregate promoting better bonding. Liquid ASA are generally added to the binder prior to mixing with the aggregates.
   b. Powder ASA are either hydrated lime or portland cement. Powder ASA are generally distributed onto the aggregates prior to drying.
5. altering aggregate gradation

Generally, when water damage susceptibility is a problem, an additive is added to the mix at three levels, and the water damage test is performed to determine the minimum amount of additive that can be used to increase the retained strength to an acceptable level. If an acceptable mix can be developed, Marshall specimens are prepared, and the mix is tested to determine whether it meets the design criteria.

9.10   Characterization of Asphalt Concrete

Tests used to characterize asphalt concrete are somewhat different from those used to characterize other civil engineering materials, such as steel, portland cement concrete, and wood. One of the main reasons for this difference is that asphalt concrete is a nonlinear viscoelastic or viscoelastoplastic material. Thus, its response to loading is greatly affected by the rate of loading and temperature. Also, asphalt pavements are typically subjected to dynamic loads applied by traffic. Moreover, asphalt pavements do not normally fail due to sudden collapse under the effect of vehicular loads, but due to accumulation of permanent deformation in the wheel path (rutting), cracking due to repeated bending of the asphalt concrete layer (fatigue cracking), thermal cracking, excessive roughness of the pavement surface, migration of asphalt binder at the pavement surface (bleeding or flushing), loss of flexibility of asphalt binder due to aging and oxidation (raveling), loss of bond between the asphalt binder and aggregate particles due to moisture (stripping), or other factors. Therefore, most of the tests used to characterize asphalt concrete try to simulate actual field conditions.

Many laboratory tests have been used to evaluate asphalt concrete properties and to predict its performance in the field. These tests are performed on either laboratory-prepared specimens or cores taken from in-service pavements. These tests measure the response of the material to load, deformation, or environmental conditions, such as temperature, moisture, or freeze and thaw cycles. Some of these tests are based on empirical relations, while others evaluate fundamental properties. All tests on asphalt concrete are performed at accurately controlled test temperatures and rates of loading, since asphalt response is largely affected by these two parameters.

The Superpave and Marshall tests used for mix design have been used to characterize asphalt concrete mixtures. Other tests are also being used, some of which are standardized by ASTM or AASHTO, while others have been used mostly for research. The following several sections discuss some of the common tests.

9.10.1 ▨ Indirect Tensile Strength

When traffic loads are applied on the pavement surface, tension is developed at the bottom of the asphalt concrete layer. Therefore, it is important to evaluate the tensile strength of asphalt concrete for the design of the layer thickness. The indirect tensile strength test is also used for determining moisture susceptibility. In this test, a cylindrical specimen 101.6 or 150 mm in diameter is typically used. A compressive vertical load is applied along the vertical diameter, using a loading device similar to that shown in Figure 9.33. The load is applied by using two curved loading strips moving with a typical rate of deformation of 51 mm/min. Tensile stresses are developed in the horizontal direction, and when these stresses reach the tensile strength, the specimen fails in tension along the vertical diameter. The test is performed at a specified temperature. With 12.5 mm loading strips, the indirect tensile strength is computed as

$$\sigma_t = \frac{2P}{\pi t D} \tag{9.21}$$

where

σ_t = tensile strength, MPa
P = load at failure, N
t = thickness of specimen, mm
D = diameter of specimen, mm

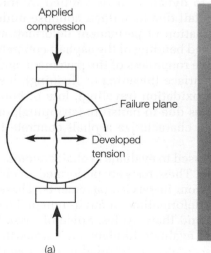

(a)   (b)

FIGURE 9.33 Indirect tension test.

Sample Problem 9.7

The following data were measured for the freeze–thaw test, AASHTO T 283. Three samples were conditioned and three were unconditioned. All the samples had a diameter of 150 mm. Determine:

a. The tensile strength of each sample in kPa.
b. The average tensile strength of the conditioned and unconditioned samples in kPa.
c. The tensile strength ratio (TSR).
d. Does the mix pass the Superpave criteria for moisture susceptibility?

	Unconditioned Samples			Conditioned Samples		
Thickness, mm	64.0	63.5	63.0	63.3	63.8	63.5
Load at failure, kN	11.8	12.6	12	6.6	7.2	7

Solution

$$\sigma_t = \frac{2P}{\pi t D}$$

	Unconditioned Samples			Conditioned Samples		
Tensile strength, kPa	782	842	808	442	479	468
Average tensile strength, kPa	811			463		

TSR = 463/811 = 57%, which is less than 80%. Mix does not pass. ∎

9.10.2 Asphalt Mixture Performance Tester

The Superpave volumetric procedure has been widely implemented, with successful results. However, the method lacks a strength test to verify the suitability of the Superpave mixes. Research completed in NCHRP Project 9-19 developed "simple performance tests" for asphalt concrete (Witczak et al., 2002), currently called the Asphalt Mixture Performance Tester, AMPT, Figure 9.34. The AMPT is not currently a requirement for Superpave mix design, but implementation studies are underway. In addition, the test method was designed to capture the material properties needed for the Mechanistic Empirical Pavement Design System, MEPDS.

The equipment and associated protocols of the AMPT tests are designed to capture the viscoelastic behavior of asphalt concrete. Tests based on measurement of dynamic modulus (for both of permanent deformation and fatigue cracking), flow time (permanent deformation), and flow number (permanent deformation) were selected for further field validation. The three tests used to obtain these parameters are the dynamic modulus test, triaxial static creep test, and triaxial repeated load permanent deformation test. All these tests use cylindrical specimens 100 mm in

FIGURE 9.34 Asphalt Mixture Performance Tester.

diameter and 150 mm high. Specimens are cored in the lab from specimens compacted using the Superpave gyratory compactor.

Dynamic Modulus Test   The dynamic modulus test in triaxial compression has been around the pavement community for many years (ASTM D3497). The test consists of applying an axial sinusoidal compressive stress to an unconfined or confined asphalt concrete cylindrical test specimen, as shown in Figure 9.35.

Assuming that asphalt concrete is a linear viscoelastic material, a complex number called the complex modulus, E^*, can be obtained from the test to define the relationship between stress and strain. The absolute value of the complex modulus, $|E^*|$,

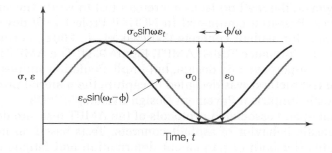

FIGURE 9.35 Stress and strain pulses for the dynamic modulus test.

is called the dynamic modulus. The dynamic modulus is mathematically defined as the peak dynamic stress σ_0 divided by the peak recoverable axial strain ε_0:

$$|E^*| = \frac{\sigma_0}{\varepsilon_0} \tag{9.22}$$

The real and imaginary portions of the complex modulus E^* can be written as

$$E^* = E' + iE'' \tag{9.23}$$

E' is generally referred to as the storage or elastic modulus component of the complex modulus and E'' is referred to as the loss or viscous modulus. The phase angle ϕ is the angle by which ε_0 lags behind σ_0. It is an indicator of the viscous properties of the material being evaluated. Mathematically, this is expressed as

$$E^* = |E^*| \cos \phi + i|E^*| \sin \phi \tag{9.24}$$

where

$$\phi = \frac{t_i}{t_p} \times 360$$

t_i = time lag between of stress and strain, s
t_p = time for a stress cycle, s
i = imaginary number

For a pure elastic material, $\phi = 0$, and the complex modulus E^* is equal to the absolute value or the dynamic modulus. For a pure viscous material, $\phi = 90°$.

The dynamic modulus obtained from this test is indicative of the stiffness of the asphalt mixture at the selected temperature and load frequency. The dynamic modulus is correlated to both rutting and fatigue cracking of asphalt concrete.

Triaxial Static Creep Test In the static compressive creep test, a total strain–time relationship for a mixture is measured in the laboratory under unconfined or confined conditions. The static creep test, using either one load–unload cycle or incremental load–unload cycles, provides sufficient information to determine the instantaneous elastic (recoverable) and plastic (irrecoverable) components (time independent), and the viscoelastic and viscoplastic components (time dependent) of the material's response. In this test, the compliance, $D(t)$ is determined by dividing the strain as a function of time $\varepsilon(t)$, by the applied constant stress σ_0:

$$D(t) = \frac{\varepsilon(t)}{\sigma_0} \tag{9.25}$$

Figure 9.36 shows typical test results between the calculated compliance and loading time. As shown, the compliance can be divided into three major zones: primary zone, secondary zone, and tertiary flow zone. The time at which tertiary flow starts is referred to as the *Flow Time*. The flow time is a significant parameter in evaluating the rutting resistance of asphalt concrete.

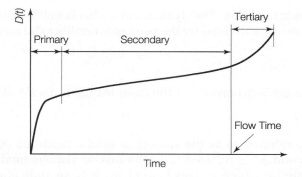

FIGURE 9.36 Compliance versus loading time for the static triaxial creep test.

Triaxial Repeated Load Permanent Deformation Test Another approach to measuring the permanent deformation characteristics of asphalt concrete is to use a repeated load test for several thousand repetitions and to record the cumulative permanent deformation as a function of the number of load repetitions. In this test, a haversine pulse load consisting of a 0.1 and 0.9 second dwell (rest) time is applied for the test duration, typically about 3 hours or 10000 loading cycles.

Results from the repeated load tests typically are presented in terms of the cumulative permanent strain versus the number of loading cycles. Figure 9.37 illustrates a typical relationship between the cumulative plastic strain and number of load cycles. In a manner similar to the triaxial static creep test, the cumulative permanent strain curve can be divided into three zones: primary, secondary, and tertiary. The cycle number at which tertiary flow starts is referred to as the *Flow Number*. In addition to the flow number, the test can provide the resilient strain and modulus, all of which are correlated to the rutting resistance of asphalt concrete.

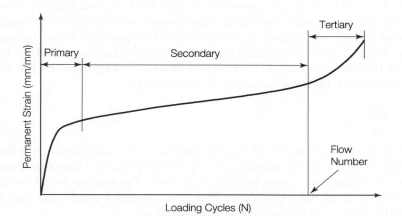

FIGURE 9.37 Typical relationship between total cumulative plastic strain and loading cycles.

9.11 Hot-Mix Asphalt Concrete Production and Construction

The production and construction of asphalt concrete highways can be described as a three-step process: production of raw materials, manufacturing asphalt concrete, and field operations. The following discussion was developed for hot-mix asphalt. Production and construction of warm-mix asphalt is similar but details of the process can vary depending on the technology used to achieve the viscosity reduction of the asphalt cement to allow for the lower temperatures compared to hot-mix asphalt.

9.11.1 Production of Raw Materials

Asphalt cement is produced at petroleum refineries and transported to the asphalt plant and placed in tanks. To maintain flow ability, the asphalt cement is kept hot, approximately 150°C, while being transported to the asphalt and while in the tanks at the asphalt plant.

Aggregates are acquired from quarries and stockpiled at the asphalt plant in the sizes needed for the asphalt mixes. As discussed in Chapter 5, good control of stockpiles is essential to limit segregation of the aggregates.

9.11.2 Manufacturing Asphalt Concrete

The manufacturing of asphalt concrete requires drying and heating the aggregates, blending them in proportions determined by the mix design, adding the required amount of asphalt cement and mixing the aggregates and asphalt cement. There are three types of plants used to produce hot mix asphalt concrete: batch, drum, and continuous. Continuous plants are only used by a small proportion of the industry so they are not discussed further.

Batch plants (Figure 9.38) are the oldest style of plant and still are widely used. Aggregates are transported from the stockpiles to the cold feed bins. Conveyors transport the aggregates from the cold feed bins to the dryer. This is a large rotating cylinder with flights which promote tumbling of the aggregates through heated air. As the aggregates are dried, dust is released into the air-flow stream of the dryer drum. This air is filtered to minimize particulate pollution of the surrounding land. The most common type filter uses a series of bags to collect the dust. The hot and dry aggregates tumble from the dryer drum to the hot elevator that lifts the aggregates to the hot sieves. The hot sieves separate the aggregates and charge the hot bins. A batch of HMA is produced by weighing out the required amount of aggregate from each hot bin into the mixer or pug mill. The aggregates are briefly mixed and then the hot asphalt cement is metered into the pug mill. The aggregates and asphalt cement are thoroughly mixed and the mixture is discharged either directly into a truck or into a silo. The entire mixing process takes about 1 minute. The capacity of the plant is controlled by the size of the pug mill. A 5-ton pug mill can produce about 300 tons per hour. The mix can be directly loaded into trucks or placed into a silo for temporary storage.

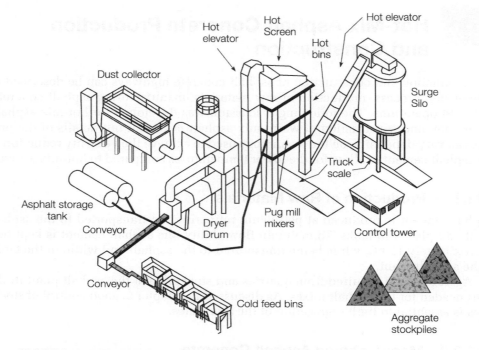

FIGURE 9.38 Batch asphalt mixing plant.

In the drum plant (Figure 9.39), a single drum is used for both drying the aggregates and mixing in the asphalt cement. Weight scales are used on the cold feed conveyor belts to monitor the amount of aggregate from each stockpile that is being fed into the drum. These weights must be adjusted for the moisture content of the stockpile. Aggregates are dried in the first two-thirds of the drum. In the last one-third, the asphalt cement is introduced. This creates a continuous process so the output from the mixing process is transported using a hot elevator into a silo. Trucks are loaded from the silo.

9.11.3 ■ Field Operations

The asphalt concrete is transported from the plant to the paving location where it is placed using an asphalt paver. Traditionally, trucks would directly load material into the paver's hopper. This has several potential problems such as the truck bumping the paver, nonuniform delivery of material, and potential for temperature segregation of the mix. In response to these problems, the industry has developed material transfer devices (Figure 9.40) that act as an interface between the trucks and the paver. The use of a material transfer device is optional but it is becoming common for paving sections where smoothness is critical such as on interstates and airfield pavements.

Asphalt pavers are designed to produce a smooth surface when operated properly. The keys to proper operation are a constant flow of material through the paver and maintaining constant paver speed.

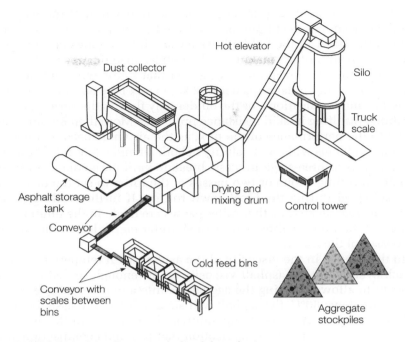

FIGURE 9.39   Drum asphalt mixing plant.

FIGURE 9.40   Material transfer devices used to transfer the asphalt concrete from the trucks to the paver.

The final step in the construction process is compaction of the pavement mat. Although the optimum percent binder is selected at 4% air voids, the specification for density of the HMA in the field allows for an air void range of 4 to 8%, usually expressed as 92 to 96% of the maximum theoretical specific gravity. A maximum of 96% G_{mm} is specified in recognition of the fact that traffic will continue to densify the pavement. If the air voids drops below 2% the mix may become unstable, resulting in bleeding and corrugations, or the binder may flow to the surface, called bleeding or flushing, resulting in skid resistance problems. More than 8% air voids is also detrimental to the performance of the mix, as the air voids become interconnected which creates problems with permeability of both air and water into the asphalt mat.

Compaction is a three-stage process: breakdown, intermediate, and finish rolling. Breakdown rolling is one or two passes of the roller to set the mat, most of the density is achieved during the breakdown rolling. If further rolling is needed to achieve the desired density, further roller passes are made in the intermediate stage. The purpose of the finish rolling is to remove roller marks left in the surface by the previous passes of the roller.

Due to the change in the viscosity of the asphalt with temperature, compaction must be achieved while the asphalt viscosity is low enough so that the asphalt acts as a lubricant to allow arranging the aggregates into a dense configuration. Usually at temperatures above 80°C, the asphalt will act as a lubricant, assisting the densification of the aggregate. At lower temperatures, the viscosity of the asphalt inhibits compaction. Most states have a specification that prohibits compaction at mat temperatures less than 80°C. If warm-mix asphalt is being used on a project most states allow a variance on the minimum compaction temperature and follow the recommendations of the manufacture of the warm-mix technology being used.

Three types of rollers are used for asphalt paving: static steel wheel, vibratory steel wheel, and pneumatic. Vibratory rollers are becoming the dominant type machine as they can be used for all stages. With the vibrators active, high forces are applied to the pavement to achieve density. The vibrators can be turned off for the finish rolling.

Quality Control During Construction   Long-term pavement performance requires proper construction of the pavement layers. Quality control for asphalt pavements generally focuses on three areas: materials, density of the mat, and smoothness. First the contractor must demonstrate that the materials specified for the job are used and that the plant can produce the mix matching the requirements set forth in the mix design for the project, called the job mix formula. Materials produced by the plant are evaluated for:

- Proper gradation of the aggregate blend
- Asphalt content specified by in the job mix formula
- Volumetric properties—samples of the plant mix are evaluated in the laboratory to evaluate air voids, voids in the mineral aggregate, and voids filled with asphalt

The key material parameter evaluated in the field is the density of the mat after compaction. Most states specify a target density of 92 to 96% of the maximum theoretical specific gravity (G_{mm}), which corresponds to 4 to 8% air voids.

The smoothness of the top layer of the pavement surface is of paramount importance to highway users. Smoothness is the lack of variations in the longitudinal profile of the pavement surface. Construction of a smooth asphalt concrete surface requires a balanced process between plant production paver speed and timely compaction.

9.12   Recycling of Asphalt Concrete

Recycling pavement materials has a long history. However, recycling became more important in the mid-1970s, after the oil embargo, due to the increase in asphalt prices. In an effort to efficiently use available resources, there was a need to recycle or reuse old pavement materials (The Asphalt Institute, 2007). The spike in petroleum prices in 2008 also heightened interest in recycling. Environmental concerns and the emphasis on sustainable design are further promoting interest in asphalt recycling. Recycling asphalt has the following advantages:

1. economic saving of about 25% of the price of materials
2. energy saving in manufacturing and transporting raw materials
3. environmental saving by reducing the amount of required new materials and by eliminating the problem of discarding old materials
4. eliminating the problem of reconstruction of utility structures, curbs, and gutters associated with overlays
5. reducing the dead load on bridges due to overlays
6. maintaining the tunnel clearance, compared with overlays

Recycling can be divided into three types: *surface recycling, central plant recycling,* and *in-place recycling.* Central plant recycling is the predominant type of recycling and offers the highest quality product, so it is featured in the following.

Central plant recycling is performed by milling the old pavement (Figure 9.41) and sending the reclaimed asphalt pavement (RAP) to a central asphalt concrete plant, where it is mixed with virgin aggregates and asphalt binder in the asphalt plant. The RAP material generally does not have the properties required for quality asphalt concrete so it is added to virgin aggregate and asphalt. The virgin materials are used to correct deficiencies in the RAP. The need to correct deficiencies in the RAP with virgin materials limits the percent RAP; generally the content of RAP ranges from 10 to 50%.

9.12.1   RAP Evaluation

Figure 9.42 (based on Newcomb et al., 2007) summarizes the material characterization process. The evaluation of the virgin aggregates and binder is the same as needed for design without RAP. In order to design an RAP mix it is necessary to know the properties of both the asphalt cement and aggregates in the RAP. Figure 9.42 is

FIGURE 9.41 Milling old pavement.

based on the assumption that the RAP material will come from a single project and be used for a single project. Frequently, this is not the case as it is a common practice for contractors to collect RAP from several projects into a single stockpile. This practice results in highly variable stockpile and quality control during production and construction will be difficult at best. To reduce the problem with variability, the contractor should develop "production" stockpiles by blending, crushing, and sorting into stockpiles of different sizes. The properties of the aggregate in each pile are evaluated following the procedures for virgin material. Evaluation of the binder in the RAP depends on the target RAP content of the mix:

RAP Content	Virgin Binder Action
<15%	Use standard grade binder
15 to 25%	Reduce binder one grade for both upper and lower temperature
>25%	Evaluated RAP binder to determine the grade for the virgin binder

9.12.2 ■ RAP Mix Design

Either of the previously described mix design methods can be used to design an RAP mix. The primary difference is in the handling of the RAP material. The recommended practice is to preheat the virgin aggregate to the mix temperature then

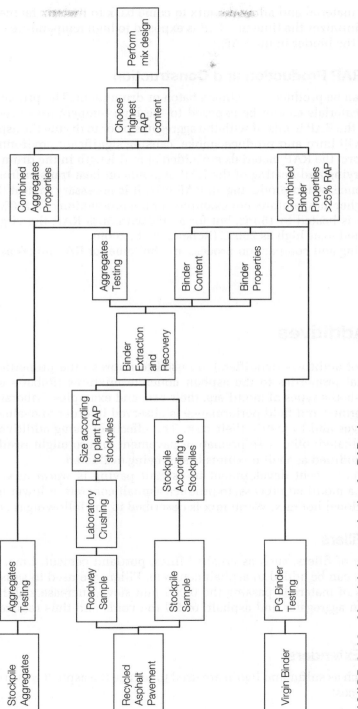

FIGURE 9.42 Material characterization for RAP mixes.

add the RAP material and allow the mix to come back to the mix temperature. This procedure minimizes the time the RAP is exposed to high temperature to reduce the hardening of the binder in the RAP.

9.12.3 ■ RAP Production and Construction

RAP mixes can be produced in either a batch or drum plant. The primary difference is the RAP materials cannot be exposed to the high temperatures used to dry the aggregates. If the RAP is mixed with the aggregates prior to drying the asphalt cement in the RAP will burn and produce smoke, causing significant environmental problems. Therefore, the RAP materials are added at mid-length in the drum at the drum mix plant. Drying and heating of the RAP depends on heat transfer from the virgin aggregate. Hence, when producing an RAP mix, it is necessary to heat the aggregate to a much higher temperature. For example, for the production of a 100% virgin mix, the aggregate is heated to 150°C, but for a mix with 50% RAP, the virgin aggregate could be heated to as high as 260°C (Roberts et al., 2009).

The placing and compaction process are the same for RAP mixes as they are for virgin mixes.

9.13   Additives

Many types of additives (modifiers) are used to improve the properties of asphalt or add special properties to the asphalt concrete mixtures (Roberts et al., 2009). Table 9.17 lists the types of modifiers, their use, and examples. Laboratory tests are usually performed and field performance is observed in order to evaluate the effect of the additives and to justify their cost. The effects of using additives should be carefully evaluated; otherwise premature pavement failure might result. The recyclability of modified asphalt mixtures is still being evaluated.

A relatively recent development in asphalt paving is *warm* mix, which uses modifiers, or a modified process, to produce asphalt concrete at lower temperatures than conventional hot mix. Warm mix is described in the following section.

9.13.1 ■ Fillers

Several types of fillers, such as crushed fines, portland cement, lime, fly ash, and carbon black, can be added to asphalt concrete. Fillers are used to satisfy gradation requirements of materials passing the 0.075 mm sieve, increase stability, improve bond between aggregates and asphalt, or fill the voids and thus reduce the asphalt required.

9.13.2 ■ Extenders

Extenders such as sulfur and lignin are used to reduce the asphalt requirements, thus reducing the cost.

9.13.3 ■ Polymer Modified Asphalt

Polymer modifiers include both the rubber and plastic category shown in Table 9.17. Rubber has been used in asphalt concrete mixture in the form of natural rubber, styrene–butadiene (SBR), styrene–butadiene–styrene (SBS), or recycled tire rubber. Rubber increases elasticity and stiffness of the mix and increases the bond between asphalt and aggregates. SBS is the most common modifier for producing PMA. Scrap rubber tires can be added to the asphalt cement (wet method) or added as crumb rubber to the aggregates (dry method).

TABLE 9.17   Asphalt Cement and Asphalt Concrete Modifiers (from Roberts et al., 2009)

Type	General Purpose or Use	Generic Examples

Filler	■ Fill voids and therefore reduce optimum asphalt content
	■ Meet aggregate gradation specifications
	■ Increase stability
	■ Improve the asphalt cement-aggregate bond
		■ Mineral filler:
		  ■ crusher fines
		  ■ lime
		  ■ portland cement
		  ■ fly ash
		■ Carbon black

Extender	■ Substituted for a portion of asphalt cement (typically between 20–35% by weight of total asphalt binder) to decrease the amount of asphalt cement required	■ Sulfur ■ Lignin
Rubber	■ Increase asphalt concrete stiffness at high service temperatures ■ Increase asphalt concrete elasticity at medium service temperatures to resist fatigue cracking ■ Decrease asphalt concrete stiffness at low temperatures to resist thermal cracking	■ Natural latex ■ Synthetic latex (e.g., polychloroprene latex) ■ Block copolymer (e.g., styrene–butadiene–styrene (SBS)) ■ Reclaimed rubber (e.g., crumb rubber from old tires)
Plastic		■ Polyethylene/polypropylene ■ Ethylene acrylate copolymer ■ Ethyl–vinyl–acetate (EVA) ■ Polyvinyl chloride (PVC) ■ Ethylene propylene or EPDM ■ Polyolefins
Rubber–plastic combinations		■ Blends of rubber and plastic

Fiber	■ Improving tensile strength of asphalt concrete
	■ Improving cohesion of asphalt concrete
	■ Permit higher asphalt content without significant increase in draindown
		■ Natural:
		  ■ Asbestos
		  ■ Rock wool
		■ Manufactured:
		  ■ Polypropylene
		  ■ Polyester
		  ■ Fiberglass
		  ■ Mineral
		  ■ Cellulose

Plastics have been used to improve certain properties of asphalt. Plastics used include polyethylene, polypropylene, ethyl–vinyl–acetate (EVA), and polyvinyl chloride (PVC). They increase the stiffness of the mix, thus reducing the rutting potential. Plastics also may reduce the temperature susceptibility of asphalt and improve its performance at low temperatures.

The introduction of Superpave and the associated Performance Grade binder specifications increased the emphasis on the use of materials that improve the long-term performance of asphalt pavements. This has lead to an increase in the use of polymer-modified asphalts (PMA). These modifiers have two primary advantages: altering the temperature–viscosity behavior of the asphalt and improving the durability of asphalt concrete. SBS is the predominant material used for modifying asphalt (Airey, 2003).

SBS blended with asphalt forms two phases, a polymer-rich phase and an asphalt-rich phase (Kraus, 2008). The polymer-rich phase swells as it absorbs aromatics from the asphalt. The absorption of the aromatics from the asphalt cement limits the amount of SBS that can be added beneficially. The amount of SBS used to modify asphalt is typically 2 to 6%. SBS modification improves both the coating of the aggregates for improved durability and the elasticity of the binder, which benefits the rutting, fatigue, and thermal cracking characteristics of the binder.

Research has demonstrated that polymer-modified binders provide measurable benefits that contribute to the performance of pavements. The improved performance has been documented in several field trials dating back over 30 years. While the improved performance would benefit all pavements, polymer modification adds approximately 10 to 15% to paving costs. Hence, most agencies limit the use of PMA to high traffic and high axle load applications where the additional costs are justified based on life-cycle cost.

9.13.4 ■ Antistripping Agents

Antistripping agents are used to improve the bond between asphalt cement and aggregates, especially for water-susceptible mixtures. Lime is the most commonly used antistripping agent and can be added as a filler or a lime slurry and mixed with the aggregates. Portland cement can be used as an alternative to lime.

9.13.5 ■ Others

Other additives, such as fibers, oxidants, antioxidants, and hydrocarbons, have been used to modify certain asphalt properties' tensile strength and stiffness.

9.14 Warm Mix

Warm-mix asphalt is the generic term for a variety of technologies that allow the producers of hot-mix asphalt pavement material to lower the temperatures at which the material is mixed and placed on the road. Reductions of 25 to 50°C are common, which offers the following benefits:

- Less energy and fuel consumption to heat the mix
- Lower greenhouse gases
- Less oxidation of the asphalt binder
- Pave at lower temperatures
- Longer haul distances are possible
- Extending the paving season since the asphalt can be compacted at lower temperatures

Warm mix technology initially developed in Europe, but the technology and implementation are developing rapidly in the United States due to the potential advantages of warm mixes. The Federal Highway Administration is working with states and the industry for the development of warm mix, with the following objectives:

- Use existing hot-mix asphalt plants and placing equipment
- Meet existing standards for hot mix asphalt specifications
- Focus on dense-graded mixes for wearing courses
- Warm mix quality equals hot-mix quality

Several warm mix technologies and producers are competing:

- Emulsion Technology
  - Evotherm—Mead Westvaco
- Mix additives
  - Aspha-min—Eurovia
  - Sasobit Sasol Int./Moore and Munger
  - Rediset—Akzo-Noble
- Materials Processing
  - WAM-Foam Shell/Kolo Veidekke/BP
  - Low Energy Asphalt Fairco
  - Green WMA Astec

The mix additive technologies modify the viscosity temperature properties of the asphalt in a nonlinear manner as shown in Figure 9.43 (Hurley and Prowell, 2005). The modifier reduces the viscosity of the asphalt cement at high temperatures, for example, the mixing temperature for the unmodified asphalt is 160°C but the modified binder has the same viscosity at a temperature of 142°C so the mixing temperature can be reduced by 18°C. The chemical used to achieve the viscosity modification can be either added at the asphalt plant or premixed into the asphalt cement by the vendor.

Warm mix can also be produced by modifying the asphalt plant to accommodate production of warm mix. One of the available technologies is the Astec Green System (www.astecinc.com/index.php), which uses carefully controlled water injection to "foam" the asphalt just before mixing with the aggregates. The water converts to steam while the asphalt is above the boiling point. In the foamed condition, the asphalt can flow around the aggregates creating the needed coating. Once the asphalt cools to below the boiling point of water, the foam collapses leaving the residual asphalt coating the aggregates. The amount of water required for the foaming process, approximately one half kilogram of water per 1000 kg of mix, is so low that the quality of the asphalt concrete is not affected.

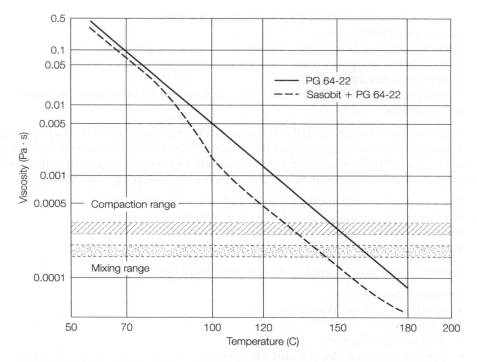

FIGURE 9.43 Temperature–viscosity chart for Sasobit warm mix modifier, Sasobit.

The substantial benefits of warm mix are such that the asphalt industry considers warm mix to be the "wave of the future" in flexible pavement construction. As with any new technology, research is needed to take the full advantage of the product. Areas that are currently being evaluated for warm mix include:

- Appropriate modifications to the mix design methods
- Plant operating procedures
- Alteration of field operations
- Quality control and assurance methods

9.15 Asphalt Sustainability

9.15.1 LEED Considerations

The National Asphalt Pavement Association (NAPA 2015) has summarized potential LEED credits in four rating categories:

- Stormwater Design: Quantity Control Stormwater Design—Porous Asphalt designed to store storm water in the structure of the pavement reducing the need for retention ponds and allowing ground water recharge.

- Heat Island Effect: Non-Roof
  - Reflective surfaces—use of chip seals and sand seals on an asphalt pavement is an economical method for improving solar reflectance
  - Open-graded asphalt—has lower heat retention than dense graded asphalt pavements; is also beneficial for noise reduction.
  - Porous pavements—evaporation of the water stored in the pavement base reduces the pavement temperature.
- Construction Waste Management: Divert from Disposal (based on weight/volume)
  - Reclaimed asphalt pavement provides a valuable commodity for the paving industry and diverts materials from landfills.
- Exceptional Performance Exceeding Expectations or Areas Not Addressed
  - Warm-mix asphalt reduces energy and emissions relative to hot mix asphalt.
  - High-RAP mixes—RAP mixes typically are 15% RAP and 85% virgin product, technology is available for using 35% or more RAP which exceeds expectations for a typical RAP mix.
  - Crum-rubber asphalt mixtures—ground tire rubber can be added to asphalt binders to produce a high-quality material from a secondary waste stream material.

9.15.2 ▨ Other Sustainability Considerations

Asphalt is unique compared to the other materials addressed in this book as the use of asphalt is heavily focused on the paving market. The LEED system is primarily focused on building projects, yet a large portion of construction is in infrastructure projects other than the building industry. Miller and Bahia (2010) presented a framework for the evaluation of the sustainability of pavements. Their recommendations include:

- Conducting energy and emissions analysis of local asphalt plants, including survey tools and audit methods.
- Refine/develop analysis tools for sustainability indicators, such as life-cycle cost analysis, into an eco-efficiency model.
- Develop a certification program for identifying and rewarding advances in sustainable pavement construction.
- Focus industry research on cold asphalt technologies that could supplement/replace hot mix for appropriate applications.

Summary

Asphalt produced from crude oil is a primary road-building material. The civil engineer is directly involved with the specification and requirements for both the asphalt cement binder and the asphalt concrete mixtures. There are several methods for grading asphalt cements. The current trend is toward the use of the performance-grading method, used in the Superpave process developed through the Strategic Highway Research Program. This grading method directly ties the binder properties to pavement-performance parameters. Similarly, the Superpave mix design method uses performance tests to evaluate the mixture characteristics relative to expected field performance. This method will continue to be refined by highway agencies. Superpave and Performance Grade binders have been adopted by all state highway agencies. Adoption for military airfields and civilian airports has been slower than for highways, but implementation is underway. Both the Corps of Engineers and the Federal Aviation Administration have specifications for the use of Superpave and Performance Grade binders.

Although the Strategic Highway Research Program is over, one of the lasting legacies of the program is the demonstration that research does improve the state of the art in asphalt materials and pavements. PMA and warm mix technologies are two of the areas that should continue to evolve.

Questions and Problems

9.1 What is the difference between tar and asphalt cement?

9.2 Discuss the main uses of asphalt.

9.3 What are the ingredients of asphalt cutbacks? What are the ingredients of asphalt emulsions? Name three uses of asphalt emulsion. Why is asphalt emulsion preferred over asphalt cutback?

9.4 Draw an asphalt concrete pavement section consisting of an asphalt concrete surface layer, aggregate base, and subgrade. Show the locations of prime coat and chip seal.

9.5 Define what is meant by temperature susceptibility of asphalt. With the help of a graph showing viscosity versus temperature, discuss the effect on the performance of asphalt concrete pavements. Are soft asphalts used in hot or cold climates?

9.6 Temperature has a large effect on the asphalt viscosity. On one graph, plot the relationship between asphalt viscosity (logarithmic) and temperature for two cases: (a) a low-temperature susceptible asphalt and (b) a high-temperature susceptible asphalt. Label all axes and relations.

9.7 Briefly discuss the chemical composition of asphalt.

9.8 What are the engineering applications of each of these tests?
a. flash point test
b. RTFO procedure
c. rotational viscometer test
d. dynamic shear rheometer test
e. penetration test.

9.9 Discuss the aging that occurs in asphalt cement during mixing with aggregates and in service. How can the different types of aging of asphalt cement be simulated in the laboratory?

9.10 Show how various Superpave tests used to characterize the asphalt binder are related to pavement performance.

9.11 Define the four methods used to grade asphalt binders. Which method is used in your state?

9.12 To select an asphalt binder for a specific location, the mean seven-day maximum pavement temperature is estimated at 55°C with a standard deviation of 2.5°C. The mean minimum pavement temperature is −9°C with standard deviation of 1.5°C. What PG grade asphalt is needed at 98% reliability?

9.13 For the following temperature conditions, calculate the proper PG grade for both 50% and 98% reliabilities (show your calculation).

Seven-day maximum pavement temperature of 48°C with a standard deviation of 2.5°C.
Minimum pavement temperature of −21°C with a standard deviation of 3°C.

9.14 As a materials engineer working for a highway department, what standard PG asphalt binder grade would you specify for each of the conditions shown in Table P9.14 (show all calculations and fill in the table)?

TABLE P9.14

	Seven-Day Maximum Pavement Temperature		Minimum Pavement Temperature		Recommended PG Grade	
Case	Mean, °C	Standard Deviation, °C	Mean, °C	Standard Deviation, °C	50% Reliability	98% Reliability
1	43	1.5	−29	2.5		
2	51	3	−18	4		
3	62	2.5	10	2		

9.15 As a materials engineer working for a highway department, what standard PG asphalt binder grade would you specify for each of the conditions shown in Table P9.15 (show all calculations and fill in the table)?

TABLE P9.15

	Seven-Day Maximum Pavement Temperature		Minimum Pavement Temperature		Recommended PG Grade	
Case	Mean, °C	Standard Deviation, °C	Mean, °C	Standard Deviation, °C	50% Reliability	98% Reliability
1	39	1	−32	3.5		
2	54	1.5	−17	2		
3	69	2	5	2.5		

9.16   What are the differences between CRS-2 and SS-1 emulsions?

9.17   Discuss how asphalt emulsions work as a binder in asphalt mixtures.

9.18   What are the components of hot-mix asphalt? What is the function of each component in the mix?

9.19   What are the objectives of the asphalt concrete mix-design process?

9.20   Why is it important to have optimum binder content in asphalt concrete? What would happen if a less-than-optimum binder content is used? What would happen if more than the optimum value is used? What is the typical range of binder content in asphalt concrete?

9.21   Explain why the strength of asphalt concrete is not necessarily the most important property of the material.

9.22   An asphalt concrete specimen has a mass in air of 1327.8 g, mass in water of 792.4 g, and saturated surface-dry mass of 1342.2 g. Calculate the bulk-specific gravity of the specimen.

9.23   As part of mix design, a laboratory-compacted cylindrical asphalt specimen is weighed for determination of bulk-specific gravity. The following numbers are obtained:

> Dry mass *in* air = 1264.7 grams
> Mass when submerged in water = 723.9 g
> Mass of saturated surface dry (SSD) = 1271.9 g

     a. What is the bulk-specific gravity of the compacted specimen (G_{mb})?
     b. If the maximum theoretical specific gravity of the specimen (G_{mm}) is 2.531, what would be the air void content of the specimen in percent?

9.24   For asphalt concrete, define

     a. air voids
     b. voids in the mineral aggregate
     c. voids filled with asphalt

9.25   An aggregate blend is composed of 59% coarse aggregate by weight (Sp. Gr. 2.635), 36% fine aggregate (Sp. Gr. 2.710), and 5% filler (Sp. Gr. 2.748). The compacted specimen contains 6% asphalt binder (Sp. Gr. 1.088) by weight of total mix and has a bulk density of 2305 kg/m^3. Ignoring absorption, compute the percent voids in total mix, percent voids in mineral aggregate, and the percent voids filled with asphalt.

9.26   An asphalt concrete mixture includes 94% aggregate by weight. The specific gravities of aggregate and asphalt are 2.65 and 1.0, respectively. If the bulk density of the mix is 2355 kg/m^3, what is the percent voids in the total mix?

9.27   After 2 years of traffic, cores were recovered from the roadway which has severe rutting and bleeding. The bulk specific gravity and the maximum theoretical specific gravity were measured on these cores and are as follows:

> Bulk specific gravity = 2.487
> Maximum theoretical specific gravity = 2.561

    a. Calculate the air voids

    b. If the design air void content was 4%, explain what effect the calculated air voids had on the rutting and bleeding noted.

9.28   An asphalt concrete specimen has the following properties:

        Asphalt content = 5.5% by total weight of mix
        Bulk specific gravity of the mix = 2.475
        Theoretical maximum specific gravity = 2.563
        Bulk specific gravity of aggregate = 2.689

    Ignoring absorption, calculate the percents VTM, VMA, and VFA.

9.29   A compacted asphalt concrete specimen has the following properties:

        Asphalt content = 5% by total weight of the mix
        Bulk specific gravity of the mix = 2.500
        Theoretical maximum specific gravity = 2.610
        Bulk specific gravity of the aggregate = 2.725

    Ignoring absorption, compute the percents VTM, VMA, and VFA.

9.30   Briefly describe the volumetric mix design procedure of Superpave.

9.31   Based on the data shown in Table P9.31, select the blend for a Superpave design aggregate structure. Assume <0.3 million ESAL and 19 mm nominal maximum aggregate size.

TABLE P9.31

	Blend		
Data	1	2	3
G_{mb}	2.451	2.465	2.467
G_{mm}	2.585	2.654	2.584
G_b	1.030	1.030	1.030
P_b	5.9	5.5	5.8
P_s	94.1	94.5	94.2
P_d	4.5	4.5	4.5
G_{sb}	2.657	2.667	2.705
H_{ini}	127	135	124
H_{des}	113	114	118

9.32   Based on the data in Table P9.32, determine the design binder content for a Superpave mix design for a 15 million ESAL. Assume a 12.5 mm nominal maximum aggregate size.

9.33   Given the data in Table P9.33, select the blend and the design binder content for a Superpave design aggregate structure for a 5 million ESAL and a 19 mm nominal maximum aggregate size.

TABLE P9.32

Data	Binder Content (%)			
	5.5	**6.0**	**6.5**	**7.0**
G_{mb}	2.351	2.441	2.455	2.469
G_{mm}	2.579	2.558	2.538	2.518
G_b	1.025	1.025	1.025	1.025
P_s	94.5	94.0	93.5	93.0
P_d	4.5	4.5	4.5	4.5
G_{sb}	2.705	2.705	2.705	2.705
h_{ini}	129	131	131	128
h_{des}	112	113	116	115

TABLE P9.33

Data	Blend		
	1	**2**	**3**
G_{mb}	2.457	2.441	2.477
G_{mm}	2.598	2.558	2.664
G_b	1.025	1.025	1.025
P_b	5.9	5.7	6.2
P_s	94.1	94.3	93.8
P_d	4.5	4.5	4.5
G_{sb}	2.692	2.688	2.665
H_{ini}	125	131	125
H_{des}	115	118	115

Data	Binder Content (%)			
	5.4	**5.9**	**6.4**	**6.9**
G_{mb}	2.351	2.441	2.455	2.469
G_{mm}	2.570	2.558	2.530	2.510
G_b	1.025	1.025	1.025	1.025
P_s	94.6	94.1	93.6	93.1
P_d	4.5	4.5	4.5	4.5
G_{sb}	2.688	2.688	2.688	2.688
h_{ini}	125	131	126	130
h_{des}	115	118	114	112

9.34 The Marshall method of mix design has been widely used by many highway agencies.

    a. What are the steps of Marshall mix design?

    b. What parameters are calculated?

    c. Show the typical graphs that are plotted after tests are completed.

    d. What is the purpose of plotting these graphs?

9.35 An asphalt concrete mixture is to be designed according to the Marshall procedure. A PG 58-28 asphalt cement with a specific gravity of 1.00 is to be used. A dense aggregate blend is to be used, with a maximum size of 19 mm and bulk specific gravity of 2.696. The theoretical maximum specific gravity of the mix is 2.470 at 5% asphalt content. Trial mixes were made, with the average results shown in Table P9.35. Using a spreadsheet program, plot the appropriate graphs necessary for the Marshall procedure, and select the optimum asphalt content using the Asphalt Institute design criteria for medium traffic. Assume a design air void content of 4% when using Table 9.16.

TABLE P9.35

Asphalt Content, % by Weight	Bulk Specific Gravity	Stability, N	Flow, 0.25 mm
4.0	2.303	7076	9
4.5	2.386	8411	10
5.0	2.412	7565	12
5.5	2.419	5963	15
6.0	2.421	4183	22

9.36 An asphalt concrete mixture is to be designed according to the Marshall procedure. A PG 64-22 asphalt cement with a specific gravity (G_b) of 1.00 is to be used. A dense aggregate blend is to be used, with a maximum aggregate size of 25 mm. and a bulk specific gravity (G_{sb}) of 2.786. The theoretical maximum specific gravity of the mix (G_{mm}), at asphalt content of 4.5%, is 2.490. Trial mixes were made with average results as shown in Table P9.36. Using a spreadsheet program, plot the appropriate six graphs necessary for the Marshall procedure and select the optimum asphalt content, using the Asphalt Institute design criteria for medium traffic (see Table 9.15). Assume a design air void content of 4% when using Table 9.16.

TABLE P9.36

Asphalt Content (P_b) (% by Weight)	Bulk Specific Gravity (G_{mb})	Stability, N	Flow, 0.25 mm
3.50	2.294	7117	8
4.00	2.396	8807	9
4.50	2.421	9475	11
5.00	2.416	7117	14
5.50	2.401	5694	20

9.37 The Marshall procedure was used to design an asphalt concrete mixture for a heavy-traffic road. Asphalt cement with a specific gravity of 1.025 is to be used. The mixture contains a 19 mm nominal maximum particle size aggregate, with bulk specific gravity of 2.654. The theoretical maximum specific gravity of the mix is 2.480 at 4.5% asphalt content. Trial mixes were made, with the average results shown in Table P9.37. Determine the optimum asphalt content using the Asphalt Institute design criteria (see Table 9.15). Assume a design air void content of 4% when using Table 9.16.

TABLE P9.37

Asphalt Content, % by Weight	Bulk Specific Gravity	Stability, kN	Flow, 0.25 mm
3.5	2.367	8.2	7
4.0	2.371	8.6	9
4.5	2.389	7.5	11
5.0	2.410	7.2	12
5.5	2.422	6.9	13

9.38 Describe the process for determining the stripping potential for an asphalt concrete mix.

9.39 Briefly discuss how the indirect tensile resilient modulus is determined in the lab.

9.40 Using the AASHTO T 283 test, three HMA specimens were freeze–thaw conditioned and three were unconditioned. All the specimens had a diameter of 150 mm. Table P9.40 shows the thickness and load at failure of each specimen. Determine:
a. The tensile strength of each specimen in MPa.
b. The average tensile strength of the conditioned and unconditioned specimens.
c. The tensile strength ratio (TSR).
d. Does the mix pass the Superpave criteria for moisture susceptibility?

TABLE P9.40

	Unconditioned Specimens			Conditioned Specimens		
Thickness, mm	64	63	63	63	64	62
Load at failure, kN	11.5	10.5	10.8	9.0	9.5	9.2

9.41 Name three methods of asphalt pavement recycling. Which one of them is the predominant method? Briefly summarize this method.

9.42 State four advantages of recycling asphalt pavement materials. Why can we not mix the RAP materials with aggregates in a conventional hot-mix asphalt concrete plant? Show the proper ways of recycling the RAP materials in the two types of hot-mix asphalt plants.

9.43 State four different asphalt modifiers that can be added to asphalt or asphalt mixtures, and indicate the effect of each.

9.44 What is the purpose of adding fly ash to asphalt concrete?

9.45 During construction of the asphalt concrete layer of asphalt pavement, cores were taken at random after compaction to detect any construction problem and to ensure that the density is within the specification limits. The target value was set at 2374 kg/m^3 and the upper and lower specification limits were set at 2385 kg/m^3 and 2363 kg/m^3, respectively. The density data shown in Table P9.45 were collected.
a. Using a spreadsheet program, create a control chart for these data showing the target value and the upper and lower specification limits. Are the density data within the specification limits?
b. Comment on any trend and possible reasons.

TABLE P9.45

Core No.	Density (kg/m³)	Core No.	Density (kg/m³)
1	2376	11	2368
2	2368	12	2372
3	2374	13	2360
4	2382	14	2366
5	2374	15	2360
6	2366	16	2355
7	2377	17	2356
8	2368	18	2350
9	2366	19	2353
10	2380	20	2352

9.46 Using online information, write a one-page, double-spaced article on how recycling asphalt pavement materials is useful in saving the environment and making material resources more sustainable. Use at least three web pages and cite them.

9.47 What is the most common polymer used for modifying asphalt cement? Briefly describe the interaction that occurs when asphalt and SBS are mixed.

9.48 List three advantages of warm mix asphalt technology.

9.49 Using Internet resources, write a paragraph on the use of porous asphalt pavements for stormwater management.

9.50 Using Internet resources, identify three technologies for producing warm mix and rank their relative sustainability with respect to energy and emissions.

9.16 References

Airey, G. D., *Rheological Properties of Styrene Butadiene Styrene Polymer Modified Road Bitumens,* Elsevier Science Ltd., 2003.

Anderson, R. M. *RAP Mix Design and Performance,* Presentation to APA Asphalt Pavement Conference, http://www.asphaltalliance.com/upload/RAP%20Mix%20Design%20&%20Performance%20-%20Mike%20Anderson.pdf, Nashville, TN, 2004.

The Asphalt Institute. *The Asphalt Handbook.* MS-4, 7th ed. Lexington, KY: The Asphalt Institute, 2007.

The Asphalt Institute. *Introduction to Asphalt.* MS-5. Lexington, KY: The Asphalt Institute, 2001.

The Asphalt Institute. *Mix Design Methods for Asphalt Concrete and Other Hot-Mix Types.* MS-2. Lexington, KY: The Asphalt Institute, 2001.

Astec Inc., Corporate literature, http://www.astecinc.com/index.php.

Chen, J. S., M. C. Liao, and M. S. Shiah, *Asphalt Modified by Styrene–Butadiene–Styrene Triblock Copolymer: Morphology and Model,* Journal of Materials in Civil Engineering, 2002.

Christensen, D. W, and R. F. Bonaquist, *Volumetric Requirements for Superpave Mix Design,* National Cooperative Highway Research Program Report 567, Washington DC, 2006.

Fernandes, M. R. S., M. M. C Forte, and L. F. M. Leite, *Rheological Evaluation of Polymer-Modified Asphalt Binders,* Materials Research, Vol. 11, No. 3. 381–386, 2008.

Goetz, W. H. and L. E. Wood. "Bituminous Materials and Mixtures." *Highway Engineering Handbook,* Section 18. New York: McGraw-Hill, 1960.

Huang, Y. H. *Pavement Analysis and Design.* 2nd ed. Upper Saddle River, NJ: Prentice Hall, 2004.

Hurley, G. C. and B. D. Prowell. Evaluation of Sasobit for Use in Warm Mix Asphalt, National Center for Asphalt Technology Report 05-06, Auburn AL, 2005.

Jansich, D. W., and F. S. Gaillard. *Minnesota Seal Coat Handbook.* Maplewood, MN: Department of Transportation, 1998.

Kluttz, R. Q. and R. Dongre. *Effect of STS Polymer Modification on the Low-Temperature Cracking of Asphalt Pavements,* Asphalt Science and Technology, editor A. M. Usmani, Marcel Dekker, Inc., NY, 1997.

Mamlouk, M. S. and J. P. Zaniewski. *Pavement Preventive Maintenance: Description, Effectiveness, and Treatments.* Special Technical Publication 1348. West Conshohocken, PA: American Society for Testing and Materials, 1998.

Miller, T. D. and H. U. Bahia. *Establishing a Framework for Analyzing Asphalt Pavement Sustainability,* International Journal of Pavement Research and Technology, Vol. 3, No. 3, 2010.

McGennis, R. B. et al. *Background of Superpave Asphalt Binder Test Methods.* Publication no. FHWA-SA-94-069. Washington, DC: Federal Highway Administration, 1994.

McGennis, R. B. et al. *Background of Superpave Asphalt Mixture Design and Analysis.* Publication no. FHWA-SA-95-003. Washington, DC: Federal Highway Administration, 1995.

NAPA, *Asphalt Pavements and the LEED Green Building System,* National Asphalt Pavement Association Brochure PS-32, http://www.asphaltroads.org/assets/_control/content/files/206086_Brochure_4.pdf, 2015.

Newcomb, D. E., E. R. Brown, and J. A. Epps. *A Practical Guide to the Design of HMA Mixtures with High Recycled Asphalt Pavement (RAP) Content,* National Asphalt Pavement Association, QIP-124, Lanham, MD, 2007.

Peterson, J. C. *Chemical Composition of Asphalt as Related to Asphalt Durability—State of the Art.* Transportation Research Record No. 999. Washington, DC: Transportation Research Board, 1984.

Prowell, B. D and E. R. Brown, Superpave Mix Design: Verifying Gradation Levels in the Ndesign Table, *National Cooperative Highway Research Program Report 573.* Washington, DC, 2007.

Roberts, F. L., P. S. Kandhal, E. R. Brown, D. Y. Lee, and T. W. Kennedy. *Hot Mix Asphalt Materials, Mixture Design, and Construction.* 3rd ed. Lanham, MD: NAPA Education Foundation, 2009.

Witczak, M. W., K. Kaloush, T. Pellinen, M. El-Basyouny, and H. Von Quintus. *Simple Performance Test for Superpave Mix Design.* NCHRP Report 465. Washington, DC: National Cooperative Highway Research Program, National Research Council, 2002.

CHAPTER

10

WOOD

Wood, because of its availability, relatively low cost, ease of use, and durability (if properly designed), continues to be an important civil engineering material. Wood is used extensively for buildings, bridges, utility poles, floors, roofs, trusses, and piles (see Figures 10.1 and 10.2). Civil engineering applications include both natural wood and engineered wood products, such as laminates, plywood, and strand board.

In order to use wood efficiently, it is important to understand its basic properties and limitations. In the United States, the Forest Service of the Department of Agriculture has broad management responsibility for the harvesting of wood from public lands and for assisting private sources with the selection of products for harvesting. This agency has produced the *Wood Handbook,* which is an excellent document describing the characteristics and properties of wood (USDA-FS, 2010).

This chapter covers the properties and characteristics of wood. In the design of a wood structure, joints and connections often limit the design elements. These are generally covered in a design class for wood construction and, therefore, are not considered in this text.

Wood is a natural, renewable product from trees. Biologically, a tree is a woody plant that attains a height of at least 6 m, normally has a single self-supporting trunk with no branches for about 1.5 m above the ground, and has a definite crown. There are over 600 species of trees in the United States.

Trees are classified as either endogenous or exogenous, based on the type of growth. Endogenous trees, such as palm trees, grow with intertwined fibers. Wood from endogenous trees is not generally used for engineering applications in the United States. Exogenous trees grow from the center out by adding concentric layers of wood around the central core. This book considers only exogenous trees.

Exogenous trees are broadly classified as deciduous and conifers, producing hardwoods and softwoods, respectively. The terms *hardwood* and *softwood* are classifications within the tree family, not a description of the woods' characteristics. In general, softwoods are softer, less dense, and easier to cut than hardwoods. However, exceptions exist such as balsa, a very soft and lightweight wood that is botanically a hardwood.

FIGURE 10.1   Wood frame used for building structural support.

FIGURE 10.2   Wooden roller coaster. (arnovdulmen/Fotolia)

Deciduous trees generally shed their leaves at the end of each growing season. Commercial hardwood production in the United States comes from 40 different tree species. Many hardwoods are used for furniture and decorative veneers, due to their pleasing grain pattern. The decorative properties of some hardwoods increase their value and cost. This makes them uneconomical for construction lumber. Again, there are some exceptions where some hardwoods are used engineered wood products that are used in construction applications.

Conifers, also known as evergreens, have needlelike leaves and normally do not shed them at the end of the growing season. Conifers grow continuously through the crown, producing a uniform stem and homogenous characteristics (Panshin & De Zeeuw, 1980). Softwood production in the United States comes from about 20 individual species of conifers. Conifers are widely used for construction. Conifers grow in large stands, permitting economical harvesting. They mature rapidly, making them a renewable resource. Table 10.1 shows examples of hardwood and softwood species (USDA-FS, 2010).

10.1   Structure of Wood

Wood has a distinct structure that affects its use as a construction material. Civil and construction engineers need to understand the way the tree grows and the anisotropic nature of wood in order to properly design and construct wood structures.

10.1.1   Growth Rings

The concentric layers in the stem of exogenous trees are called *growth rings* or *annual rings,* as shown in Figure 10.3a. The wood produced in one growing season constitutes a single growth ring. Each annual ring is composed of earlywood, produced by rapid growth during the spring, and latewood from summer growth. Latewood consists of dense, dark, and thick-walled cells producing a stronger structure than earlywood, as shown in Figure 10.3b. The predominant physical features of the tree stem include the *bark, cambium, wood,* and *pith,* as shown in Figure 10.4. The bark is the exterior covering of the tree and has an outer and an inner layer. The outer layer is dead and corky and has great variability in thickness, dependent on the species and age of the tree. The inner bark layer is the growth layer for bark but is not part of the wood section of the tree. The cambium is a thin layer of cells situated between the wood and the bark and is the location of all wood growth.

The wood section of the tree is composed of *sapwood* and *heartwood.* Heartwood is often darker colored and occurs at the center of the cross section and is surrounded by sapwood. Sapwood functions as a storehouse for starches and as a pipeline to transport sap. Generally, faster growing species have thick sapwood regions. In its natural state, sapwood is not durable when exposed to conditions that promote decay. Heartwood is not a living part of the tree. It is composed of cells that have been physically and chemically altered by mineral deposits. The heartwood

TABLE 10.1 Major Sources of Hardwood and Softwood Species by Region

Hardwoods			Softwoods		
Western	Northern and Appalachia	Southern	Western	Northern	Southern
Ash	Ash	Red alder	Incense cedar	Northern white cedar	Atlantic white cedar
Basswood	Aspen	Oregon ash	Port Orford cedar	Balsam fir	Bald cypress
American Beech	Basswood	Aspen	Douglas fir	Eastern hemlock	Fraser fir
Butternut	Buckeye	Black cottonwood	Western hemlock	Fraser fir	Southern pine
Cottonwood	Butternut	California black oak	Western larch	Jack pine	Eastern red cedar
Elm	American beech	Big leaf maple	Lodgepole pine	Eastern white pine	
Hackberry	Birch	Paper birch	Ponderosa pine	Eastern red cedar	
Pecan hickory	Black cherry	Tan oak	Sugar pine	Eastern spruces	
True hickory	American chestnut*		Western white pine	Tamarack	
Honey locust	Cottonwood		Western red cedar		
Black locust	Elm		Red wood		
Magnolia	Hackberry		Engelmann spruce		
Soft maple	True hickory		Sitka spruce		
Red oak	Honey locust		Yellow cedar		
White oak	Black locust				
Sassafras	Hard maple				
Sweetgum	Soft maple				
American Sycamore	Red oak				
Tupelo	White oak				
Black walnut	American sycamore				
Black willow	Black walnut				
Yellow poplar	Yellow poplar				

*Chestnut is no longer harvested, but lumber from salvage timbers is available.

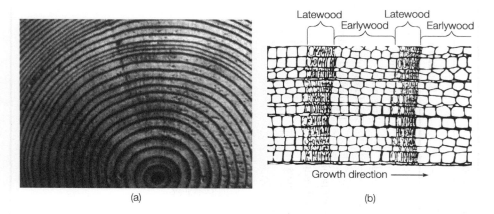

Latewood   Latewood
Earlywood   Earlywood

Growth direction ⟶

(a)   (b)

FIGURE 10.3   Cross section of a typical tree stem: (a) annual rings (Courtesy of American Forest & Paper Association, Washington, D.C.) and (b) earlywood and latewood.

provides structural strength for the tree. Also, the heartwood of some species is decay resistant due to the presence of extractives. Most of the species used in the United States do not have decay-resistant heartwood. A few species, such as redwood and cedar, are recognized as having durable heartwood. The *Wood Handbook* (USDA-FS, 2010) lists species based on the decay resistance of heartwood. The extractives present in some heartwood can also protect the wood from termites.

The pith is the central core of the tree. Its size varies with the tree species, ranging from barely distinguishable to large and conspicuous. The color ranges from blacks to whitish, depending on the tree species and locality. The pith structure can be solid, porous, chambered, or hollow.

10.1.2 ■ Anisotropic Nature of Wood

Wood is an anisotropic material in that it has different and unique properties in each direction. The three axis orientations in wood are longitudinal, or parallel to the grain; radial, or cross the growth rings; and tangential, or tangent to the growth rings, as illustrated in Figure 10.4. The anisotropic nature of wood affects physical and mechanical properties such as shrinkage, stiffness, and strength.

The anisotropic behavior of wood is the result of the tubular geometry of the wood cells. The wood cells have a rectangular cross section. The centers of the tubes are hollow, whereas the ends of the tubes are tapered. The length-to-width ratio can be as large as 100. The long dimension of the majority of cells is parallel to the tree's trunk. However, a few cells, in localized bundles, grow radially, from the center to the outside of the trunk. The preponderance of cell orientation in one direction gives wood its anisotropic characteristics. The hollow tube structure is very efficient in resisting compressive and tensile stresses parallel to its length but readily deforms when loaded on its side. Also, fluctuations in moisture contents expand or contract the tube walls but have little effect on the length of the tube.

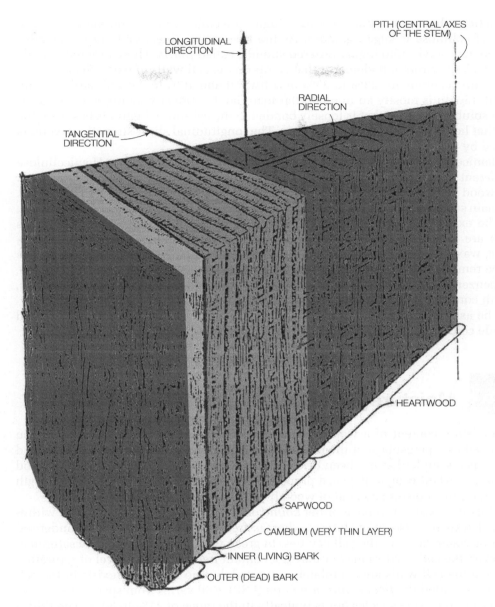

FIGURE 10.4   Main structural features of a typical tree stem.

10.2   Chemical Composition

Wood is composed of cellulose, lignin, hemicellulose, extractives, and ash-producing minerals. Cellulose accounts for approximately 50% of the wood substance by weight (USDA-FS, 2010). The exact percent is species dependent. It is a linear polymer (aliphatic carbon compound) having a high molecular weight. The main building block of cellulose is sugar: glucose. As the tree grows, linear cellulose molecules arrange themselves into highly ordered strands, called *fibrils.* These ordered strands form the large structural elements that compose the cell walls of wood fibers.

Lignin accounts for 23% to 33% of softwood and 16% to 25% of hardwood by weight. Lignin is mostly an intercellular material. Chemically, lignin is an intractable, insoluble, material that is loosely bonded to the cellulose. Lignin is basically the glue that holds the tubular cells together. The longitudinal shear strength of wood is limited by the strength of the lignin bounds.

Hemicelluloses are polymeric units made from sugar molecules. Hemicellulose is different from cellulose in that it has several sugars tied up in its cellular structure. Hardwood contains 20% to 30% hemicellulose and softwood averages 15% to 20%. The main sugar units in hardwood and softwood are xylose and monnose, respectively.

The extractives compose 5% to 30% of the wood substance. Included in this group are tannins and other polyphenolics, coloring matters, essential oils, fats, resins, waxes, gums, starches, and simple metabolic intermediates. These materials can be removed with simple inert neutral solvents, such as water, alcohol, acetone, and benzene. The amount contained in an individual tree depends on the species, growth conditions, and time of year the tree is harvested.

The ash-forming materials account for 0.1% to 3.0% of the wood material and include calcium, potassium, phosphate, and silica.

10.3 Moisture Content

The moisture content of a wood specimen is the weight of water in the specimen expressed as a percentage of the oven-dry weight of the wood. An oven-dried wood sample is a sample that has been dried in an oven at 100°C to 105°C until the wood attains a constant weight. Physical properties such as weight, shrinkage, and strength depend on the moisture content of wood (Panshin & De Zeeuw, 1980).

Moisture exists in wood as either *bound* or *free water.* Bound water is held within the cell wall by adsorption forces, whereas free water exists as either condensed water or water vapor in the cell cavities. In green wood, the cell walls are saturated. However, the cell cavities may or may not contain free water. The level of saturation at which the cell walls are completely saturated, but no free water exists in the cell cavities, is called the *fiber saturation point* (FSP). FSP varies from species to species and within the same species but is typically in the range of 21% to 32%. The FSP is of great practical significance because the addition or removal of moisture below the FSP has a large effect on practically all physical and mechanical properties of wood, whereas above the FSP, the properties are independent of moisture content.

When the moisture content of wood is above the fiber saturation point, the wood is dimensionally stable. However, moisture fluctuations below the FSP always result in dimensional changes. Shrinkage is caused by loss of moisture from the cell walls, and conversely, swelling is caused by the gain of moisture in the cell walls.

Figure 10.5 shows that the changes in wood dimensions vary from one direction to another. Dimensional changes in the radial direction are generally one-half the change in the tangential direction. Swelling and shrinkage in the longitudinal direction is minimal, typically 0.1% to 0.2% for a change in the moisture content from the FSP to oven dry. This anisotropy of dimensional changes of wood causes warping, checking, splitting, and structural performance problems, as discussed in more detail later in the chapter. It is also the reason that the sawing pattern of boards affects the amount of distortion when subjected to changes in moisture. Note that the shrinkage values shown in Figure 10.5 are related to the green dimensions at or above the FSP.

For engineering calculations, the FSP is commonly assumed to be 30%. This is easier to remember and more useful than expecting the engineer to find and use species-specific values. The same applies for shrinkage calculations. Tangential shrinkage can be assumed to be 8% and radial shrinkage 4%, relative to the green dimensions. For structural lumber where growth ring or orientation is not known in advance and can be reasonably expected to vary from piece to piece, a value of 6% is commonly chosen for shrinkage from an FSP (30%) to oven-dry (0%). Thus, shrinkage could be estimated as 6% shrinkage in 30% change in moisture content, or 1% shrinkage per 5% change in moisture content, relative to the green dimensions. This is the rule of thumb recommended by the American Institute of Timber Construction (AITC) for calculating shrinkage of a glued laminated (glulam) beam, which consists of several laminations of varying growth ring orientations.

The moisture content in wood varies depending on air temperature and humidity. However, the natural change of moisture content is a slow process, so as atmospheric conditions change, the moisture content in wood tends to adjust to conditions near the average. The moisture content for the average atmospheric conditions is the *equilibrium moisture content* (EMC). The wood handbook includes values of the EMC as a function of temperature and humidity. The EMC ranges from less than 1%, at temperatures greater than 55°C and 5% humidity, to over 20% at temperatures less than 27°C and 90% humidity.

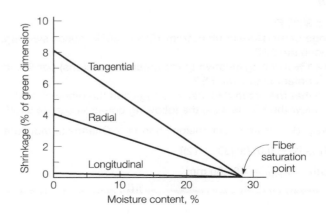

FIGURE 10.5   Relationship between shrinkage and moisture content.

Sample Problem 10.1

A 250-mm wide red spruce plank is cut at the mill at a moisture content of 50% in such a way that the width is in the tangential direction of annual rings. What is the new width as the moisture content changes to 15%? The fiber saturation point of red spruce is 27%, and it shrinks 7.8% (of the green dimension) in drying from FSP to oven dry in the tangential direction.

Solution

Shrinkage does not occur above FSP. Decreasing the moisture content from 27% to 15% causes shrinkage. From the data given, red spruce shrinks 7.8% for a 27% change in moisture content (from FSP to zero moisture content).
Rate of shrinkage $= 7.8/27 = 0.289\%$ length for each 1% reduction in moisture content.
Note that this shrinkage rate is related to the green dimension. Assuming that the green dimension is X_{FSP} and the dimension at any moisture content below the FSP is X_{MC}, the following equation can be used.

$X_{MC} = X_{FSP} (1 - $ Rate of shrinkage $\times$ Change in moisture content$)$

$X_{15} = 250 [1 - 0.00289 (27 - 15)] = 241.33$ mm

Sample Problem 10.2

A glulam beam is manufactured with a depth of 450 mm and a 12% moisture content. Assuming common values for estimating dimensional changes, compute the new depth of the beam if the moisture content is increased to 40%.

Solution

Assume a 30% FSP.
The change of moisture content from 12% to 30% causes swelling. Swell does not occur above the FSP.
Assume a 1% swelling (relative to the green dimensions) per 5% increase in moisture content below the FSP.
Assuming that the green dimension is X_{FSP} and the dimension at any moisture content below the FSP is X_{MC}, the following equation can be used.

$X_{MC} = X_{FSP} (1 - $ Rate of dimension change $\times$ Change in moisture content$)$

$450 = X_{FSP} [1 - 0.01/5*(30 - 12)]$

New depth $= X_{FSP} = 466.80$ mm

10.4 Wood Production

A vast industry has developed to harvest and process wood. Wood is harvested from forests as *logs*. They are transported to saw mills, where they are cut into dimensional shapes to produce a variety of products for engineering applications:

1. *Dimension lumber* is wood from 50 mm to 125 mm thick, sawn on all four sides. Common shapes include 50 × 100, 50 × 150, 50 × 200, 50 × 250, 50 × 300, 100 × 100 and 100 × 150.[1] These sizes refer to the rough-sawn dimensions of the lumber in inches. The rough-sawn lumber is surfaced to produce smooth surfaces; this removes 6.5 mm per side. For dimensions 200 mm and larger, 9.5 mm is removed per side. For example, the actual dimensions of a 50 × 100 are 37 mm by 87 mm Dimension lumber is produced in lengths of 2.4 m to 7.2 m in 0.6 m increments. Dimension lumber is typically used for studs, sill and top plates, joists, beams, rafters, trusses, and decking.

2. *Heavy timber* is wood sawn on all four sides; common shapes include 150 × 150 and 200 × 200 and larger. Heavy timber includes *Beams* and *Stringers* (subjected to bending) and *Posts* and *Timbers* (used as posts or columns). As with the case of dimension lumber, these sizes specify rough-sawn dimensions in inches. Heavy timbers are used for heavy frame construction, landscaping, railroad ties, and marine construction.

3. *Round stock* consists of posts and poles used for building poles, marine piling, and utility poles.

4. *Engineered wood* consists of products manufactured by bonding together wood strands, veneers, lumber, and other forms of wood fiber to produce a larger and

FIGURE 10.6 Frames made of structural glued laminated timber (glulam). (Courtesy of the American Institute of Timber Construction)

[1]The current standards for dimension lumber and heavy timber standards were implemented in 1970. When remodeling older structures, the dimension of the existing lumber must be measured.

FIGURE 10.7   An arch bridge made of structural glued laminated timber (glulam). (Courtesy of the American Institute of Timber Construction)

integral composite unit (Figures 10.6 and 10.7). These products are engineered and tested to have specific mechanical responses to loads. Structural engineered wood products include the following:

- structural panels including plywood, oriented strand board, and composite panels,
- structural glued laminated timber (glulam),
- structural composite lumber, and
- composite structural members.

5. Specialty items are milled and fabricated products to reduce on-site construction time, includes lattice, handrails, spindles, radius edge decking, and turned posts.

10.4.1 ▦ Cutting Techniques

Sawn wood production includes the following steps:

- Sawing into desired shape
- Seasoning
- Surfacing
- Grading
- Preservative treatment (optional)

Surfacing (planing) of the wood surface, to produce a smooth face, can be done before or after drying. Post drying surfacing is superior, because it removes small defects developed during the drying process. When surfacing is done before seasoning, the dimensions are slightly increased to compensate for shrinkage during seasoning.

The harvested wood is cut into lumber and timber at saw mills using circular saws, band saws, or frame saws. The most common patterns for sawing a log are *live (plain) sawing, quarter sawing,* and *combination sawing,* as shown in Figure 10.8 (Levin, 1972).

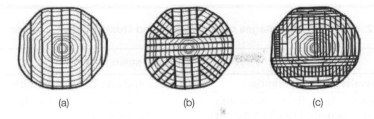

FIGURE 10.8 Common log sawing patterns: (a) live (plain) sawing, (b) quarter sawing, and (c) combination sawing.

The angle the growth rings make with the face of the board (i.e., the angle between the growth ring and the saw blade) produces three categories of board cuts, as illustrated in Figure 10.9:

1. Flat-sawn, 45° or less
2. Rift-sawn, 45° to 80°
3. Quarter-sawn (vertical- or edge-sawn), 80° to 90°

These cut types affect the surface appearance of the board but do not largely affect the structural integrity of the board.

The sawing pattern selected depends on the cross section of the tree, the capability of the mill, and the desired product. Plain sawing is rapid and economic, whereas quarter sawing maximizes the amount of vertical-sawn cuts. Some of the advantages of the different sawing patterns are summarized in Table 10.2 (USDA-FS, 2010).

10.4.2 ▨ Seasoning

Green wood, in living trees, contains from 30% to 200% moisture by the oven-dry weight. Seasoning removes the excess moisture from wood. For structural wood, the recommended moisture content varies from 7% in the dry southwestern states to 14% in the damp coastal regions. However, as it leaves the mill, framing lumber typically has an average moisture content of 15%.

Wood is seasoned by air and kiln drying. Air drying is inexpensive but slow. The green lumber is stacked in covered piles to dry. These piles of lumber are made of successive layers of boards separated by 25 mm strips so that air can flow between the layers. The time required for drying varies with the climate and temperature of the area. Normally, three to four months is the maximum air drying time used in the United States. Air drying is complete when the moisture content of the wood is in equilibrium with the air humidity. The optimum moisture may not be achievable through air drying alone.

FIGURE 10.9 Types of board cut: (a) flat sawn, (b) rift sawn, and (c) quarter-sawn (vertical- or edge-sawn).

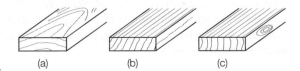

TABLE 10.2   Relative Advantages of Plain-Sawn and Quarter-Sawn Lumber

Plain-sawn	Quarter-sawn
Shrinks and swells less in thickness	Shrinks and swells less in width
Surface appearance less affected by round or oval knots, compared to effect of spike knots in quarter-sawn boards; boards with round or oval knots not as weak as boards with spike knots	Cups, surface checks, and splits less in seasoning and in use
Shakes and pitch pockets, when present, extend through fewer boards	Raised grain caused by separation in annual rings does not become as pronounced
Figure patterns resulting from annual rings and some other types of figure brought out more conspicuously	Figure patterns resulting from pronounced rays, interlocked grain, and wavy grain are brought out more conspicuously
Less susceptible to collapse in drying	Does not allow liquids to pass through readily in some species
Costs less because it is easy to obtain	Holds paint better in some species
	Sapwood appears in boards at edges and its width is limited by the width of the log

After air drying, the lumber may be kiln dried. A kiln is a large oven where all variables can be closely monitored. Drying temperatures in a kiln range from 20°C to 50°C, typically requiring 4 to 10 days. Care must be taken to slowly reduce the moisture content of wood. Drying too rapidly can result in an increase in cracking and warping. Dried lumber will take on moisture again if exposed to water; therefore, care must be used when storing and transporting wood. Structural lumber is not typically air-dried prior to kiln drying.

Since wood shrinks at different rates in the different directions (see Figure 10.5), the combined effects of radial and tangential shrinkage can distort the shape of wood pieces. Figure 10.10 shows types of distortion that could happen in flat, square and round pieces after seasoning since tangential shrinkage is about twice as great as radial.

10.5   Lumber Grades

The final step in lumber production involves grading the lumber according to quality. Typically, lumber is graded according to the characteristics that affect strength, durability, or workability. The most common grade-reducing qualities of lumber are knots, checks, pitch pockets, shakes, and stains. Due to the high degree of natural variability within lumber, it is nearly impossible to develop an exact, uniform set of grading standards. As a result, grading techniques and standards can, and do, vary among organizations.

FIGURE 10.10   Distortion of flat, square, and round pieces as affected by direction of growth rings after seasoning. (USDA-FS, 2010)

The American Lumber Standards Committee (ALSC) in Germantown, MD, certifies or accredits lumber grading agencies and rules writing agencies. The ALSC lumber program currently has 30 accredited independent third-party agencies headquartered throughout the United States and Canada. The accredited agencies in the lumber program operate under one or more of the sets of grading rules published by the following seven rule writing agencies:

- Northeastern Lumber Manufacturers Association (NELMA), Cumberland Center, Maine
- Northern Softwood Lumber Bureau (NSLB), Cumberland Center, Maine
- Redwood Inspection Service (RIS), Novato, California
- Southern Pine Inspection Bureau (SPIB), Pensacola, Florida
- West Coast Lumber Inspection Bureau (WCLIB), Tigard, Oregon
- Western Wood Products Association (WWPA), Portland, Oregon
- National Lumber Grades Authority (NLGA), New Westminster, BC, Canada

Each of these agencies writes standards and specifications for particular species or combinations of species that are produced within their operating region. For example, the SPIB provides the grading rules for all species of Southern Pine, the WWPA governs grading of most western species, and the NLGA provides grading rules for all grades produced in Canada.

10.5.1 ▦ Hardwood Grades

Hardwood can be graded for either appearance or structural applications. For appearance purpose, the National Hardwood Lumber Association bases the grading of hardwood on the amount of usable lumber in each piece of standard length lumber. The inspection is performed on the poorest side of the material and the grade is based on the number and size of clear "cuttings" that can be produced from a given piece of lumber. Cuttings must have one face clear of strength-reducing imperfections and the other side must be sound. Based on "cutting" quantity, the wood is given a classification of *Firsts, Seconds, Selects,* and *Common* (No. 1, No. 2, No. 3A, or 3B), with Firsts being the best. Frequently, Firsts and Seconds are grouped into one grade of Firsts and Seconds (FAS).

For civil engineering applications, appearance grades are less important than structural grades. Other grading systems exist and may use some of the same terminologies; that is, No. 1 may be used to describe structural grade or an appearance grade in different products.

10.5.2 ▨ Softwood Grades

Softwood for structural applications is graded by visual inspection and is machine stress graded. The purpose of grading is to ensure that all lumber graded to a specific grade designation has at least the minimum mechanical or load-carrying capability with respect to critical design parameters. Under machine stress grading, each piece of wood is subjected to X-ray to measure density, which is correlated to strength and stiffness. Alternatively, the wood piece is subjected to a bending stress and is graded based on the mechanical response, as shown in Table 10.3. The grade designation identifies the minimum extreme fiber bending stress, tensile stress parallel to grain, compressive stress parallel to grain, and modulus of elasticity of the wood.

For visual classification, the basic mechanical properties of the wood are determined by testing small clear wood specimens. These results are then adjusted for allowable defects and characteristics for each class of wood. Unlike stress graded lumber, visual stress grade properties are defined for each species of softwood. Table 10.4 is an example of design values for grades of Eastern White Pine. These design values are only for dimension lumber. Different design values apply for beams and stringers or posts and timbers with sizes of 125 mm and larger.

Visual grade designations include *Yard, Structural,* or *Factory and Shop,* with subgrades of Select, Select B, Select C, and No. 1, No. 2, and No. 3 commons, appearance, and studs. Other commonly used ratings are *Construction, Standard,* or *Utility* and combinations such as No. 2&BTR (Number 2 and Better) and STD&BTR. Not all grades are used for all species of wood.

Yard lumber frequently refers to some specialty stress grades of lumber such as those used for light structural framing. Structural lumber typically comprises pieces 50 to 125 mm thick and is graded according to its intended use. Grading categories include light framing, joists and planks, beams and stringers, and posts and timbers. Factory and shop lumber includes siding, flooring, casing, shingles, shakes, and finish lumber.

TABLE 10.3 Sample of Design Values for Mechanically Graded Dimension Lumber of Softwood for Structural Applications[1,4]

Grade Designation	Design Values,[2] MPa				
	Bending[3]	Tension Parallel to Grain	Compression Parallel to Grain	Modulus of Elasticity	Minimum Modulus of Elasticity
900f-1.0E	6.21	2.4	7.24	6895	3516
1650f-1.3E	11.38	7.0	11.7	8963	4551
1950f-1.5E	13.44	9.5	12.4	10342	5240
2250f-1.7E	15.51	12.1	13.3	11721	5929
2400f-2.0E	16.55	13.3	13.6	13790	7033
2850f-2.3E	19.65	15.9	14.8	15858	8067
3000f-2.4E	20.68	16.5	15.2	16547	8412

[1] Courtesy of American Wood Council, Washington, D.C.

[2] Stresses apply to lumber used at 19% maximum moisture content. When lumber is designed for use where the moisture content will exceed 19% for an extended period of time, the values shown herein shall be multiplied by certain *wet service* factors.

[3] Bending values are applicable to lumber loaded on edge. When loaded flatwise, these values may be increased by multiplying by certain *flat use* factors.

[4] For a complete list of grade designations and more detailed design values see reference (American Wood Council, 2012).

10.6 Defects in Lumber

Lumber may include defects that affect its appearance, its mechanical properties, or both. These defects can have many causes, such as natural growth of the wood, wood diseases, animal parasites, too rapid seasoning, or faulty processing. Common defect types are shown in Figure 10.11.

Knots are branch bases that have become incorporated into the wood of the tree trunk or another limb. Knots degrade the mechanical properties of lumber, affecting the tensile and flexural strengths.

Shakes are lengthwise separations in the wood occurring between annual rings. They develop prior to cutting the lumber and could be due to heavy winds.

Wane is bark or other soft material left on the edge of the board or absence of material.

Sap Streak is a heavy accumulation of sap in the fibers of the wood, which produces a distinctive streak in color.

TABLE 10.4 Example of Design Values for Visually Graded Dimension Lumber (50 mm – 100 mm thick) of Eastern White Pine[1,4]

Grade Designation	Size Classification	Design Values,[2] MPa						
		Bending[3]	Tension Parallel to Grain	Shear Parallel to Grain	Compression Perpendicular to Grain	Compression Parallel to Grain	Modulus of Elasticity	Minimum Modulus of Elasticity
Select Structural		8.60	3.96	0.93	2.41	8.27	8274	3034
No. 1	50 mm & wider	5.34	2.41	0.93	2.41	6.90	7584	3034
No. 2		3.96	1.90	0.93	2.41	5.69	7584	3034
No. 3	50 mm & wider	2.41	1.03	0.93	2.41	3.28	6205	2275
Stud		3.10	1.38	0.93	2.41	3.62	6205	2275
Construction		4.65	2.07	0.93	2.41	7.24	6895	2551
Standard	50 mm–100 mm wide	2.59	1.21	0.93	2.41	5.86	6205	2275
Utility		1.21	0.52	0.93	2.41	3.79	5516	2000

[1] Courtesy of American Wood Council, Washington, D.C.

[2] Stresses apply to lumber used at 19% maximum moisture content. When lumber is designed for use where the moisture content will exceed 19% for an extended period of time, the values shown herein shall be multiplied by certain *wet service* factors.

[3] Bending values are applicable to lumber loaded on edge. When loaded flatwise, these values may be increased by multiplying by certain *flat use* factors.

[4] For a complete list of grade designations and more detailed design values see reference (American Wood Council, 2012).

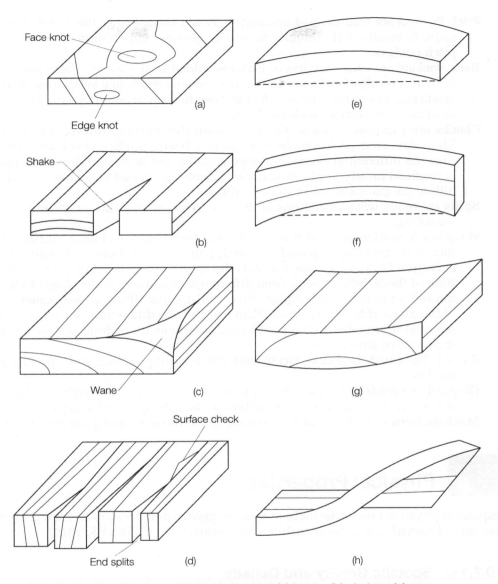

FIGURE 10.11   Common defects in lumber: (a) knots, (b) shakes, (c) wanes, (d) checks and splits, (e) bowing, (f) crooking, (g) cupping, and (h) twisting.

Reaction Wood is abnormally woody tissue that forms in crooked stems or limbs. Reaction wood causes the pith to be off center from the neutral axis of the tree. It creates internal stresses which can cause warping and longitudinal cracking.

Pitch Pockets are well-defined openings between annual rings that contain free resin. Normally, only Douglas fir, pines, spruces, and western larches have pitch pockets.

Bark Pockets are small patches of bark embedded in the wood. These pockets form as a result of an injury to the tree, causing death to a small area of the cambium. The surrounding tree continues to grow, eventually covering the dead area with a new cambium layer.

Checks are ruptures in wood along the grain that develop during seasoning. They can occur on the surface or end of a board. Surface checking results from the differential shrinkage between radial and tangential directions and is confined mostly to planer surfaces. Cracks due to end checking normally follow the grain and result in end splitting.

Splits are lengthwise separations of the wood caused by either mishandling or seasoning.

Warp is a distortion of wood from the desired true plane (see Figure 10.10). The four major types of warp are *bow, crook, cup,* and *twist.* Bow is a longitudinal curvature from end to end. Crook is the longitudinal curvature side to side. Both of these defects result from differential longitudinal shrinkage. Cup is the rolling of both edges up or down. Twist is the lifting of one corner out of the plane of the other three. Warp results from differential shrinkage, differential drying due to the production environment, or from the release of internal tree stress.

Raised, Loosened, or Fuzzy Grain may occur during cutting and dressing of lumber.

Chipped or Torn Grain occurs when pieces of wood are scooped out of the board surface or chipped away by the action of the cutting and planing tools.

Machine Burn is an area that has been darkened by overheating during cutting.

10.7 Physical Properties

Important physical properties include specific gravity and density, thermal properties, and electrical properties (Panshin & De Zeeuw, 1980; USDA-FS, 2010).

10.7.1 Specific Gravity and Density

Specific gravity of wood depends on cell size, cell wall thickness, and number and types of cells. Regardless of species, the substance composing the cell walls has a specific gravity of 1.5. Because of this consistency, specific gravity is an excellent index for the amount of substance a dry piece of wood actually contains. Therefore, specific gravity, or density, is a commonly cited property and is an indicator of mechanical properties within a clear, straight-grained wood.

The dry density of wood ranges from 160 kg/m³ for balsa to 1335 kg/m³ for black ironwood. The majority of wood types have densities in the range of 300 to 700 kg/m³.

10.7.2 ■ Thermal Properties

Thermal conductivity, specific heat, thermal diffusivity, and coefficient of thermal expansion are the four significant thermal properties of wood.

Thermal Conductivity Thermal conductivity is a measure of the rate at which heat flows through a material. The reciprocal of thermal conductivity is the thermal resistance (insulating) value (R). Wood has a thermal conductivity that is a fraction of that of most metals and three to four times greater than common insulating materials. The thermal conductivity ranges from 0.06 W/(mK) for balsa to 0.17 W/(mK) for rock elm. Structural woods average 0.12 W/(mK) as compared to 200 W/(mK) for aluminum and 0.04 W/(mK) for wool. The thermal conductivity of wood depends on several items including (1) grain orientation, (2) moisture content, (3) specific gravity, (4) extractive content, and (5) structural irregularities such as knots.

Heat flow in wood across the radial and tangential directions (with respect to the growth rings) is nearly uniform. However, heat flow through wood in the longitudinal direction (parallel to the grain) is 2.0 to 2.8 times greater than in the radial direction.

Moisture content has a strong influence on thermal conductivity. When the wood is dry, the cells are filled with air and the thermal conductivity is very low. As the moisture content increases, thermal conductivity increases. As the moisture content increases from 0% to 40%, the thermal conductivity increases by about 30%.

Because of the solid cell wall material in heavy woods, they conduct heat faster than light woods. This relationship between specific gravity and thermal conductivity for wood is linear. Also affecting the heat transfer in wood are increases in extractive content and density (i.e., knots), which increase thermal conductivity.

Specific Heat Specific heat of a material is the ratio of the quantity of heat required to raise the temperature of the material one degree to that required to raise the temperature of an equal mass of water one degree. Temperature and moisture content largely control the specific heat of wood, with species and density having little to no effect. When wood contains water, the specific heat is increased because the specific heat of water is higher than that of dry wood. However, the value of specific heat for the wet wood is higher than just the sum of the specific heats for the wood and water combined. This increase in specific heat beyond the simple sum is due to the wood–water bonds absorbing energy. An increase in temperature increases the energy absorption of wood and results in an increase in the specific heat.

Thermal Diffusivity Thermal diffusivity is a measure of the rate at which a material absorbs heat from its surroundings. The thermal diffusivity for wood is much smaller than that of other common building materials. Generally, wood has a thermal diffusivity value averaging 0.006 mm/s, compared with steel, which has a thermal diffusivity of 0.5 mm/s. It is because of the low thermal diffusivity that wood does not feel hot or cold to the touch, compared with other materials. The small thermal conductivity, moderate density, and moderate specific heat contribute to the low value of thermal diffusivity in wood.

Coefficient of Thermal Expansion The coefficient of thermal expansion is a measure of dimensional changes caused by a temperature variance. Thermal expansion coefficients for completely dry wood are positive in all directions. For both hard and soft woods, the longitudinal (parallel to the grain) coefficient values range from 0.009 to 0.0014 mm/m/°C. The expansion coefficients are proportional to density and therefore are five to ten times greater across the grain than those parallel to it.

When moist wood is heated, it expands due to thermal expansion and then shrinks because of the loss of moisture (below the fiber saturation point). This combined swelling and shrinking often results in a net shrinkage. Most woods, at normal moisture levels, react in this way.

10.7.3 ▦ Electrical Properties

Air-dry wood is a good electrical insulator. As the moisture content of the wood increases, the resistivity decreases by a factor of three for each 1% change in moisture content. However, when wood reaches the fiber saturation point, it takes on the resistivity of water alone.

10.8 Mechanical Properties

Knowing the mechanical properties of wood is a prerequisite to a proper design of a wood structure. Typical mechanical properties of interest to civil and construction engineers include modulus of elasticity, strength properties, creep, and damping capacity.

10.8.1 ▦ Modulus of Elasticity

The typical stress–strain relationship of wood is linear up to a certain limit, followed by a small nonlinear curve after which failure occurs, as shown in Figure 10.12. The modulus of elasticity of wood is the slope of the linear portion of the representative

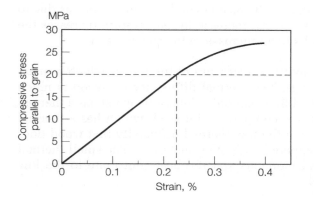

FIGURE 10.12 Typical stress–strain relationship for wood.

stress–strain curve. The stress–strain relationship of wood varies within and between species and is affected by variation in moisture content and specific gravity. Also, since wood is anisotropic, different stress–strain relationships exist for different directions. The moduli of elasticity along the longitudinal, radial, and tangential axes are typically different.

10.8.2 ▒ Strength Properties

Strength properties of wood vary to a large extent, depending on the orientation of grain relative to the direction of force. For example, the tensile strength in the longitudinal direction (parallel to grain) is more than 20 times the tensile strength in the radial direction (perpendicular to grain). Also, tensile strength parallel to grain is larger than the compressive strength in the same direction for small, clear specimens. In full size structural members, the tensile strength controls the capacity of most pieces. Also, tension is more sensitive to strength-reducing characteristics such as knots and slope of grain. Common strength properties for wood include modulus of rupture in bending, compressive strength parallel and perpendicular to the grain, tensile strength parallel to the grain, and shear strength parallel to the grain. Some of the less common strength properties are torsion, toughness, and fatigue strength.

10.8.3 ▒ Load Duration

Wood can support higher loads of short duration than sustained loads. Under sustained loads, wood continues to deform. The design values of material properties contemplate fully stressing the member to the tabulated design values for a period of 10 years and/or the application of 90% of the full maximum load continuously throughout the life of the structure. If the maximum stress levels are exceeded, the structure can deform prematurely.

Generally, a load duration of 10 years is used for design. Typical load duration factors that would be applied to the values in Tables 10.3 and 10.4 (except for compression perpendicular to the grain) are shown in Figure 10.13. For example, if a floor joint is designed for a temporary stage that is used only for a one-day presentation, the allowable bending fiber stress can be increased by 33% over the allowable stress for a normal application.

10.8.4 ▒ Damping Capacity

Damping is the phenomenon in which the amplitude of vibration in a material decreases with time. Reduction in amplitude is due to internal friction within the material and resistance of the support system. Moisture content and temperature largely govern the internal friction in wood. At normal ambient temperatures, an increase in moisture content produces a proportional increase in internal friction up to the fiber-saturation point. Under normal conditions of temperature and moisture content, the internal friction in wood (parallel to the grain) is 10 times that of structural metals. Because of these qualities, wood structures dampen vibrations more quickly than metal structures of similar design.

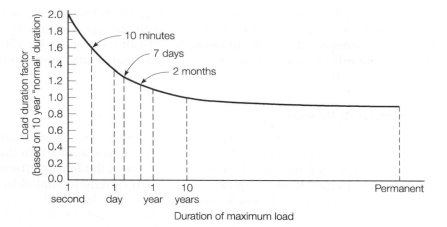

FIGURE 10.13   Adjustment factors for load duration for working-stress wood design based on 10-year normal duration. Reprinted with permission of American Wood Council, Leesburg, VA (2015)

10.9 Testing to Determine Mechanical Properties

Standard mechanical testing methods for wood are designed almost exclusively to obtain data for predicting performance. To achieve reproducibility in the testing environment, specifications include methods of material selection and preparation, testing equipment and techniques, and computational methods for data reduction. Standards for testing wood and wood composites are published by ASTM, the U.S. Department of Commerce, the National Standard Institute (NSI), and various other trade associations, such as the Western Woods Product Association.

Due to the many variables affecting the test results, it is of primary importance to correctly select the specimen and type of test. There are two main testing techniques for establishing strength parameters: the testing of timbers of structural sizes and the testing of representative, small, clear specimens.

Testing of structural-size timbers provides relationships among mechanical and physical properties, working stress data, correlations between environmental conditions, wood imperfections, and mechanical properties. The primary purposes for testing small, clear specimens are to obtain the mechanical properties of various species and provide a means of control and comparison in production activities. Testing of structural-size members is more important than testing small, clear specimens since the design values are more applicable to the actual size members. ASTM D198 presents the testing standards for full-size tests, whereas ASTM D143 presents the testing standards for small, clear wood specimens.

Tests for structural-size lumber include (ASTM D198):

- Flexure (bending)
- Compression (short column)
- Compression (long column)
- Tension

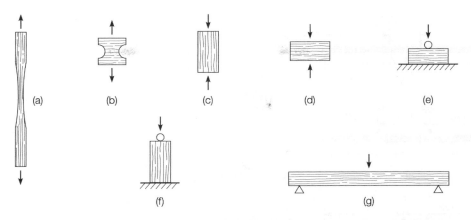

FIGURE 10.14   Test specimens of wood: (a) tension parallel to grain, (b) tension perpendicular to grain, (c) compression parallel to grain, (d) compression perpendicular to grain, (e) hardness perpendicular to grain, (f) hardness parallel to grain, and (g) bending. (© Pearson Education, Inc. Used by permission)

Tests for small-clear wood specimens include (ASTM D143):

- Static bending (flexure)
- Impact bending
- Compression parallel to the grain
- Compression perpendicular to the grain
- Tension parallel to the grain
- Tension perpendicular to the grain

Figure 10.14 shows a schematic of test specimens of wood tested in tension, compression, bending, and hardness.

For both full-size lumber and small, clear specimens, the bending test is more commonly used than the other tests.

10.9.1 ▨ Flexure Test of Structural Members (ASTM D198)

The flexure test is performed on members in sizes that are usually used in structural applications. The test specimen may be solid wood such as 50 mm × 100 mm, laminated wood, or a composite construction of wood or of wood combined with plastics or metals. The test method is intended primarily for beams of rectangular cross section but is also applicable to beams of round and irregular shapes, such as round posts, I-beams, or other special sections. The member is supported near its ends and is subjected to two-point, third-point, or center-point loading as shown in Figure 10.15, depending on the purpose of the test. Figure 10.16 illustrates a two-point loading flexure test on a glulam structural member.

The span of the beam varies depending on the member size. Specimens with a depth-to-width ratio of three or greater require lateral support to avoid instability. Loading is continued until complete failure or an arbitrary terminal load has been

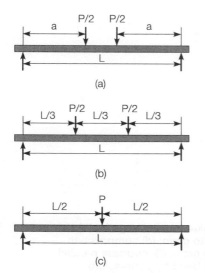

(a)

(b)

(c)

FIGURE 10.15 Methods of loading a wood beam: (a) two-point loading, (b) third-point loading, and (c) center-point loading.

FIGURE 10.16 Two-point loading flexure test on a glulam structural member. (Courtesy of the American Institute of Timber Construction)

reached. The load and the deflection of the neutral axis of the beam at the center of the span are recorded during the test. From the load-deflection data, the following data are noted: load and deflection at first failure, at maximum load, and at points of sudden change. Properties that can be obtained from the test include fiber stress at proportional limit, modulus of rupture, modulus of elasticity, and maximum shear stress.

The results of this test can be used in several applications, such as the development of grading rules and specifications, development of working stresses, and checking the influence of imperfections on the mechanical properties of structural members.

Sample Problem 10.3

A third-point bending test was performed on a 100 × 150 mm wood lumber according to ASTM D198 procedure with a span of 1.8 m and the 150 mm side is positioned vertically. A strain gauge is glued at the center point between the supports. The strain gauge is zeroed at the start of the test when the force is zero. A strain of 0.00456 is measured when the combined load is 6.672 kN and the load–strain curve is in the linear range. Determine the flexural modulus of elasticity. If the maximum combined load on both loading bearings was 22.722 kN, calculate the modulus of rupture.

Solution
The actual cross section dimensions are 87 mm × 137 mm

To determine modulus:

Reaction at each support = 6672/2 = 3336 N

Bending moment centered between the loads

M = 3336 × 0.6 = 2002 N·m

c = 1/2 of the specimen height = 68.5 mm

$$I = \frac{137^3 \times 87}{12} = 18.6 \times 10^6 \text{ mm}^4$$

Stress centered between the loads

σ = Mc/I = 2002 × 68.5/(18.6 × 10^6) = 0.007373 GPa = 7.373 MPa

$E = \sigma/\varepsilon$ = 7.373/0.00456 = 1617 MPa = 1.617 GPa

To determine modulus of rupture:

$$\text{Reaction at each support at failure} = \frac{22722}{2} = 11361 \text{ N}$$

Bending moment at the center third of span at

failure = 11361 × 0.6 = 6817 N·m

$$\text{Modulus of rupture} = \frac{Mc}{I} = \frac{6817 \times 68.5}{18.6 \times 10^6} = 0.02511 \text{ GPa} = 25.11 \text{ MPa}$$

10.9.2 ▓ Flexure Test of Small, Clear Specimen (ASTM D143)

The test is performed on either 50 × 50 × 760 mm or 25 × 25 × 410 mm specimens. The specimen is supported near its ends and is subjected to center-point loading. For large specimens, a span of 710 mm is used and the load is applied at a rate of 2.5 mm/min. For the small specimens, a span of 360 mm is used and the load is applied at a rate of 1.3 mm/min Load–deflection data are recorded to or beyond the maximum load. Within the proportional limit, readings are taken to the nearest 0.02 mm. After the proportional limit, deflection readings are usually measured with a dial gauge, to the limit of the gauge, usually 25 mm. Load and deflection of the first failure, the maximum load, and points of sudden change are recorded. The failure appearance is described as either brash or fibrous. Brash indicates an abrupt failure and fibrous indicates a failure showing splinters. The modulus of rupture is calculated from the results of the test.

Sample Problem 10.4

A static center-point bending test was performed on a 50 × 50 × 760 mm wood sample according to ASTM D143 procedure (span between supports = 710 mm) If the maximum load was 2.67 kN, calculate the modulus of rupture.

Solution

$$\text{Modulus of rupture} = \frac{Mc}{I}$$

where

M = bending moment at maximum load

c = 1/2 of the specimen height

I = moment of inertia of the specimen cross section

$$\text{Reaction at each support at failure} = \frac{2.67}{2} = 1.335 \text{ kN}$$

$$\text{Bending moment at the center at failure} = 1.335 \times \left(\frac{710}{2}\right) = 473.9 \text{ N} \cdot \text{m}$$

$$I = \frac{(0.05)(0.05)^3}{12} = 5.21 \times 10^{-7} \text{ m}^4$$

$$\text{Modulus of rupture} = \frac{473.9 \times 0.025}{5.21 \times 10^{-7}} = 22.75 \times 10^6 \text{ Pa} = 22.75 \text{ MPa}$$

10.10 Design Considerations

Measurement of wood properties in the lab does not reflect all the factors which affect behavior of the material in engineering applications. For design of wood structures, the strength properties given in Tables 10.3 and 10.4 must be adjusted for the following (National Design Specification® for Wood Construction, 2005):

- Load duration
- Wet service
- Temperature
- Beam stability
- Size
- Volume (glulam only)
- Flat use
- Repetitive member (lumber only)
- Curvature (glulam only)
- Column stability
- Bearing area

10.11 Organisms that Degrade Wood

Wood can experience degradation due to attack of fungi, bacteria, insects, or marine organisms.

10.11.1 Fungi

Most forms of decay and sap stains are the result of fungal growth. Fungi need four essential conditions to exist: food, proper range of temperature, moisture, and oxygen. Fungi feed on either the cell structure or the cell contents of woody plants, depending on the fungus type. The temperature range conducive for fungal growth is from 5°C to 40°C. Moisture content above the fiber saturation point is required for fungal growth. Fungi require oxygen for respiration. Fungi attack produces *stains* and/or *decay damage.*

To protect against fungal attack, one of the four essential conditions for growth needs be removed. The most effective protection measure is to keep the wood dry by correct placement during storage and in the structure. Fungi growth can also be prevented by treating the wood fibers with chemical poisons through a pressure treatment process.

Construction procedures that limit decay in buildings include the following:

1. Building with dry lumber that is free of incipient decay and excessive amounts of stains and molds
2. Using designs that keep the wood components dry
3. Using a heartwood from decay-resistant species or pressure-treated wood in sections exposed to above-ground decay hazards
4. Using pressure-treated wood for components in contact with the ground.

10.11.2 Insects

Beetles and *termites* are the most common wood-attacking insects. Several types of beetles, such as bark beetles, attack and destroy wood. Storage of the logs in water or a water spray prevents the parent beetle from boring. Quick drying or early removal of the bark also prevents beetle attack. Damage can be prevented by proper cutting practices and dipping or spraying with an appropriate chemical solution.

Termites are one of the most destructive insects that attacks wood. The annual damage attributed to termites exceeds losses due to fires. Termites enter structures through wood that is close to the ground and is poorly ventilated or wet. Prevention is partially achieved by using pressure-treated wood and otherwise prohibiting insect entry into areas of unprotected wood through the use of screening, sill plates, and sealing compounds.

10.11.3 ▪ Marine Organisms

Damage by marine boring organisms in the United States and surrounding oceans is principally caused by shipworms, pholads, *Limnoria,* and *Sphaeroma.* These organisms are almost totally confined to salt or brackish waters.

10.11.4 ▪ Bacteria

Bacteria cause "wet wood" and "black heartwood" in living trees and a general degradation of lumber. Wet wood is a water-soaked condition that occupies the stem centers of living trees and is most common in poplar, willows, and elms. Black heartwood has characteristics similar to those of wet wood, in addition to causing the center of the stem to turn dark brown or black.

Bacterial growth is sometimes fostered by prolonged storage in contact with soils. This type of bacteria activity produces a softening of the outer wood layers, which results in excessive shrinkage when redried. Bacterial attack does not pose a significant problem to common structural wood species.

10.12   Wood Preservation

The susceptibility of wood to deterioration varies from one location to another depending on moisture condition, temperature, and wood species. Figure 10.17 shows five wood deterioration zones in the United States, ranging from low to severe (Southern Forest Products Association, 2014). Untreated wood can deteriorate when four conditions required for decay and insect occur: high moisture, favorable temperature, oxygen, and food source (wood fibers). If any one of these conditions is removed, deterioration cannot occur. Treating wood with preservatives eliminates wood fiber as a food source, which can reduce or eliminate decay.

Petroleum-based solutions, and waterborne oxides (salts) are the principal types of wood preservatives. The degree of preservation achieved depends on the type of preservative, the degree of penetration, and the amount of the chemical retained within the wood. Effective preservatives must be applied under pressure to increase penetration into the wood.

Deterioration Zones

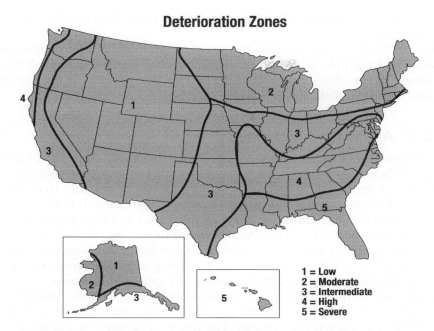

1 = Low
2 = Moderate
3 = Intermediate
4 = High
5 = Severe

FIGURE 10.17   Potential for deterioration of wood used in contact with the ground (American Wood Protection Association Standard U1-15, Commodity Specification D: Poles).

10.12.1 ▪ Petroleum-Based Solutions

Coal-tar creosote, petroleum creosote, creosote solutions, and pentachlorophenol solutions are some of the oil-based preservatives. These preservatives are very effective, but some are environmentally sensitive. They are commonly used where a high degree of environmental exposure exists and human contact is not a concern. Applications include utility poles, railroad ties, and retaining walls.

10.12.2 ▪ Waterborne Preservatives

The typical solutes used in waterborne preservative mixtures are ammoniacal copper arsenate, chromated copper arsenate (CCA), and ammoniacal copper zinc arsenate. The advantages of the waterborne preservative over the oil based are cleanliness and its ability to be painted. The disadvantage of some of these treatments is their removal by leaching when exposed to moist conditions over long periods of time. These preservatives are also environmentally sensitive and must be applied under carefully controlled conditions. The level of potential danger to humans from contacting wood pressure treated with CCA is controversial. Trade groups are supportive of the product, but several other agencies see potential health effects. By 2003, the wood preservation industry agreed to stop using arsenic-based preservatives for products intended for residential use or direct human contact. CCA can still be used for commercial applications and permanent foundations. Although CCA was removed from use in some applications, the EPA is not calling for the removal of it from existing structures.

The most common replacement preservatives are ammoniacal copper quat, or ACQ, followed by copper azole and borate. Borate may be used in home foundation sill plates and other "dry" applications, but borate-treated wood is not appropriate for outdoor uses (Forintek, 2002).

10.12.3 ■ Application Techniques

Superficial treatments are generally not effective. Pressure-treated wood has greater resistance to degradation than surface-treated wood. The preservative is forced into the structure of the wood. Pressure treatment does not thoroughly treat the entire cross section for most species but provide a "shell" of treatment. For this reason, it is recommended that structural members be fabricated as much as possible before treatment in order not to expose untreated wood by cutting, drilling holes, and so on. When fabrication of a treated member is unavoidable, the American Wood Preservative Association (AWPA) recommends treating cuts and holes with a liberal application of field-applied preservative, such as copper naphthenate.

With proper treatment, decay can be eliminated for an extended period of time. Some vendors of pressure-treated wood provide a lifetime warranty for their products when in direct contact with the ground. The key to ensuring long life is the amount of preservative retained in the wood. Minimum retention requirements for different treatments and applications are published by different agencies (e.g., Southern Forest Products Association, 2014).

10.12.4 ■ Construction Precautions

A few issues should be recognized during the design and construction with pressure-treated materials. Care should be taken to avoid inhaling sawdust during cutting of the wood. Hands should be washed after handling pressure treated wood and before consuming food or beverages. Clothes should be washed separately from other items. Waste material must be disposed of properly; it should not be burned.

Some pressure treatments promote corrosion of hangers and other fasteners. Historically, galvanization was adequate. However, there is some evidence that the new preservatives are even more corrosive. The higher metal corrosion rates associated with ACQ-treated wood have raised concerns with the federal Consumer Product Safety Commission (CPSC) and a Bay Area district attorney who issued a consumer alert (e-builder, 2005).

> "CPSC is recommending consumers use stainless steel brackets and fasteners in conjunction with ACQ-treated lumber," said commission spokesman Scott Wolfson. The CPSC is considering whether it needs to study the corrosion issue further, based on information from the connector industry and Contra Costa County. That county's district attorney, Bob Kochly, warned in a recent consumer alert that wood treated with ACQ and copper azole "may result in serious and premature corrosion . . . especially in wet or moist conditions unless stainless steel connectors are used."

10.13 Engineered Wood Products

Engineered wood includes a wide variety of products manufactured by bonding together wood strands, veneers, lumber, or other forms of wood fibers to produce large and integral units. These products are "engineered" to produce specific and consistent mechanical behavior and thus have consistent design properties. There is a wide variety of engineered wood products produced for many applications. Only the products used in structural applications are considered here.

An engineered product consists of wood stock material glued together with an appropriate adhesive. These are predominantly wood materials, so they are liable to the same concerns as natural wood products with respect to the effects of moisture and decay. The wood stock may consist of veneers, strands, or dried lumber as shown in Table 10. 5.

Veneer-based materials consist of wood plies that are glued together. A ply is a thin sheet of wood. To produce plies, logs are saturated by storage in ponds, water vats, steam vats, or water sprays. Prior to processing, the logs are moved into a boiling water bath. Next, they are debarked and sectioned into the desired width. The segments are rotated in a giant lathe and peeled into continuous sheets of veneer, or sliced, as shown in Figure 10.18. These segments of veneer are trimmed and combined into continuous rolls. Each roll of veneer is seasoned, dried to the desired moisture content. The plies, cut from the veneer rolls, are assembled, glued, and pressed. The grains of the plies are alternated at 90° for plywood and they are arranged parallel for laminated veneer lumber (LVL). Plywood is made in sheets, whereas LVL is made in structural shapes.

Strands have uniform thickness, with a combination of length and width dimensions up to 150 mm long and 25 mm wide for panel products and may be up to

TABLE 10.5 Classification of Engineered Wood Products

Wood Stock	Class of Product	
	Structural Panels/Sheets (sheathing, flooring)	**Structural Shapes (beams, columns, headers)**
Veneer	Plywood	Laminated veneer lumber (LVL)
Strands	Flakeboard	Oriented strand lumber
	Waferboard—random orientation of strands	Laminated strand lumber
	Oriented Strand Board—strands oriented, panels made in plys	Parallel strand lumber (Parallam)
Composite	COM-PLY—Wood fiber core with veneer exterior	N/A
Dried lumber	N/A	Glued laminated lumber (Glulam)

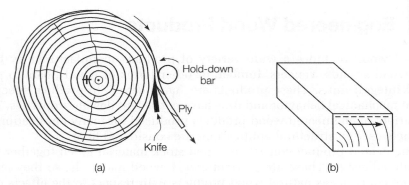

FIGURE 10.18 Cutting of plies for plywood and glulam: (a) rotary cutting and (b) slicing. (© Pearson Education, Inc. Used by permission)

600 mm long for structural shapes. Plies of strands are made by gluing together the strands to form plies. Structural panels are made by gluing together plies in three or five layers. These panels fall into a general class of material called flakeboard. Originally, the strands had a random orientation, producing waferboard. In the 1970s, recognition of the advantages of orientating the strands prior to gluing, then forming panels by using plies with alternating orientation of strands, led to the development of oriented strandboard (OSB). This is the dominant strand product used in the market today. It's structural characteristics compare favorably with plywood. Strandboard can also be made into structural shapes, such as beams.

Dried lumber products can be glued together (laminated) to form structural shapes to meet a wide variety of applications. Typically, either southern pine or western species of softwoods are used. The individual boards are referred to as lams, which are glued together to form the glulam product. Lams of southern pine are typically 35 mm; Western species lams are typically 38 mm. The widths of glulam products are typically 63 to 275 mm, although virtually any size member can be produced.

Engineered wood products can be broadly classified as either structural panels/sheets or structural shapes as shown in Table 10.5. Panels are primarily used for sheathing, whereas structural shapes are used as beams, columns, etc. Sheet products and structural shape products can be combined to produce a variety of composite structural members, such as I-joists and T-sections.

10.13.1 ▓ Structural Panels/Sheets

Structural panel products include plywood and oriented strand board (OSB). Plywood is manufactured as a composite of veneer plies, with an alternating orientation of the grain. OSB is manufactured as a composite of strands; the strands are oriented and pressed into sheets. The OSB panel consists of three to five sheets glued together, with alternating orientation of the strands. Panels may be used directly as sheathing, roofs, sides, and floors. Sheet panels can be cut and bonded into structural components, such as glued or nailed box beams, stressed skin panels, and structural insulated panels.

Conventional wood-based composite products are typically made with a thermosetting or heat-curing resin or adhesive that binds the wood fibers together. Adhesives are chosen based on their suitability for the particular product. Factors considered are the materials to be bonded, moisture content at time of bonding, mechanical property, durability, and cost (USDA-FS, 2010). Commonly used resin-binder systems for panels include the following:

Phenol–Formaldehyde (PF) Used for materials designed for exterior exposure, including plywood, OSB, and siding.

Urea–Formaldehyde (UF) typically used for manufactured products where dimensional uniformity and surface smoothness are of primary concern, such as particle board and medium density fiberboard designed for interior applications.

Melamine–Formaldehyde (MF) used for decorative laminates, paper treating, and paper coating.

Isocyanate as diphenylmethane di-isocyanate, commonly used for OSB. Highly toxic during the manufacturing process.

UF and PF systems are expected to continue to be the dominant wood adhesives, unless their use is curtailed due to concerns about the hazards of formaldehyde products or interruptions to the supply and manufacturing of the material. Research is underway to develop adhesives from renewable resources.

Structural panels are manufactured to meet requirements based on either the composition of the material or the stress rating of the panels. The National Institute of Standards and Technology, NIST, publishes these standards. PS 1-95 specifies the composition of structural plywood. This standard specifies wood species, grades of veneer, what repairs are permissible, and how the repairs may be made. NIST standard PS 2-92 is a performance-based standard for structural composite panels. Figure 10.19 shows typical grade stamps for plywood and OSB (USDA-FS, 2010). Figures 10.19(a) and (b) are grade stamps for plywood showing (1) conformance of plywood to product standards, (2) nominal panel thickness, (3) grades of face and back veneers or grade name based on panel use, (4) performance-rated panel standard, (5) recognition as a quality assurance agency by the National Evaluation Service, NES, (6) exposure durability classification, (7) span rating, which refers to maximum allowable roof support spacing and maximum floor joist spacing, and (8) panel size for spacing. Figure 10.19(c) and (d) demonstrates the typical grade stamps for OSB. The ratings are similar to those for plywood, with the exception that OSB does not have a grade rating for the face and back veneer material.

The mechanical properties of plywood are rated based on the orientation of the strength axis relative to the orientation of the supports, as shown in Figure 10.20 (Williamson, 2002). The differences in the strength properties are due to the orientation of the grain in the layers of the plywood. Similar layers occur in OSB; however, the layers are neither defined nor specified. The mechanical properties of structural panels are rated based on the orientation of the load, as shown in Figure 10.21, and several other factors (Williamson, 2002). Table 10.6 summarizes the mechanical properties

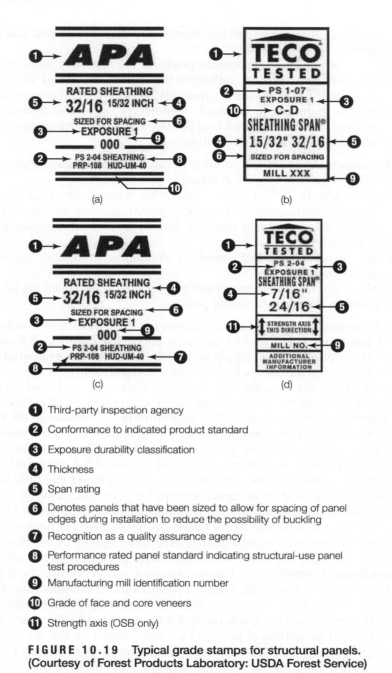

1 Third-party inspection agency

2 Conformance to indicated product standard

3 Exposure durability classification

4 Thickness

5 Span rating

6 Denotes panels that have been sized to allow for spacing of panel edges during installation to reduce the possibility of buckling

7 Recognition as a quality assurance agency

8 Performance rated panel standard indicating structural-use panel test procedures

9 Manufacturing mill identification number

10 Grade of face and core veneers

11 Strength axis (OSB only)

FIGURE 10.19 Typical grade stamps for structural panels. (Courtesy of Forest Products Laboratory: USDA Forest Service)

of plywood and OSB used in structural sheathing applications (USDA-FS, 2010). As with natural wood products, there are several design adjustment factors that must be considered in the selection of the structural panel for a specific application.

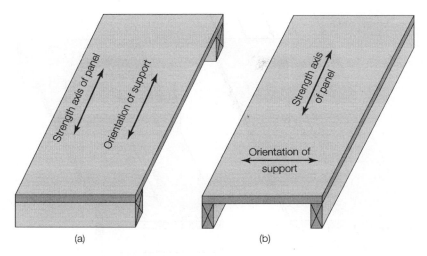

FIGURE 10.20   Typical plywood panel: (a) strength axis parallel to orientation of support and (b) strength axis perpendicular to orientation of support.

10.13.2 ■ Structural Shapes

Three classes of engineered wood products are used to manufacture structural shapes, based on the type of wood stock: laminated veneer, strand, and glue laminated. The laminated and strand products are classified as structural composite lumber (SCL). All types of SCL products can be used in place of sawn lumber for many applications. LVL is used for scaffold planks and in flanges of prefabricated I-joists. Both

TABLE 10.6   General Properties of Plywood and OSB in Structural Sheathing Applications

Property	Plywood	OSB	ASTM Test Method
Linear hygroscopic expansion (30–90%), %	0.15	0.15	
Linear thermal expansion m/m/°C	6.12×10^{-6}	6.12×10^{-6}	
Flexure—modulus of rupture, MPa	20.7 – 48.3	20.7 – 27.6	D 3043
Flexure—modulus of elasticity, GPa	6.89 – 13.1	$4.8 – 8.3 \times 10^3$	
Tensile strength, MPa	10.3 – 27.6	6.9 – 10.3	D 3500
Compressive strength, MPa	20.7 – 34.5	10.3 – 17.2	D 3501
Shear through thickness—strength, MPa	4 – 7.6	6.9 – 10.3	D 2719
Shear through thickness—modulus, GPa	0.47×0.76	1.24×2.00	D 3044
Shear in plane of plies—strength, MPa	1.7 – 2.1	1.4 – 2.1	D 2718
Shear in plane of plies—modulus, GPa	0.14×0.21	0.14×0.34	D 2718

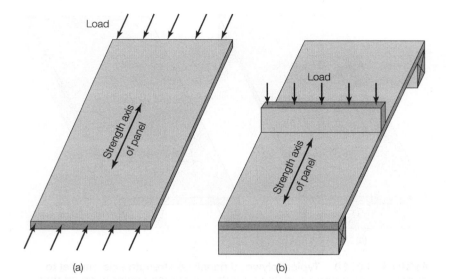

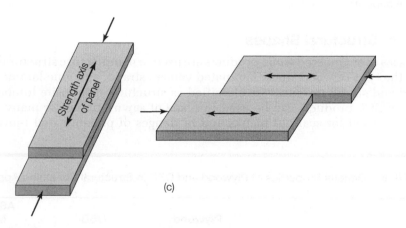

FIGURE 10.21 Load orientation conditions for determining mechanical properties of plywood: (a) axial, (b) bending, and (c) shear.

LVL and parallel strand lumber (PSL) are used for headers and major load-carrying elements. The strand products are used for band joists in floor construction and as substitutes for studs and rafters in wall and roof construction (USDA-FS, 2010).

Laminated Veneer Lumber LVL was developed in the 1940s to produce high-strength parts for aircraft. It is currently a widely accepted construction material. Veneer sheets, 2.5 to 3.2 mm thick are glued with phenol–formaldehyde adhesive to form billets, which are 0.6 to 1.2 m wide by 38 mm thick. The veneer sheets are all aligned with the grain parallel to the length of the member. The end joint between sheets is either staggered or overlapped to minimize any strength-reducing effect. Continuous presses are used to form sheets with lengths limited only by

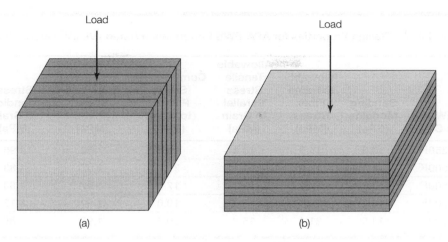

FIGURE 10.22 Axis convention used for defining loading conditions for laminated veneer lumber: (a) edge use and (b) flat use.

handling capability. Alternatively, the billets may be manufactured in 2.4 m lengths and enjoined with either scarf or finger joints to form longer pieces.

The veneer used to manufacture LVL must be carefully selected for uniformity. Visual grading may be used, but ultrasonic testing is commonly used to identify and sort acceptable sheets. The ultrasonic testing can rapidly detect flaws in the veneer, such as knots and splits that would limit the strength of the finished product. Since the quality of each veneer layer is controlled, the variations in product properties are less than for natural products.

LVL is manufactured to meet specific stress grade requirements. Due to the ply construction, the orientation of the member and load must be considered when selecting properties. Figure 10.22 identifies the common terminology used to describe loading conditions for testing and design (Williamson, 2002). Table 10.7 is a portion of the design property table in the APA-EWS performance standard PRL-501, which covers the required properties for stress-graded LVL (APA-EWS, 2000). There are several footnotes associated with the information in this table; therefore, the cited reference should be consulted before using this information for design. The footnotes deal with modifications to the design values as a function of load conditions and duration. It should be noted that the values in Table 10.7 are for design and are therefore conservative. Table 10.8 provides characteristic test values, which are much higher than the design values for all properties other than elastic modulus in bending.

Parallel and Oriented Strand Lumber As with OSB, strands of wood can be glued together into structural members. Depending on the length of strand and degree of orientation, products are classified as parallel strand lumber, laminated strand lumber (LSL), or oriented strand lumber (OSL).

Parallel strand lumber is a composite of wood strand elements, with wood fibers oriented primarily along the length of the member. The smallest dimension of the strand must be less than 6.4 mm and the average length of the strand must be

T A B L E 1 0 . 7   Design Properties for APA-EWS Performance Rated LVL (APA-EWS, 2000)

APA-EWS Stress Class	Bending Modulus (GPa)	Allowable Extreme Fiber Stress (MPa)	Allowable Tensile Stress Parallel to Grain (MPa)	Compressive Stress Parallel to Grain (MPa)	Edgewise Shear Stress (MPa)	Stress Perpendicular to Grain (MPa)
1.5E-2250F	10.3	15.5	10.3	13.4	1.52	3.96
1.8E-2600F	12.4	17.9	11.7	37.2	1.97	4.83
1.9E-2600F	13.1	17.9	11.7	17.6	1.97	4.83
2.0F-2900F	13.8	20.0	13.1	19.0	1.97	5.17
2.1E-3100F	14.5	21.4	15.2	20.7	1.97	5.86

T A B L E 1 0 . 8   Characteristic Test Values for APA-EWS Performance Rated LVL (APA-EWS, 2000)

APA-EWS Stress Class	Bending Modulus (GPa)	Allowable Extreme Fiber Stress (MPa)	Allowable Tensile Stress Parallel to Grain (MPa)	Compressive Stress Parallel to Grain (MPa)	Edgewise Shear Stress (MPa)	Compressive Stress Perpendicular to Grain (MPa)
1.5E-2250F	10.3	32.58	21.72	25.55	4.79	6.62
1.8E-2600F	12.4	37.65	24.61	31.44	6.21	8.07
1.9E-2600F	13.1	37.65	24.61	33.41	6.21	8.07
2.0F-2900F	13.8	37.65	27.51	36.03	6.21	8.65
2.1E-3100F	14.5	37.65	31.85	39.30	6.21	9.79

150 times the smallest dimension. Typically, the strands are 3 mm thick, 19 mm wide, and 0.6 m long. A waterproof structural adhesive, such as phenol–resorcinol–formaldehyde, bonds the strands together. The strands are carefully oriented, coated with adhesive, and run through a continuous press to achieve a high density, producing billets that are typically 0.28 m by 0.48 m. Length is limited only by handling capability. The billets are sawn into the desired dimensions (USDA-FS, 2010).

Currently, parallel strand lumber is a proprietary material with a single producer, marketed under the trade name Parallam (Trusjoist, 2003). The mechanical design properties of this product, as reported by the manufacturer, are given in Table 10.9.

LSL and OSL are extensions of the technology used to produce oriented strand board. They are manufactured using similar processes, with the primary difference being in the length of the strands—approximately 300 mm and 150 mm for LSL and OSL, respectively. As shown in Table 10.9, the mechanical properties of LSL are

TABLE 10.9 Mechanical Properties of Parallel Strand and Laminated Strand Lumber (Trusjoist)

Property	Parallel Strand Lumber	Laminated Strand Lumber
Shear modulus of elasticity (GPa)	0.9	
Modulus of elasticity (GPa)	13.8	10.3
Flexural stress (MPa)	20.0	15.5
Tension stress (MPa)	14.0	
Compression perpendicular to grain (MPa)	5.2	3.3
Compression parallel to grain (MPa)	20.0	13.4
Horizontal shear parallel to grain (MPa)	2.0	2.8

somewhat less than those of PSL. The mechanical properties of OSL are less than those of LSL (USDA-FS, 2010).

Glued–Laminated Timbers   Laminated timbers are composed of two or more layers of dimensional lumber glued together with the grain of all layers laid parallel. Almost any species can be used when its mechanical and physical properties are suited for the design requirements. However, softwoods, such as Douglas fir and southern pine, are most commonly used for laminated structural timbers.

Glued–laminated wood, commonly called "glulam," is used for structural beams and columns as shown in Figure 10.23. In such cases, glued-laminated wood is preferred over large sawn timbers for many reasons, including the following:

- ease of manufacturing of large structural members from standard commercial sized lumber
- the opportunity to design large members that vary in cross section along their length, as required by the application
- specialized design to meet architectural appeal, opportunity to use lower grades of wood within the less stressed areas of the member
- minimization of checking and other seasoning defects associated with curing large one-piece members

Assembly of the laminates into full-depth members is a critical stage in manufacturing. Laminations are planed to strict tolerances before gluing to ensure that the final assembly will be rectangular and that pressure will be applied evenly. Prequalified adhesives, typically phenol–resorcinol, are applied to the gluing face. The laminations are laid up and clamped as an assembly while the glue cures. The adhesive is allowed to cure at room temperature for 6 to 24 hours, or, in some newer processes, radio-frequency curing is used to reduce the time to a few minutes (USDA-FS, 2010).

Glued–laminated wood is designed and produced to have consistent and uniform properties. Due to the laminated structured, the glulam can be laid up to optimize the use of the wood products. For example, glulams designed as bending members are laid up with higher quality wood at the edges of the beam and lower quality

(a)   (b)

FIGURE 10.23 Examples of glued-laminated wood.

wood in the center, as shown in Figure 10.24. Bending members used for cantilevers or continuous members over multiple spans have a balanced layup, since both the top and bottom of the beam carry tensile loads. However, glulams designed for simple spans, which are designed to have only tension at the bottom and compression at the top of the beam, have unbalanced designs, with the higher quality wood at the bottom of the member. This is an unbalanced design and these members are carefully labeled to minimize the risk of installing the beam in an incorrect orientation.

Glulams are produced in four appearance grades:

Framing grade members match the width of conventional framing for use as door and window headers. Poor appearance will typically slow glue squeeze out and may have some unevenness of laminations.

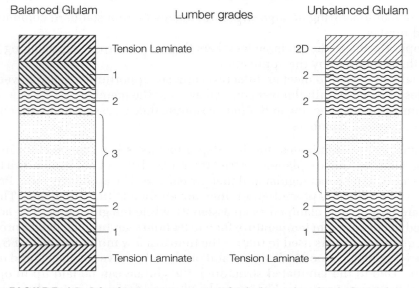

FIGURE 10.24 Layup arrangement for balanced and unbalanced glulams.

Industrial grade is suitable where the glulam will not be exposed to view and appearance is not a concern.

Architectural grade is suitable where appearance is an important requirement. Poor appearance will typically slow glue squeeze out and may have some unevenness of laminations.

Premium grade is the highest grade and is used when appearance is of the utmost importance.

Appearance grades do not modify design values, grades of lumber used, or other provisions governing the manufacture or use of glued–laminate timber.

Glulams are specified by stress class, based on minimum values for their flexural strength and elastic modulus, as shown in Table 10.10 (APA-EWS, 2004). The design values in this table are for members with four or more laminations. The stress grade

TABLE 10.10   Stress Classes of Glulams (APA-EWS, 2004)

| | Extreme Fiber in Bending | | | | |
Stress Class	Tension Zone Stressed in Tension (MPa)	Compression Zone Stressed in Tension[1] (MPa)	Compression Perpendicular to Grain (MPa)	Shear Parallel to Grain[3] (MPa)	Modulus of Elasticity (GPa)
16F-1.3E	11.0	6.37	2.17	1.34	8.96
20F-1.5E	13.8	7.58	2.93	1.45	10.34[5]
24F-1.7E	16.5	10.0	3.45	1.45	11.72
24F-1.8E	16.5	10.0[2]	4.48	1.83[4]	12.41
26F-1.8E	17.9	13.4	4.48	1.83[4]	13.10
28F-1.8E	19.3	15.9	5.10	2.07	14.48[6]
30F-2.1E SP[7]	20.7	16.5	5.10	2.07	14.48[6]
30F-2.1E LVL[8]	20.7	20.7	3.52[9]	2.07	14.48

[1]For balanced layups, bending stress when compression zone is stressed in tension shall be equal to the bending stress when tension zone is stressed in tension for the stress class. Designer shall specify when balanced layup is required.

[2]Negative bending stress is permitted to be increased to 12.8 MPa for Douglas fir and to 13.4 MPa for southern pine for specific combinations. Designer shall specify when these increased stresses are required.

[3]For non-prismatic members, notched members, and members subjected to impact or cyclic loading, the design value for shear shall be multiplied by a factor of 0.72.

[4]Shear Parallel to Grain (Horizontal) = 2. 07 MPa for glulam made of southern pine.

[5]Modulus of Elasticity may be increased to 12.4 GPa for glulam made of Canadian spruce-fir-pine or Eastern spruce.

[6]Modulus of Elasticity = 13.8 GPa for members with more than 15 laminations except for 30F hybrid LVL glulam.

[7]Limited to a maximum of 152 mm except for hybrid LVL glulam.

[8]Requires the use of an outermost LVL lamination on the top and bottom.

[9]Compressive perpendicular to grain stress can be increased to the published value for the outer most LVL lamination.

classification does not designate the species used or whether the wood in the laminate is visually graded or stress graded. Species may be specified in combination with these stress classes to obtain the required design properties. The presence of a stress classification does not ensure that the material is available. In addition, higher design values for some properties may be obtained by specifying a specific combination of laminations. Design values and layup information for specific combinations can be obtained from the American Institute of Timber Construction (AITC) or the APA–The Engineered Wood Association. Table 10.11 presents a partial list of allowable design stresses for glulams softwood members stressed primarily in bending (American Wood Council, 2012). It should be noted that there are many adjustment factors and footnotes associated with this table, and before use of this information, the reader should consult the cited reference (American Wood Council, 2012). Table 10.11 addresses design values only for glulams softwood members stressed primarily in bending. Similar tables are available in the cited reference for design values for glulams softwood members stressed primarily in axial tension or compression and for glulams hardwood timber.

10.13.3 ■ Composite Structural Members

As with natural wood products, glued products can be combined in a variety of ways to produce structural elements. Typical examples of manufactured composite members (Figure 10.25) are:

> I-beams
> Box beams
> Stressed skinned panels
> Structural insulated panels

For example, I-beams are produced by gluing flange elements to an oriented strand board web, as shown in Figure 10.25(b). The flange elements may be either dimensional lumber or, more commonly, LVL.

As manufactured products, the properties of the elements are frequently specified based on the performance of the member, rather than being based on the individual material components. For example, I-joists are specified based on the spacing of the joists and the depth of the member. Standards for composite structural members may be developed by national organizations, such as ASTM, or by trade associations concerned with the proper use of the products they promote.

10.14   Wood Sustainability

10.14.1 ■ LEED Considerations

The LEED process for wood is very straightforward. If 50% of the wood is sourced from a supplier who is certified by the Forest Stewardship Council, FSC, then one credit is earned. If 95% of the wood is from an FSC supplier, an additional credit can be earned for exceeding the minimum. In large building projects the amount of wood

TABLE 10.11 Reference Design Values for Structural Glued–Laminated Softwood Timber (Members Stressed Primarily in Bending) (American Wood Council, 2012)

| Stress Class | Bending (Loaded Perpendicular to Wide Faces of Laminations) | | | | | | |
| | Extreme Fiber in Bending | | | | | | |
	Tension Zone Stressed in Tension (MPa)	Compression Zone Stressed in Tension (MPa)	Compression Perpendicular to Grain (MPa)	Shear Parallel to Grain (MPa)	Modulus of Elasticity (GPa)	Modulus of Elasticity for Beam and Column Stability (GPa)
16F-1.3E	11.0	6.38	2.17	1.34	8.96	4.62
20F-1.5E	13.8	7.58	2.93	1.45	10.34	5.38
24F-1.7E	16.5	10.0	3.45	1.45	11.72	6.07
24F-1.8E	16.5	10.0	4.48	1.83	12.41	6.41
26F-1.9E	17.9	13.4	4.48	1.83	13.10	6.76
28F-2.1E SP	19.3	15.9	5.10	2.07	14.48	7.52
30F-2.1E SP	20.7	16.5	5.10	2.07	14.48	7.52

(Continued)

TABLE 10.11 (Continued)

	Bending (Loaded Parallel to Wide Faces of Laminations)					Axially Loaded			Fasteners
Stress Class	Extreme Fiber in Bending (MPa)	Compression Perpendicular to Grain (MPa)	Shear Parallel to Grain (MPa)	Modulus of Elasticity (GPa)	Modulus of Elasticity for Beam and Column Stability (GPa)	Tension Parallel to Grain (MPa)	Compression Parallel to Grain (MPa)	Modulus of Elasticity (GPa)	Specific Gravity for Fastener Design
16F-1.3E	5.52	2.17	1.17	7.58	3.93	4.65	6.38	8.27	0.42
20F-1.5E	5.52	2.17	1.28	8.27	4.27	5.00	6.38	8.96	0.42
24F-1.7E	7.24	2.17	1.28	8.96	4.62	5.34	6.89	9.65	0.42
24F-1.8E	10.00	3.86	1.59	11.03	5.72	7.58	11.03	11.72	0.5
26F-1.9E	11.03	3.86	1.59	11.03	5.72	7.93	11.03	11.72	0.5
28F-2.1E SP	11.03	4.48	1.79	11.72	6.07	8.62	12.07	11.72	0.55
30F-2.1E SP	12.07	4.48	1.79	11.72	6.07	8.62	12.07	11.72	0.55

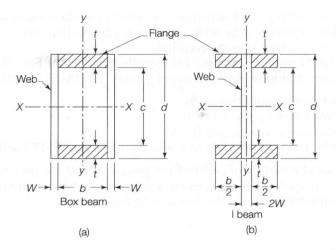

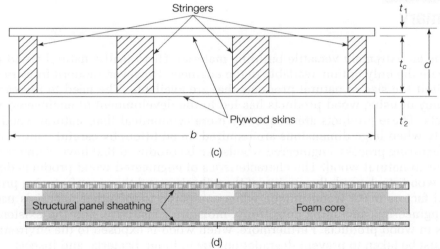

FIGURE 10.25   Composite structural members.

used relative to the project size is frequently a small percent of the project cost, so specifying wood from an FSC supplier, which may carry a premium price, may have little impact on the project cost.

Under the latest LEED v4 criteria, there is an option to use a "USGBC-approved equivalent". However, the process for determining equivalency has not been developed.

10.14.2 ▨ Other Sustainability Considerations

Wood is a renewable resource when the forests are properly managed. Although certification of forest products is growing, much of the forests are not under certification programs. Globally only 8% of the natural forest areas and 13% of the managed forests produce certified products (USDA-FS, 2010). Certification is predominantly in North America, approximately 30%, and Europe, more than 50%. Less than 0.1% of the forests in Africa and Asia are certified.

While the LEED program accepts the certification of the FSC there are many other organizations that have certification programs (USDA-FS, 2010):

- Sustainable Forestry Initiative (SFI)
- American Tree Farm System (ATFS)
- Canadian Standards Association (CSA)
- Programme for the Endorsement of Forest Certification (PEFC) schemes

The existence of a variety of certification programs and the percent of the wood industry that participate in these programs suggests that the sustainability movement has substantially impacted the forest industry.

Summary

Wood is an extremely versatile building material. Historically, natural wood products were the only option available to the engineer. However, modern forestry practices limit the size of natural products that are available. The need to increase the efficiency of using wood products has led to the development of engineered wood products. These products are frequently more economical than natural wood, particularly when large dimensions are required. In addition, by careful control of the manufacturing process, engineered woods can be produced that have characteristics superior to natural wood. The characteristics of engineered wood products depend on the wood stock used, the quality of the adhesive, and the manufacturing process. Several factors make wood unique when compared to the other materials used in civil engineering, including anisotropy, moisture sensitivity, and the existence of defects in wood products. Furthermore, when wood is exposed to the environment, care must be taken to prevent degradation due to fungi, bacteria, and insects.

Questions and Problems

10.1   What are the two main classes of wood? What is the main use of each class? State the names of two tree species of each class.

10.2   What is the difference between earlywood and latewood? Describe each.

10.3   A simple lab test for specific gravity, G, on two samples of lumber indicate that sample A has G = 0.4 and sample B has G = 0.5 Based on this information alone, which wood sample would you choose as a structural member for your construction project? Briefly explain why.

10.4 Discuss the anisotropic nature of wood. How does this phenomenon affect the performance of wood?

10.5 Briefly describe the chemical composition of wood.

10.6 The moisture content of wood test was performed according to ASTM D4442 procedure and produced the following data:

> Mass of specimen in the green condition = 317.5 g
> Mass of oven-dry specimen = 203.9 g
> Calculate the moisture content of the given wood.

10.7 What is the fiber saturation point? What is the effect of the fiber saturation point on the shrinkage of wood in the different directions? How does this phenomenon affect the properties of lumber?

10.8 A stud had dimensions of 38 mm × 89 mm × 2.438 m and a moisture content of 150% when it was prepared. After seasoning, the moisture content was reduced to 7%. If the tangential, radial, and longitudinal directions of the grains are on the same order as the dimensions indicated above, what are the dimensions of the seasoned stud if the moisture-shrinkage relation follows Figure 10.5?

10.9 A wood pole with a diameter of 250 mm. has a moisture content of 5%. The fiber saturation point (FSP) for this wood is 30%. The wood shrinks or swells 1% (relative to the green dimensions) in the radial direction for every 5% change in moisture content below FSP.
   a. What would be the percent change in the wood's diameter if the wood's moisture is increased to 55%?
   b. Would the wood swell or shrink?
   c. What would be the new diameter?

10.10 A glulam beam with a manufactured depth of 250 mm at 14% moisture content. Assuming common values for estimating dimensional changes, compute the new depth of the beam if the moisture content is increased to 35%?

10.11 Wood is cut at sawmills into a variety of products, with different sizes and shapes for engineering applications. What are these products?

10.12 Construction lumber can be cut from the tree using one of two methods or a combination of them. Name these two methods and show a sketch of each. What is the main advantage of each method?

10.13 Why are the actual dimensions of lumber different from the nominal dimensions? Explain.

10.14 What are the factors considered in grading lumber?

10.15 State five different imperfections that may be found in lumber, and briefly define them.

10.16 Draw a graph to show the typical stress–strain curve for wood. On the graph, show the modulus of elasticity. State three different factors that affect this relationship.

10.17 Compute the modulus of elasticity of the wood species whose stress–strain relationship is shown in Figure 10.12. Compare the results with the typical values shown in Table 1.1 in Chapter 1 and comment about the results.

10.18 What is the typical load duration used in designing wood structures? If a wood beam is designed for use at a one-week event only, should the designer increase or decrease the allowable fiber stress relative to the allowable stress used for normal applications? How much increase or decrease?

10.19 To evaluate the mechanical properties of wood, structural-size specimens or small, clear specimens can be tested. Which technique is more important? Why? What is the most common test used in each technique?

10.20 A center-point bending test was performed on a 50 mm × 100 mm wood lumber according to ASTM D198 procedure with a span of 1.2 m and the 102 mm side is positioned vertically. If the maximum load was 1068 kN and the corresponding deflection at the mid-span was 61 mm, calculate the modulus of rupture and the apparent modulus of elasticity. See Experiment No. 30 for equations.

10.21 A 100 mm × 100 mm wood lumber was subjected to bending with a span of 1.5 m until failure by applying a load in the middle of its span. The load and the deflection in the middle of the span were recorded as shown in Table P10.21.
   a. Using a computer spreadsheet program, plot the load–deflection relationship.
   b. Plot the proportional limit on the graph.
   c. Calculate the modulus of rupture (flexure strength).

TABLE P10.21

Load (N)	Deflection (mm)
0	0
431	3406
867	5189
1517	6744
3247	8344
6005	9235
8443	9990
10235	10470
13149	11476
15475	12276
17748	13282
19750	14082

10.22 A wood specimen was subjected to bending until failure by applying a load in the middle of its span. The specimen has a cross section of 25 mm × 25 mm (actual dimensions) and a span of 350 mm between the simple supports. The load and the deflection in the middle of the span were recorded as shown in Table P10.22.
   a. Using a computer spreadsheet program, plot the load–deflection relationship.
   b. Plot the proportional limit on the graph.
   c. Calculate the modulus of rupture (flexure strength).

TABLE P10.22

Load (N)	Deflection (mm)
0	0
36	3.78
71	5.77
125	7.49
267	9.27
494	10.3
694	11.1
881	11.9
1081	12.8
1272	13.6
1459	14.8
1624	15.6

10.23 A wood specimen having a square cross section of 50 mm × 50 mm (actual dimensions) was tested in bending by applying a load at the middle of the span, where the span between the simple supports was 700 mm The deflection under the load was measured at different load levels as shown in Table P10.23.
   a. Using a computer spreadsheet program plot the load–deflection relationship.
   b. Plot the proportional limit on the graph.
   c. Calculate the modulus of rupture (flexure strength).
   d. Does the modulus of rupture computed in (c) truly represent the extreme fiber stresses in the specimen? Comment on the assumptions used to compute the modulus of rupture and the actual response of the wood specimen.

10.24 A wood specimen was prepared with actual dimensions of 25 mm × 25 mm × 150 mm and grain parallel to its length. Displacement was measured over a 100 mm gauge length. The specimen was subjected to compression parallel to the grain to failure. The load–deformation results are as shown in Table P10.24.

TABLE P10.23

Load (N)	Deflection (10⁻³ mm)
0	0
445	709
890	1412
1335	2113
1779	2825
2224	3556
2669	4234
3114	4935
3559	5644
4003	6353
4448	6995
4893	7996
5338	9131
5783	10287
6228	11902 (failure)

TABLE P10.24

Load (N)	Displacement (mm)
0	0
31	0.30
45	1.73
387	4.17
2358	4.57
7584	5.28
12740	5.99
16859	6.81
20489	7.62
23745	8.23
22757	9.14
19875	9.75
19265	10.5

a. Using a computer spreadsheet program, plot the stress–strain relationship.
b. Calculate the modulus of elasticity.
c. What is the failure stress?

10.25 A pine wood specimen was prepared with actual dimensions of 50 mm ×
50 mm × 250 mm and grain parallel to its length. The deformation was measured over a gauge length of 200 mm. The specimen was subjected to compression parallel to the grain to failure. The load–deformation results are as shown
in Table P10.25.

TABLE P10.25

Load (Kn)	Deformation (mm)
0	0
8.9	0.457
17.8	0.597
26.7	0.724
35.6	0.838
44.5	0.965
53.4	1.118
62.3	1.270
71.2	1.422
80.1	1.588
89.0	1.765
97.9	1.956
106.8	2.159
111.3	2.311

a. Using a computer spreadsheet program, plot the stress–strain
relationship.
b. Calculate the modulus of elasticity.
c. What is the failure stress?

10.26 A wood specimen was prepared with dimensions of 25 mm × 25 mm × 100 mm
and grain parallel to its length. The specimen was subjected to compression
parallel to the grain to failure. The load (P) versus deformation (ΔL) results are
as shown in Table P10.26. Using a spreadsheet program, complete the table
by calculating engineering stress (σ) and engineering strain (ε). Determine the
toughness of the material (u_t) by calculating the area under the stress–strain
curve, namely,

$$u_t = \int_0^{\varepsilon^f} \sigma \, d\varepsilon$$

where ε_f is the strain at fracture. This integral can be approximated numerically using a trapezoidal integration technique:

$$u_t = \sum_{i=1}^{n} u_i = \sum_{i=1}^{n} \frac{1}{2}(\sigma_i + \sigma_{i-1})(\varepsilon_i - \varepsilon_{i-1})$$

TABLE P10.26

Observation No.	P (kN)	ΔL >(mm)	σ (MPa)	ε (m/m)	u_j (MPa)
0	0	0			N/A
1	3.20	0.51			
2	7.65	1.22			
3	12.23	1.93			
4	16.86	2.74			
5	20.49	3.56			
6	23.75	4.17			
7	27.45	5.08			
8	28.83	5.69			
9	24.02	6.43			
					$u_t =$

10.27 A short round wood column with an actual diameter of 250 mm is to be constructed. If the failure stress is 29.6 MPa, what is the maximum load that can be applied to this column, using a factor of safety of 1.3?

10.28 For the purpose of designing wood structures, laboratory-measured strength properties are adjusted for application conditions. State five different application conditions that are used to adjust the strength properties.

10.29 What are three types of organisms that attack wood?

10.30 Untreated wood can deteriorate when four conditions required for decay and insects occur. States these four conditions. How do preservatives reduce or eliminate wood decay?

10.31 What are the two types of preservatives that can be used to protect wood from decay? How are these preservatives applied?

10.32 What are the main types of engineered wood products?

10.33 What are the main advantages of engineered wood products over natural-timber members?

10.34 In addition to the LEED system, several other organizations have certification programs for wood. Using Internet resources, summarize some of these certification programs.

10.15 References

American Institute of Timber Construction (AITC), http://www.aitc-glulam.org/.

American Wood Council. *National Design Specification for Wood Construction.* American Wood Council, Washington, D.C., 2012.

American Wood Protection Association. *2014 APWA Book of Standards*, Birmingham, AL, 2014.

APA-EWS, PRL-501. *Performance Standard for APA EWS Laminated Veneer Lumber.* Tacoma, WA: APA-The Engineered Wood Association, 2000.

APA-EWS, *Glulam Design Properties and Layup Combinations.* Tacoma, WA: APA-The Engineered Wood Association, 2004.

e-builder. http://www.ebuild.com/guide/resources/product-news.asp? ID = 81435, 2005.

Forintek. *Borate-Treated Wood for Construction.* Forintek Canada Corp. http://cwc.ca/wp-content/uploads/durabilitybytreatment-BorateTreatedWood.pdf, 2002.

Levin, E., ed. *The International Guide to Wood Selection.* New York, NY: Drake Publishers, Inc, 1972.

Panshin, A. J. and C. De Zeeuw. *Textbook of Wood Technology.* 4th ed. New York, NY: McGraw-Hill Book Company, 1980.

Southern Forest Products Association. *Pressure Treated Southern Pine.* Southern Forest Products Association, Metairie, LA, 2014.

Trus Joist Engineering Wood Products, Parallam Plus PSL Headers, Beams, and Columns. Boise, ID, #TJ-7102 Specifier's Guide, a Weyerhaeuser Business. 2014.

USDA-FS. *Wood Handbook—Wood as an Engineering Material.* Gen. Tech. Rep. FPL-GTR-190. Madison, WI: USDA, Forest Service, Forest Products Laboratory, Centennial Edition. http://www.woodweb.com/Resources/wood_eng_handbook/wood_handbook_fpl_2010.pdf, 2010.

Van Vlack, L. H. *Materials for Engineering: Concepts and Applications,* Reading, MA: Addison-Wesley, 1982.

Van Vlack, L. H. *Elements of Materials Science and Engineering,* 6th ed. Reading, MA: Addison Wesley, 1989.

Williamson, T. G. (ed.). *APA Engineered Wood Handbook.* New York, NY: McGraw-Hill Book Company, 2002.

C H A P T E R

COMPOSITES

The need for materials with properties not found in conventional materials, combined with advances in technology, has resulted in combining two or more materials to form what are called composite materials. These materials usually combine the best properties of their constituents and frequently exhibit qualities that do not even exist in their constituents. Strength, stiffness, specific weight, fracture resistance, corrosion resistance, wear resistance, attractiveness, fatigue life, temperature susceptibility, thermal insulation, thermal conductivity, and acoustical insulation can all be improved by composite materials. These improved properties could have a significant effect on the sustainability of the structure, including higher durability, better performance, reducing life-cycle cost, lower cross section, and more friendly environment.

Of course, not all these properties are improved in the same composite, but typically a few of these properties are improved. For example, materials needed to build aircraft and space vehicles must be light, strong, and stiff and must exhibit high resistance to abrasion, impact, and corrosion. Fiber-reinforced polymer (fiber-reinforced plastic, in a layman term) is an example of a composite material that is very useful for civil engineers. It is strong, stiff, and corrosion resistant and can be used to make concrete reinforcing rebars to eliminate the corrosion problem of steel rebars (Figure 11.1). These combinations of properties are formidable and typically cannot be found in a conventional material.

Composite materials have been used throughout history, with differing levels of sophistication. For example, straw was used to strengthen the mud bricks in ancient civilizations. Swords and armor were constructed with layers of different materials to obtain unique properties. Portland cement concrete, which combines paste and aggregate with different properties to form a strong and durable construction material, has been used for many years. In recent years, fiber-reinforced concrete has been used as a building material that is strong in both tension and compression. The automobile industry has been using composite metals to build lightweight vehicles that are strong

FIGURE 11.1   FRP reinforcing bars used for bridge abutment. (Courtesy of West Virginia University Constructed Facilities Center)

and impact resistant. Recently, a new generation of composites, such as polymer and ceramic composites, revolutionized the material industry and opened new horizons for civil and construction engineering applications.

Although several definitions of composites exist, it is generally accepted that a composite is a material that has two or more distinct constituent materials or phases. The constituents of a composite typically have significantly different physical properties, and thus the properties of the composite are noticeably different from those of the constituents. This definition eliminates many multiphase materials that do not have distinct properties, such as many alloys with components that are similar.

There are a number of naturally formed composites, such as wood—which consists of cellulose fibers and lignin—and bone, which consists of protein collagen and mineral apatite. However, in this chapter, we discuss artificially made composites only.

Composite materials can be classified as *microscopic* or *macroscopic,* as shown in Figure 11.2. The distinction between microscopic and macroscopic depends on the type of properties being considered. This distinction seems arbitrary, but normally microscopic composites include *fibers* or *particles* in sizes up to a few hundred microns. On the other hand, macroscopic composites could have constituents of much larger size, such as aggregate particles and rebars in concrete.

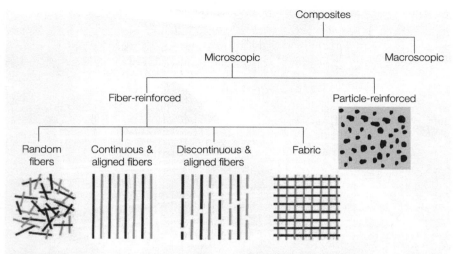

FIGURE 11.2 A classification scheme for composite materials.

Alternatively, composites can be classified by function (Barbero, 1988):

Reinforcement	• Continuous long fibers • Unidirectional fiber orientation • Bidirectional fiber orientation (woven, stitched, mat, etc.) • Random orientation (continuous strand mat)	• Discontinuous fiber • Random orientation (chopped strand mat) • Preferential orientation (oriented strand board)	• Particles and whiskers • Random orientation • Preferential orientation
Laminate	• Unidirectional	• Laminate—some layers with different orientation	
Hybrid Structure	• Different Materials in Layers	• Different reinforcement in a layer	

11.1 Microscopic Composites

Many microscopic composite materials consist of two constituent phases: a continuous phase, or *matrix,* and the *dispersed phase* or *reinforcing phase,* which is surrounded by the matrix. In most cases, the dispersed phase is harder and stiffer than the matrix. The properties of the composite depend on the properties of component phases, their relative properties, and the geometry of the dispersed phase, such as the particle shape, size, distribution, and orientation.

As indicated in Figure 11.2, microscopic composites fall into two basic classes: *fiber-reinforced* and *particle-reinforced.* This classification is based on the shape of the dispersed phase. Fibers can be either random, continuous and aligned, discontinuous and aligned, or fabric. The mechanism of strengthening microscopic composites varies for different classes and different sizes and orientations of the dispersed shape. The fiber-reinforced polymer (FRP) is a common composite consisting of a polymer reinforced with either fibers or fabrics. FRP composite has been used in many engineering applications.

Since the applied load needs to be transmitted to the fibers by the matrix, the fiber length is important. A minimum fiber length is needed for effectively strengthening and stiffening the composite. A critical fiber length (L_c) is needed for full effectiveness of the fibers, which depends on fiber diameter (d), the tensile strength of the fiber (σ_f), and the fiber–matrix bond strength (or the shear yield strength of the matrix, whichever is smaller) (τ_c) as shown in Equation 11.1 (Callister, 2006, 2008).

$$L_c = \frac{\sigma_f\, d}{2\tau_c} \tag{11.1}$$

When the fiber length is equal to, or larger than, the critical length (L_c), the fibers reach their full effectiveness in strengthening the composite. Also, when the length of the fiber is larger than $15 \times L_c$, the fibers are generally called continuous, otherwise they are called discontinuous.

11.1.1 ▨ Fiber-Reinforced Composites

Fiber-reinforced microscopic composites include fibers dispersed in a matrix such as metal, polymer, or ceramic. Fibers have a very high strength-to-diameter ratio, with near crystal-sized diameters. Because of the very small diameter of the fibers, they are much stronger than the bulk material. For example, a glass plate fractures at stresses of 10 to 20 kPa, yet glass fibers have strengths of 3 to 5 GPa or more. Fibers are much stronger than the bulk form because they have fewer internal defects.

Fibers can be classified on the basis of their diameter and character as whiskers, fibers, and wires. Whiskers are very thin single crystals, have extremely large length-to-diameter ratios, have a high degree of crystalline perfection, and, consequently, are extremely strong. Fibers have larger diameters than whiskers, while wires are even larger. Whiskers are not commonly used for reinforcement because of their high cost, poor bond with many common matrix materials, and the difficulty of incorporating them into the matrix.

Common fiber types include (Figure 11.3):

- Steel fibers
- Glass fibers
- Synthetic fibers such as polypropylene, carbon, nylon, polyethylene, and poly vinyl alcohol (PVA)
- Natural fibers such as wood, sisal, and coconut

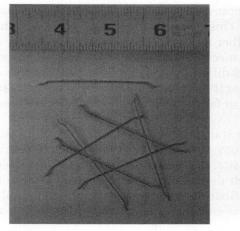

(a)

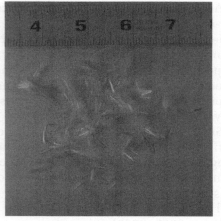

(b)

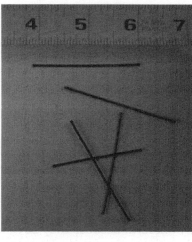

(c)

FIGURE 11.3 Common fibers used in fiber-reinforced composites: (a) steel, (b) glass, and (c) polypropylene.

Table 11.1 shows the strength characteristics of some common fibers. Because of their low cost and high strength, glass fibers are the most common of all reinforcing fibers for polymer matrix composites. Glass fibers are commercially available in several forms suitable for different applications. Common glass fibers include veils, rovings (continuous fibers), and mats (Figure 11.4). A strand consists of about 30 or 40 fibers twisted together to form a ropelike length.

TABLE 11.1   Materials and Mechanical Properties of Some Fibers (Callister, 2006, 2008)

Material	Specific Gravity	Tensile Strength, GPa	Elastic Modulus, GPa
Aramid (Kevlar)	1.4	3.5	130
E-Glass	2.5	3.5	72
Graphite	1.4	1.7	255
Nylon 6,6	1.1	1.0	4.8
Asbestos	2.5	1.4	172

One of the common terms for a fiber-reinforced composite is fiberglass. Fiberglass is simply a composite consisting of glass fibers, either continuous or discontinuous, contained within a plastic matrix. Typical fiberglass applications include aircraft, automobiles, boats, storage containers, water tanks, sporting equipment, and flooring.

FRP composites have been used to make laminates. A lamina is a single layer containing continuous fibers in the plane of the layer as shown in Figure 11.5(a). The fibers in the lamina could be in one direction or two directions (mostly orthogonal), either woven or knitted. The lamina is anisotropic and its mechanical properties are influenced by the orientation of the fibers. The lamina is the usual building block for design and fabrication of composite structures. A laminate, on the other hand, is a stack of laminas, usually with alternating or varying principal directions as shown in Figure 11.5(b). The laminas in the laminate are arranged in a particular way to provide nearly identical properties along the laminate plane.

FRP laminates have been used as structural components of various structures and also in different types of repair of aged or partially damaged structures. Repair

FIGURE 11.4   Common continuous glass fibers (veil, roving, and mat). (Courtesy of Creative Pultrusions, Inc.)

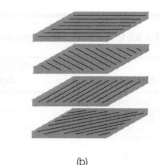

FIGURE 11.5 FRP laminates
and panels: (a) lamina and
(b) laminate.   (a)   (b)

applications include deflection reduction, reduction of stresses in steel reinforcement, crack width reduction, and structural modifications such as removal of walls, columns, or slab sections.

11.1.2 ■ Particle-Reinforced Composites

Particle-reinforced composites consist of particles dispersed in a matrix phase. The strengthening mechanism of particle-reinforced composites varies with the size of the reinforcing particles. When the size of the particles is about 0.01 to 0.1 μm, the matrix bears most of the applied load, whereas the small dispersed particles hinder or impede the motion of dislocations. An example of dispersed-reinforced composite is thoria-dispersed nickel, in which about 3% of thoria (ThO_2) is finely dispersed in a nickel alloy to increase its high-temperature strength. On the other hand, when the particles are larger than 1 μm, particles act as fillers to improve the properties of the matrix phase and/or to replace some of its volume, since the filler is typically less expensive. Here, the matrix retains movement in the vicinity of the particle. Thus, the applied load is shared by the matrix and dispersed phases. The stronger the bond between the dispersed particles and the matrix, the larger is the reinforcing effect. An example of particle reinforcing is adding fillers to polymers to improve tensile and compressive strengths, abrasion resistance, toughness, dimensional and thermal properties, and other properties.

11.1.3 ■ Matrix Phase

Typically, the matrix used in most microscopic composites is polymer (plastic), ceramic, or metal. The matrix binds the dispersed materials (particles or fibers) together, transfers loads to them, and protects them against environmental attack and damage due to handling.

Polymer matrices have the advantages of low cost, easy processability, good chemical resistance, and low specific gravity. The shortcomings of polymers are their low strength, low modulus, low operating temperatures, and low resistance to prolonged exposure to ultraviolet light and some solvents. Modern resins have minimized these shortcomings. Common polymers used in fiber-reinforced composites include epoxy, polyester, and vinylester.

Ceramic matrices can be either non-oxide or oxide. Non-oxide matrices are mostly silicon carbide, carbon or mixtures of silicon carbide and silicon. Oxide matrices consist of alumina, zirconia, mullite, or other alumino-silicates. Ceramic matrices are commonly used in kiln lining, gas turbines, and bone replacement.

Metal matrices have high strength, high modulus, high toughness and impact resistance, relative insensitivity to temperature changes, and high resistance to high temperatures and other severe environmental conditions. However, metals have high density and high processing temperatures, due to their high melting points. Metals also may react with particles and fibers and they are vulnerable to attack by corrosion (Agarwal and Broutman, 2006). The metals most commonly used as the matrix phase in composites are aluminum and titanium alloys.

11.1.4 ▨ Fabrication

Fabrication of microscopic composites includes the merging of the matrix and dispersed material into a product with minimum air voids. Several methods have been used to fabricate the composites. The selection of the fabrication process typically is based on the chemical nature of the matrix and the dispersed phases, the shape and strength requirements, and on the temperature required to form, melt, or cure the matrix. Figure 11.6 illustrates fabrication of structural shape fiber-re-inforced composites by using the pultrusion process. Pultrusion is an automated process for manufacturing fiber-reinforced composite materials into continuous, constant-cross-section profiles.

11.1.5 ▨ Civil Engineering Applications

Microscopic composites have been used in many civil and construction engineering applications in the past several decades (Bakis, 2002). In fact, composite materials compete with, and in many cases are preferred over, conventional building materials. Composites are used by civil engineers as structural shapes in buildings and

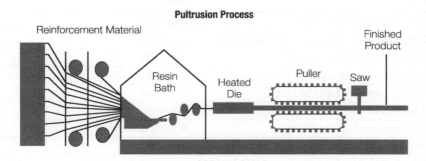
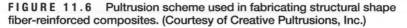

FIGURE 11.6   Pultrusion scheme used in fabricating structural shape fiber-reinforced composites. (Courtesy of Creative Pultrusions, Inc.)

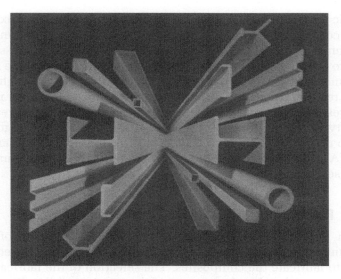

FIGURE 11.7   Structural shapes made of fiberglass composites. (Courtesy of Creative Pultrusions, Inc.)

other structures and can replace steel and aluminum structural shapes (Figure 11.7). Table 11.2 provides an example of physical properties of FRP composite round rods and bars.

FRP rebars can also be used for concrete reinforcement instead of steel rebars. Composites have been used for bridge decks, tanks, industrial flooring, trusses and joists, walkways and platforms, wind turbines, waste treatment plants, handrailings, plastic pipes, utility and light poles, door and window panels and frames, and electrical enclosures. Figure 11.8 shows cables of a suspension long-span bridge retrofit

TABLE 11.2   An Example of Physical Properties of Fiber-Reinforced Composite Round Rods and Bars (Creative Pultrusions, Inc., 2006)

Property	Value
Tensile strength (ASTM D638)	690 MPa
Tensile modulus of elasticity (ASTM D638)	41.4 GPa
Flexural strength (ASTM D790)	690 MPa
Compressive strength (ASTM D695)	41 MPa
Izod impact strength (ASTM D256)	2.1 kJ/m
Barcol hardness (ASTM D2583)	50
Water absorption (ASTM D570)	0.25% (maximum)
Specific gravity (ASTM D792)	2.0
Coefficient of thermal expansion (ASTM D696)	1.7×10^{-6} m/m/°C

FIGURE 11.8   Retrofit bridge cables made of carbon
fiber reinforced polymer. (Penobscot Narrows Bridge and
Observatory, Maine) (Fotolia/Chris Soucy)

with FRP composite. Note that the FRP bridge cables can carry very large stresses
and resist corrosion. Figure 11.9 shows some civil engineering applications made of
FRP composites. FRP composites can also be used to strengthen and wrap columns
and bridge supports that are partially damaged by earthquakes and other environmental factors (Grace, 2002). In Figures 11.10 and 11.11, the structure is wrapped
with FRP composites flexible sheets to provide confinement in order to strengthen
the structural members. In Figure 11.12, an old bridge is coated with a FRP composite material for confinement and strengthening. Figure 11.13 shows an old concrete
bridge that has been strengthened by adding FRP composite bars to carry the tension
and increase the stability of the bridge.

Fiber-reinforced concrete is another composite material that has been used by
civil engineers in various structural applications. Different types of fibers, such as
separate fibers, chopped-strands, or rovings, can be used to reinforce the concrete. If
separate fibers or chopped-strands are used, they are mixed with the fresh concrete
in a random order. In such a case, fibers hinder or impede the progression of cracks
in concrete. Figure 11.14 shows a scanning electron micrograph of concrete mortar
mixed with about 3-mm-long carbon fibers at a volume fraction of 12%. Fiber rovings, on the other hand, are placed in the direction in which the tension is applied
in the structural member. In this case, fibers carry the tensile stresses. In general,
fibers increase the tensile and flexure strength of concrete so that a more efficient
structural member can be designed. Table 11.3 shows typical ranges of physical
properties of glass fiber-reinforced concrete at 28 days. Research has shown that
glass fiber-reinforced concrete offers two to three times the flexural strength of
unreinforced concrete. Moreover, the material under increasing load does not
fail abruptly but yields gradually. This gradual yielding occurs because fibers
are stronger than the matrix and, therefore, arrests cracks. Therefore, instead of a

(a)   (b)

(c)   (d)

FIGURE 11.9 FRP composites used in civil engineering applications. (Courtesy of the Constructed Facilities Center, West Virginia University)

FIGURE 11.10   Wrapping an old structure with glass fiber-reinforced polymer composite. (Courtesy of R.J. Watson)

FIGURE 11.11   Wrapping columns of a bridge with FRP composites.

(a)

(b)

FIGURE 11.12 Woodland viaduct over metro north railroad in Westchester County, NY: (a) reinforcing with FRP composite and (b) finished structure. (Courtesy of R.J. Watson)

FIGURE 11.13
Strengthening a concrete bridge with FRP composite bars.

FIGURE 11.14
Scanning electron micrograph of concrete mortar mixed with carbon fibers.

TABLE 11.3   Typical Ranges of Physical Properties at 28 Days of Glass Fiber-Reinforced Concrete (Neal, 1977)

Property	Value
Flexural strength	21–32 MPa
Tensile strength	7–11 MPa
Compressive strength	50–79 MPa
Impact strength	10–25 kN/m
Elastic modulus	10.5–20.5 GPa
Density	1.70–2.10 Mg/m^3

worsening of the first crack that occurs in the concrete, more cracks are developed elsewhere, and failure finally occurs when fibers pull out or break (Neal, 1977).

Entrained air in concrete can also be considered as a component in a microscopic composite material. Entrained air increases the durability of concrete since it releases internal stresses caused by the freezing of water within the concrete. For the same water-to-cement ratio, however, air bubbles reduce the concrete strength by about 20%. Since entrained air also improves the workability of fresh concrete, the water-to-cement ratio can be reduced to compensate for some of the strength reduction.

11.2 Macroscopic Composites

Macroscopic composites are used in many engineering applications. Because macroscopic composites are relatively large, how the load is carried and how the properties of the composite components are improved vary from one composite to another. Common macroscopic composites used by civil and construction engineers include plain portland cement concrete, steel-reinforced concrete, asphalt concrete, and engineered wood such as glued–laminated timber, and structural strand board.

11.2.1 ▒ Plain Portland Cement Concrete

Plain portland cement concrete is a composite material consisting of cement paste and aggregate particles with different physical and mechanical properties, as discussed in Chapter 7 (Figure 11.15). Aggregate particles in concrete act as a filler material, since it is cheaper than the portland cement. In addition, since cement

FIGURE 11.15 Cross section of portland cement concrete showing cement paste and aggregate particles.

paste shrinks as it cures, aggregate increases the volume stability of the concrete. When the concrete structure is loaded, both cement paste and aggregate share the load. Both the strength of aggregate particles and the bond between the aggregate and cement paste play an important role in determining the strength of the concrete composite, which is limited by the weaker of the two. The bond between cement paste and aggregate is affected by roughness and absorption of the aggregate particles, as well as by other physical and chemical properties of aggregate.

11.2.2 ▓ Reinforced Portland Cement Concrete

Steel-reinforced concrete can be viewed as a composite material, consisting of plain concrete and steel rebars, as shown in Figure 11.16. Since concrete has a very low tensile strength, which is typically ignored in designing concrete structures, steel rebars are usually placed in areas within the structure that are subjected to tension. When the concrete structure is loaded, the concrete carries compressive stresses and steel carries tensile stresses. In such cases, steel allows concrete to be used as structural members carrying tension. Steel rebars are also used in areas subjected to compression, such as columns, to share the load support. In such cases, steel reduces the required cross-sectional area of the compressive member, since the compressive strength of steel is larger than that of concrete. Steel reinforcing is also used in prestressed concrete, where the reinforcement is prestressed under tension so that the concrete remains under compression even when it is externally loaded. In such cases, a smaller cross-sectional area of the concrete member is required. Steel rebars can also be used to control cracking in concrete due to temperature change. For example, concrete pavement is sometimes reinforced by placing longitudinal and transverse steel bars at the midheight of the concrete slab. In this case, when the concrete shrinks due to reduction in ambient temperature, many cracks will develop; these cracks are uniformly distributed within the pavement section, but each crack will be tight. Typically, tight cracks are not harmful to concrete pavement, since they transfer the load from one side of the crack to the other by interlocking. In all applications of steel reinforced concrete, the bond between the rebars and the concrete is important in order to allow the composite to work as one unit. Therefore, bars have a deformed surface to prevent slipping between steel and concrete.

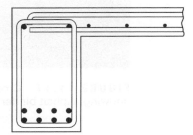

FIGURE 11.16 An example of steel-reinforced concrete beam and slab details.

11.2.3 ■ Asphalt Concrete

Asphalt concrete used in pavements is another composite material as shown in Figure 11.17. It consists of two materials with distinct properties, as presented in Chapter 9. Asphalt concrete consists of approximately 95% aggregate and 5% asphalt binder, by weight. When the traffic loads are applied on the asphalt concrete composite, most of the compressive stresses are supported by the aggregate-to-aggregate contact. The asphalt acts as a binder that prevents slipping of aggregate particles relative to each other. When tensile stresses are applied due to bending of the asphalt concrete layer or due to thermal contraction, the aggregate particles are supported by the asphalt binder. One important property of asphalt is that it gets soft at high temperatures and brittle at low temperatures, whereas aggregate does not change its properties with temperature fluctuation. It is important, therefore, to properly select the asphalt grade that will perform properly within the temperature range of the region in which it is being used. Also, since aggregate represents a major portion of the mixture, it is important to use aggregate with proper gradation and other properties. The asphalt binder content must be carefully designed in order to ensure that aggregate particles are fully coated, without excessive lubrication. When the asphalt concrete mixture is appropriately designed and compacted, it should last for a long time without failure.

11.2.4 ■ Engineered Wood

Engineered wood is manufactured by bonding together wood strands, veneers, or lumber with different grain orientations to produce large and integral units. Since engineered wood consists of components of the same material, it does not

FIGURE 11.17 Cross section of asphalt concrete showing asphalt binder and aggregate particles.

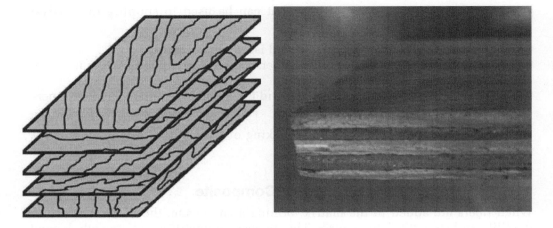

FIGURE 11.18 Plywood with alternating grain orientation.

qualify as a composite according to our definition. However, engineered wood is presented in this chapter because it follows a strengthening mechanism similar to that of composites. Since wood has anisotropic properties due to the existence of grains, engineered wood produces specific and consistent mechanical behavior and thus has consistent design properties. For example, alternating the grain orientation of the plies of plywood (see Figure 11.18) provides nearly identical properties along the length and width and provides resistance to dimensional change under varying moisture conditions. The plywood composite has about one-tenth of the dimensional change of solid lumber under any temperature or moisture condition. As discussed in Chapter 10, engineered wood products include plywood, oriented strand boards, composite panels, glued–laminated timber (glulam), laminated veneer lumber, parallel strand lumber, oriented strand lumber, and wood I-joists.

11.3 Properties of Composites

The properties of composite materials are affected by the component properties, volume fractions of components, type and orientation of the dispersed phase, and the bond between the dispersed phase and the matrix. The properties of the composite can be viewed as the weighted average of the properties of the components (Shackelford, 2014). Equations can be derived to estimate the composite properties under certain idealized material properties, loading patterns, and geometrical conditions. Assumptions that can be used to simplify the analysis include the following:

- Each component has linear, elastic, and isotropic properties.
- A perfect bond exists between the dispersed and matrix phases without slipping.
- The composite geometry is idealized and the loading pattern is parallel or perpendicular to reinforcing fibers.

Properties that are affected by the making of the composite include ductility, strength, and modulus of elasticity.

11.3.1 ▓ Ductility and Strength of Composite

When fibers are added to the matrix forming a composite, the fibers increase the ductility and strength of the matrix. The increase in ductility and strength depends on the fiber properties, volume fraction, and orientation. This concept is useful in many engineering applications. It allows for smaller structural components to carry higher loads and producing safer structures that do not catastrophically collapse when overloaded.

One of the structural applications that utilizes this phenomenon is adding dispersed steel fibers to concrete. Since concrete is brittle, it has low toughness and it fails at a low strain level. When steel fibers are added to the plain concrete and the concrete starts to crack the fibers continue to carry the load and the structure does not suddenly fail as demonstrated in Figure 11.19. Steel fiber-reinforced concrete can be used in industrial floors, highway pavements, airport runways, pipes, shotcrete, and precast segments. Adding steel fibers to plain concrete has the following benefits:

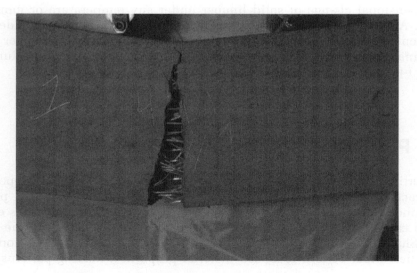

FIGURE 11.19 Fiber bridging cracks in concrete. (Courtesy of Barzin Mobasher)

- High-energy absorption (toughness)
- High flexural fatigue resistance
- Cracking resistance/redistribution of cracking
- Redistribution of moments
- Increased impact resistance

Figure 11.20 shows stress–strain relationships of plain concrete mixed with short fibers with different amounts and subjected to compression (Johnson, 1974). As discussed in Chapter 1, toughness of a material is measured as the area under the stress–strain curve up to fracture. Figure 11.20 shows the fiber-reinforced concrete is more ductile than plain concrete. It also shows that increasing the fiber content in the composite increases the ductility. The figure also shows that fibers slightly increase the compressive strength of the concrete. Similar results are obtained if the concrete is subjected to tension or bending.

11.3.2 ■ Modulus of Elasticity of Composite

When fibers are added to the matrix forming a composite the fibers changes the modulus of elasticity of the matrix depending on the fiber properties, volume fraction, and orientation.

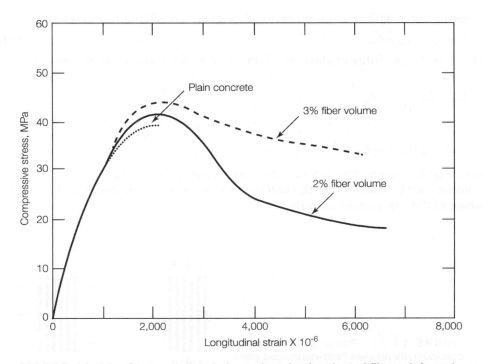

FIGURE 11.20   Stress–strain relations of randomly oriented fiber-reinforced concrete with different fiber contents. (Johnson, 1974)

Loading Parallel to Fibers When load is applied to a continuous and aligned fiber-reinforced composite parallel to the fibers, as seen in Figure 11.21(a), both matrix and fiber phases will deform equally. Thus, the strains of both phases will be the same (known as an *isostrain condition*) and are given by

$$\varepsilon_c = \varepsilon_m = \varepsilon_f = \varepsilon \tag{11.2}$$

where

ε = total strain
ε_c = composite strain
ε_m = matrix strain
ε_f = fiber strain

Also, the force applied to the composite F_c is the sum of the force carried by the matrix F_m and the force carried by the fibers F_f:

$$F_c = F_m + F_f \tag{11.3}$$

Thus,

$$\sigma_c A_c = \sigma_m A_m + \sigma_f A_f \tag{11.4}$$

where

σ_i = stress of component i
A_i = area of component i

Replacing σ with $E\varepsilon$ for each material, we can write Equation 11.4 as

$$E_c \varepsilon A_c = E_m \varepsilon A_m + E_f \varepsilon A_f \tag{11.5}$$

where E is the modulus of elasticity. Canceling ε and dividing by A_c, we have

$$E_c = E_m \frac{A_m}{A_c} + E_f \frac{A_f}{A_c} \tag{11.6}$$

or

$$E_c = \nu_m E_m + \nu_f E_f \tag{11.7}$$

where ν is the volume fraction of each component and $\nu_m + \nu_f = 1$.

Equation 11.7 shows that the composite's modulus of elasticity is the weighted average of the component moduli.

FIGURE 11.21 Patterns of loading continuously aligned fiber-reinforced composites: (a) loading parallel to fibers and (b) loading perpendicular to fibers.

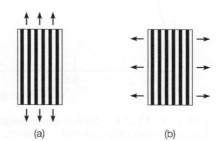

(a)   (b)

The share of the load carried by the fibers can be determined as follows:

$$\frac{F_f}{F_c} = \frac{\sigma_f A_f}{\sigma_c A_c} = \frac{E_f \varepsilon A_f}{E_c \varepsilon A_c} = \frac{E_f}{E_c} \nu_f \tag{11.8}$$

Sample Problem 11.1

Calculate the modulus of elasticity of fiberglass under isostrain condition if the fiberglass consists of 70% E-glass fibers and 30% epoxy by volume. Also, calculate the percentage of load carried by the glass fibers. The moduli of elasticity of the glass fibers and the epoxy are 70.5 and 6.9 GPa, respectively. If a longitudinal stress of 60 MPa is applied on the composite with a cross-sectional area of 300 mm², what is the load carried by each of the fiber and the matrix phases? What is the strain sustained by each of the fiber and the matrix phases?

Solution

From Equation 11.7,

$$E_c = (0.3)(6.9) + (0.7)(70.5) = 51.4 \text{ GPa}$$

From Equation 11.8,

$$\frac{F_f}{F_c} = \frac{70.5}{51.4}(0.7) = 0.96 = 96\%$$

Therefore, 96% of the load is carried by the fibers.

Total load carried by the composite $F_c = A_c \sigma_c = (300 \text{ mm}^2)(60 \text{ MPa}) = 18000 \text{ N}$

Load carried by the fibers $= (0.96)(18000) = 17280 \text{ N}$

Load carried by the matrix $= (0.04)(18000) = 720 \text{ N}$

$$\varepsilon_c = \varepsilon_m = \varepsilon_f = \frac{\sigma_c}{E_c} = \frac{60}{51400} = 0.001 \text{ m/m}$$

Equation 11.7 can be generalized to cover other composite material properties as a function of the properties of the components as

$$X_c = \nu_m X_m + \nu_f X_f \tag{11.9}$$

where X is a property such as Poisson's ratio, thermal conductivity, electrical conductivity, or diffusivity.

Loading Perpendicular to Fibers   When load is applied to a continuous and aligned fiber-reinforced composite perpendicular to the fibers [Figure 11.21(b)], both matrix and fiber phases will be subjected to the same stress (isostress condition). In other words,

$$\sigma_c = \sigma_m = \sigma_f = \sigma \tag{11.10}$$

The elongation of the composite in the direction of the applied stress is the sum of the elongations of the matrix and fibers:

$$\Delta L_c = \Delta L_m + \Delta L_f \tag{11.11}$$

Dividing Equation 11.11 by the composite length L_c in the stress direction gives

$$\frac{\Delta L_c}{L_c} = \frac{\Delta L_m}{L_c} + \frac{\Delta L_f}{L_c} \tag{11.12}$$

Assuming that the fibers are uniform in thickness, the cumulative length of each component in the direction of the stress is proportional to its volume fraction. Thus,

$$L_m = \nu_m L_c \tag{11.13}$$

and

$$L_f = \nu_f L_c \tag{11.14}$$

Substituting the values of L_c from Equations 11.13 and 11.14 in Equation 11.12 yields

$$\frac{\Delta L_c}{L_c} = \frac{\nu_m \Delta L_m}{L_m} + \frac{\nu_f \Delta L_f}{L_f} \tag{11.15}$$

Since $\varepsilon = \dfrac{\Delta L}{L}$, Equation 11.15 can be rewritten as

$$\varepsilon_c = \nu_m \varepsilon_m + \nu_f \varepsilon_f \tag{11.16}$$

Replacing ε with $\dfrac{\sigma}{E}$ gives

$$\frac{\sigma}{E_c} = \nu_m \frac{\sigma}{E_m} + \nu_f \frac{\sigma}{E_f} \tag{11.17}$$

or

$$\frac{1}{E_c} = \frac{\nu_m}{E_m} + \frac{\nu_f}{E_f} \tag{11.18}$$

Equation 11.18 can be rewritten as

$$E_c = \frac{E_m E_f}{\nu_m E_f + \nu_f E_m} \tag{11.19}$$

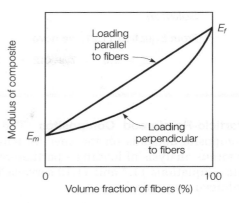

FIGURE 11.22   Modulus of elasticity of the composite versus fiber volume fraction.

As with Equation 11.9, Equation 11.19 can be generalized to cover other composite material properties as a function of the properties of the components as

$$X_c = \frac{X_m X_f}{\nu_m X_f + \nu_f X_m}$$ (11.20)

where X is a property such as Poisson's ratio, thermal conductivity, electrical conductivity, or diffusivity.

The moduli in Equations 11.7 and 11.19 can be plotted as functions of the volume fraction of the fiber, as shown in Figure 11.22. Clearly, the fibers are more effective in raising the modulus of the composite when loading is parallel to fibers than when loading is perpendicular to fibers.

Randomly Oriented Fiber Composites   Unlike continuously aligned fiber composites, the mechanical properties of randomly oriented fiber composites are isotropic. The modulus of elasticity of randomly oriented fiber composites falls between the moduli of loading parallel to fibers and perpendicular to fibers. To estimate the modulus of elasticity of randomly oriented fiber composites, Equation 11.7 can be rewritten as

$$E_c = \nu_m E_m + K \nu_f E_f$$ (11.21)

where K is a fiber efficiency parameter (Callister, 2006, 2008). For fibers randomly and uniformly distributed within three dimensions in space, K has a value of 0.2. Similarly, Equation 11.21 can be generalized to cover other composite material properties as a function of the properties of the components.

Sample Problem 11.2

A fiberglass composite consists of epoxy matrix reinforced with randomly oriented and uniformly distributed E-glass fibers. The moduli of elasticity of the glass fibers and the epoxy are 65 and 7 GPa, respectively. If the volume percentage of fibers is 33%, and the fiber efficiency is 0.2, calculate the modulus of elasticity of the fiberglass.

Solution

From Equation 11.21, we have

$$E_c = 0.67 \times 7 + 0.2 \times 0.33 \times 65 = 9.0 \text{ GPa}$$

■

Particle-Reinforced Composites The analysis of loading a particle-reinforced composite depends on the specific nature of the dispersed and matrix phases. A rigorous analysis of loading a particle-reinforced composite can become quite complex. Equations 11.7 and 11.19 serve as upper and lower bounds for the particle-reinforced properties.

11.4 Composites Sustainability

11.4.1 ■ LEED Considerations

The composites materials industry does not appear to have the same level of organization that is evident for the other materials covered in this book. American Composites Manufactures Association and the International Institute for FRP in Construction were identified but neither of these organizations have information about LEED on their web sites. There are several individual manufactures that represent their products as meeting LEED criteria. For example, Crane Composites (2010), a manufacture of composite wall and ceiling panels, provides the following concerning their product and LEED:

- The panels contain no recycled material, so they do not qualify for LEED credits in this area.
- Tenant Space, Long Term Commitment, the product warranty is 10 years which supports the LEED credit for a minimum of a 10-year occupancy commitment.
- Building Reuse Crane claims that the 10 year plus life of the product may qualify for this credit.
- Regional Materials is a possible credit if the project is within the LEED limit for regional materials
- Low-Emitting is a possible credit as the Crane panels and adhesive do not produce measureable volatile organic compounds.

11.4.2 ■ Other Sustainability Considerations

Lee and Jain (2009) state that based on energy and material resources that the argument for FRP is "questionable," for example, plastics are produced from nonrenewable petrochemicals. To improve the sustainability of the FRP, they recommend developing industry practices that focus on:

- Reducing use of natural resources which are in limited supply
- Minimizing energy during manufacturing
- Replace or reduce use of hazardous and polluting materials
- Use of waste and recycled materials for FRP composites

- Produce longer lasting and upgradable FRP products
- Minimizing life-cycle cost.

Even with these limitations, FRP can play a significant role in sustainable design and engineering. For example, the Center for Integration of Composites into Infrastructure at West Virginia University is working to develop accelerated adoption of polymer composites into the infrastructure. Other examples of the case for FRP as a sustainable material are described by Rizkalla et al. (2013), who presents several examples of FRP use in construction projects with the claim that the durable and structurally efficient precast products result in sustainable structures. Mara et al. (2014) and Hota and Liang (2011) make similar cases for the use of FRP infrastructure projects.

Summary

Combining different materials to produce a composite that has properties superior to the component materials has been practiced since ancient times. In fact, many of the conventional materials currently used in civil engineering are composites, including portland cement concrete, reinforced concrete, asphalt concrete, and engineered woods. Composites are generally classified as either fiber or particle reinforced, depending on the nature of the dispersed phase material. The properties of composites depend on the characteristics of the component materials, the bonding between the dispersed and matrix phases, and the orientation of the dispersed phase. FRPs have been used as structural components of various civil engineering structures and also in different types of repair of aged or partially damaged structures.

Questions and Problems

11.1 What is a composite material? List some composite materials that you use in your daily life.

11.2 List five different advantages of composite materials over conventional materials.

11.3 Define microscopic composites. What are the two phases of microscopic composites?

11.4 What are the two types of microscopic composites? Show the mechanism for strengthening of each type.

11.5 Why are fibers much stronger than the bulk material? Give an example of a material that is relatively weak in the bulk form and very strong in the fiber form.

11.6 Compare the desired properties of the matrix and the fiber phases of the fiber-reinforced composite.

11.7 Name three functions of the matrix phase in fiber-reinforced composites. State the reason for the need for a strong bond between the fibers and the matrix.

11.8 An epoxy is randomly reinforced with Kevlar fibers with 0.04 mm diameter, 15 mm length, 3.5 GPa ultimate tensile strength, and 50 MPa shear strength. Does this fiber length fully strengthen the composite? What is the minimum fiber length that would make the composite continuously reinforced?

11.9 What are the functions of aggregate used in portland cement concrete?

11.10 How is the load supported by asphalt concrete in the cases of tension and compression? Under what conditions is the asphalt concrete layer subjected to tension?

11.11 Briefly describe why engineered wood is stronger and has better properties than natural wood.

11.12 What are the benefits of adding dispersed steel fibers to plain concrete?

11.13 Getting measurements from Figure 11.20, determine approximate values of the following items for plain concrete, concrete with 2% steel fibers, and concrete with 3% steel fibers:

a. Modulus of elasticity
b. Ultimate strength
c. Strain at failure
d. Toughness. Curves may be approximated with a series of straight lines.
e. Comment on the effects of increasing the steel fiber content on items a, b, c, and d above.

11.14 Three 150 mm × 300 mm concrete cylinders with randomly oriented steel fiber contents of 0, 2, and 3% by weight, respectively. After curing for 28 days, the specimens were subjected to increments of compressive loads until failure. The load versus deformation results were as shown in Table P11.14.

TABLE P11.14

Specimen No.	1	2	3
Fiber Content (%)	0	2	3
Deformation (mm)		Load (kN)	
0	0	0	0
0.31	565	565	565
0.61	716 (failure)	756	792
0.91		641	756
1.22		440	690
1.52		378	641
1.83		351 (failure)	605 (failure)

Assuming that the gauge length is the whole specimen height, determine the following:

a. The compressive stresses and strains for each specimen at each load increment.
b. Plot stresses versus strains for all specimens on one graph.

    c. The modulus of elasticity for each specimen.

    d. The ultimate strength for each specimen.

    e. The strain at failure for each specimen.

    f. Toughness. Curves may be approximated with a series of straight lines.

    g. Comment on the effects of increasing the fiber content on the following:

        i.   Modulus of elasticity

        ii. Ultimate strength

        iii. Ductility

        iv. Toughness

11.15 A fiber-reinforced polymer composite under isostrain condition consists of 35% fibers and 65% polymer by volume. The moduli of elasticity of the fibers and the polymer are 260 GPa and 3.4 GPa, respectively, and the Poisson's ratios of the fibers and the polymer are 0.25 and 0.41, respectively. Calculate the following:

    a. modulus of elasticity of the composite.

    b. percentage of load carried by the fibers

    c. Poisson's ratio of the composite

11.16 Repeat Problem 11.15 for 45% fibers by volume.

11.17 Repeat Problem 11.15 under isostress condition.

11.18 Calculate the modulus of elasticity of carbon–epoxy composite under isostrain condition if the composite consists of 40% carbon fibers and 60% epoxy by volume. Also, calculate the percentage of load carried by the carbon fibers. The moduli of elasticity of the carbon fibers and the epoxy are 350 and 3.5 GPa, respectively.

11.19 Repeat Problem 11.18 for 35% carbon fibers by volume.

11.20 Repeat Problem 11.18 under isostress condition.

11.21 An aligned and continuous fiber reinforced composite consists of fibers with an elastic modulus of 290 GPa and a matrix with an elastic modulus of 3.5 GPa. If the fibers are required to carry 92% of the total load in the longitudinal direction (isostrain condition), determine the volume fraction of the fibers and matrix needed to accomplish this task. Also, determine the composite modulus of elasticity for this case.

11.22 A fiberglass composite consists of epoxy matrix reinforced with randomly oriented and uniformly distributed E-glass fibers. The moduli of elasticity of the glass fibers and the epoxy are 70 and 6 GPa, respectively. Calculate the modulus of elasticity of the fiberglass if the volume percentage of fibers is (a) 25%, (b) 50%, and (c) 75%. Plot a graph showing the relationship between the modulus of elasticity of the fiberglass and the percent of fibers. Comment on the effect of the percent of glass fibers on the modulus of elasticity of fiberglass.

11.23 A short reinforced concrete column is subjected to a 1000 kN axial compressive load. The moduli of elasticity of plain concrete and steel are 25 and 207 GPa, respectively, and the cross-sectional area of steel is 2% of that of the reinforced concrete. Considering the column as a structural member made of a composite material and subjected to load parallel to the steel rebars, calculate the following:

a. the modulus of elasticity of the reinforced concrete
b. the load carried by each of the steel and plain concrete
c. the minimum required cross-sectional area of the column given that the allowable compressive stress of plain concrete is 20 MPa and that the allowable compressive stress of plain concrete will be reached before that of steel.

11.24 An FRP composite includes 40% continuous and aligned Aramid fibers by volume. The moduli of elasticity of the Aramid fibers and the epoxy resin matrix are 130 and 3.9 GPa, respectively.
a. Compute the modulus of elasticity of the composite in the longitudinal and transverse directions.
b. If the composite cross-sectional area is 150 mm^2 and a stress of 40 MPa is applied in the longitudinal direction, compute the load carried by the fibers and the matrix.
c. Determine the strain sustained by each of the phases when stress in part (b) is applied.

11.25 The first parts of the stress–strain relationships of the components of a randomly oriented composite are shown in Figure P11.25. Estimate the volume percentage of fibers in the composite assuming a fiber efficiency parameter (K) of 0.2.

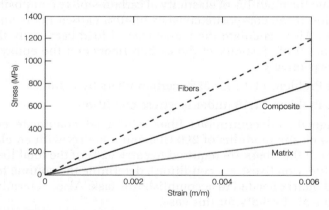

FIGURE P11.25

11.26 A circular FRP composite rod with continuous and aligned fibers has a 75 mm diameter. The rod is simply supported with a span of 1.2 m and is subjected to three-point bending with a 1.2 kN center load.
a. If the maximum allowable center deflection cannot exceed 0.5 mm, what is the minimum modulus of the composite in order to meet this requirement?
b. If the FRP has a modulus as obtained in Part a and the moduli of the matrix and the fibers are 3.5 GPa and 138 GPa, respectively, what is minimum percent of fibers that meets this requirement?

11.27 Using online information, write a one-page, double-spaced article on the use of fiber-reinforced polymers in civil engineering applications. Use at least two web sites and cite them.

11.5 References

Agarwal, B. D. and L. J. Broutman. *Analysis and Performance of Fiber Composites,* 3rd ed. New York: John Wiley and Sons, 2006.

Bakis, C. E., "Fiber-Reinforced Polymer Composites for Construction—State-of-the-Art Review," American Society of Civil Engineers, Journal of Composites for Construction, May 2002.

Callister, W. D., Jr. *Materials Science and Engineering—an Introduction.* 7th ed. New York: John Wiley and Sons, 2006.

Callister, W. D., Jr. and D. G. Rethwisch. *Fundamentals of Materials Science and Engineering—an Integrated Approach.* 3rd ed. New York: John Wiley and Sons, 2008.

Crane Composites, *Sustainability at Crane Composites,* Company Brochure, https://www.kamcoboston.com/store/redirector.asp?documentID=259&whereFrom=Product_View, Channahon IL, 2010.

Creative Pultrusions, Inc. *Creative Pultrusions Design Guide.* Alum Bank, PA: Creative Pultrusions, Inc, 2006.

Grace, N. F. *Materials for the Future-Innovative Use for FRP Materials for Construction.* Detroit, MI: The Engineering Society of Detroit, Technology Century, 2002.

Grace, N. F., F. C. Navarre, R. B. Nacy, W. Bonus, and L. Collavin. "Design-Construction of Bridge Street Bridge—First CFRP Bridge in the United States," *PCI Journal* 47(5): 20–35, September/October 2002. http://www.pci.org/pdf/journal/bridge_street.pdf.

Hota, G, and R. Liang, Advanced Fiber Reinforced Polymer Composites for Sustainable Civil Infrastructures, International Symposium on Innovation and Sustainability of Structures in Civil Engineering, Xiamen University, China, 2010.

Johnson, C.D., "Steel Fiber Reinforced Mortar and Concrete," A review of mechanical properties. In fiber reinforced concrete ACI-SP44-Detroit, 1974.

Lee, L. S. and Jain, R., The Role of FRP Composites in a Sustainable World, http://link.springer.com/article/10.1007/s10098-009-0253-0#page-1, 2009.

Mallick, P. K. *Fiber-Reinforced Composites—Materials, Manufacturing, and Design,* 3rd ed. New York: Marcel Dekker, 2007.

Mara, V., R. Haghani, and P. Harryson, *Bridge Decks of Fibre Reinforced Polymer (FRP): A Sustainable Solution*, Construction and Building Materials 50, 2014.

Mobasher, B., *Mechanics of Fiber and Textile Reinforced Cement Composites*, CRC Press, 2012.

Neal, W. "Glass Fiber Reinforced Concrete—A New Composite for Construction." *The Construction Specifier,* March, 1977. Washington, D.C.: Construction Specifications Institute, 1977.

Rizkalla, S. H., G. Lucier, D. Lunn, L. Sennour, President, H. Gleich, and J. Carson, Innovative Use of FRP for Sustainable Precast Structures, FACADE TECTONICS Journal: Number 8: 2013.

Shackelford, J. E. *Introduction to Materials Science for Engineers,* 8th ed. Upper Saddle River, NJ: Prentice Hall, 2014.

APPENDIX

LABORATORY MANUAL

The laboratory tests discussed in this appendix can be performed as a part of the civil and construction engineering materials course. More tests are included in this manual than typically can be performed in one semester. The extent of tests provided gives the instructor flexibility to choose the appropriate tests. In order for the students to get the most benefit from the laboratory sessions, tests should be coordinated with the topics covered in the lectures.

This laboratory manual summarizes the main components of each test method. Students are encouraged to read the corresponding ASTM or AASHTO test methods for the detailed laboratory procedures. The ASTM and AASHTO standards are usually available in college libraries.

In many cases, the time available for a laboratory session is not sufficient to permit the performance of the complete test, as specified in the ASTM or AASHTO procedure. Therefore, the instructor may limit the number of specimens or eliminate some portions of the test. However, the student should be aware of the complete test procedure and the specification requirements.

In some cases, different experiments are used to obtain the same material properties, such as the air content of freshly mixed concrete. In such cases, the instructor may require the specific test used by the state or the test for which equipment is available.

It is recommended that the instructors would show the students video clips before running the experiments in the lab. Video clips are available online for many of the tests used in this book.

Typically a laboratory report is required from the student after each laboratory session. The format of the report can vary, depending on the requirements of the instructor. A suggested format may include the following items:

- Title of the experiment
- ASTM or AASHTO designation
- Purpose
- Significance and use
- Test materials
- Main pieces of equipment
- Summary of test procedure and test conditions
- Raw data sheets—may be placed in an appendix
- Test results and analysis
- Comments, conclusions, and recommendations. Any deviations from the standard test procedure should be reported and justified.

Experiment No. 1
Introduction to Measuring Devices

▓ ASTM Designation

There is no ASTM procedure for the main portion of this session. The information at the end of Chapter 1 can be used as a reference. Some discussion on precision and bias can be helpful; this is included in ASTM C670 (Practice for Preparing Precision and Bias Statements for Test Methods for Construction Materials).

▓ Purpose

To introduce the students to common measuring devices such as dial gauges, LVDTs, strain gauges, proving rings, and extensometers. An introduction to precision and bias can also be included.

▓ Apparatus

The instructor may demonstrate one or more of the following items:
- A few dial gauges with different ranges and sensitivities (Figures A.1 and 1.26)
- LVDT (Figures 1.28–1.30) and necessary attachments such as power supply, signal conditioner, voltmeter, display device, and calibration device
- Extensometer (Figures A.2, 1.27, 1.31, and 1.33)
- Strain gauge (Figure 1.32) and necessary attachments
- Proving ring (Figures A.3 and 1.34)
- Load cell (Figures A.4 and 1.35)

▓ Calibration

The instructor may require the students to calibrate one or more measuring devices, such as a proving ring or an LVDT. Static loads can be used to calibrate the proving ring to develop a

FIGURE A.1   Examining the sensitivity and range of a dial gauge.

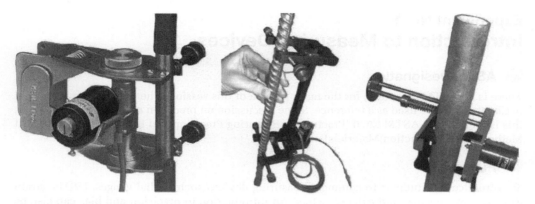

FIGURE A.2 Extensometers. (Courtesy of Instron®)

FIGURE A.3 Calibrating a proving ring using a known weight.

relation between force and the reading of the proving ring (Figure A.3). To calibrate the LVDT, a calibration device equipped with a micrometer, such as that shown in Figures A.5 and 1.29 is used. A relation is developed between voltage and displacement of the LVDT.

Requirements

- Brief description of the demonstrated device(s) including the use, components, theory, and sensitivity
- Calibration table, graph, and equation for each device calibrated

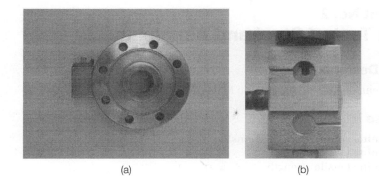

(a)   (b)

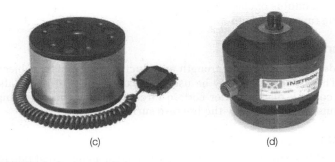

(c)   (d)

FIGURE A.4 Load cells.

FIGURE A.5 Calibrating an LVDT with a micrometer using a calibration device.

Experiment No. 2
Tension Test of Steel and Aluminum

▪ ASTM Designation

ASTM E8—Tension Testing of Metallic Materials

▪ Purpose

- To determine stress–strain relationship
- To determine yield strength
- To determine tensile strength
- To determine elongation and reduction of cross-sectional area
- To determine modulus of elasticity
- To determine rupture strength

▪ Significance and Use

This test provides information on strength and ductility for metals subjected to a uniaxial tensile stress. This information may be useful in comparison of materials, alloy developments, quality control, design under certain circumstances, and detecting nonuniformity and imperfections, as indicated by the fracture surface.

▪ Apparatus

- A testing machine capable of applying tensile load at a controlled rate of deformation or load. The testing machine could be either mechanical or closed-loop electrohydraulic. The machine could be equipped with a dial gauge to indicate the load or could be connected to a chart recorder or computer to record load and deformation.
- A gripping device, used to transmit the load from the testing machine to the test specimen and to ensure axial stress within the gauge length of the specimen
- An extensometer with an LVDT or dial gauge used to measure the deformation of the specimen
- Caliper to measure the dimensions of the specimen

▪ Test Specimens

Either plate-type or rounded specimens can be used, as shown in Figures A.6 and 3.15. Specimen dimensions are specified in ASTM E8.

▪ Test Procedure

1. Mark the gauge length on the specimen, either by slight notches or with ink.
2. Place the specimen in the loading machine (Figure 3.16).
3. Attach the extensometer to the specimen (Figure A.7).
4. Set the load reading to zero, then apply load at a rate less than 690 MPa/min. Unless otherwise specified, any convenient speed of testing may be used up to half of the specified yield strength or yield point, or one quarter of the specified tensile strength, whichever is smaller. After the yield strength or yield point has been determined, the strain rate may be increased to a maximum of 0.5 mm/mm of the gauge length per minute.

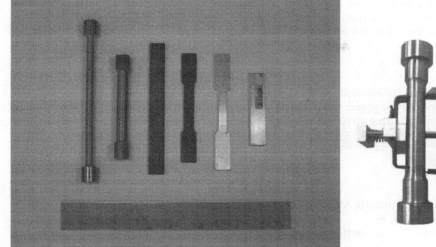

FIGURE A.6 Rounded and plate-type steel and aluminum tension test specimens.

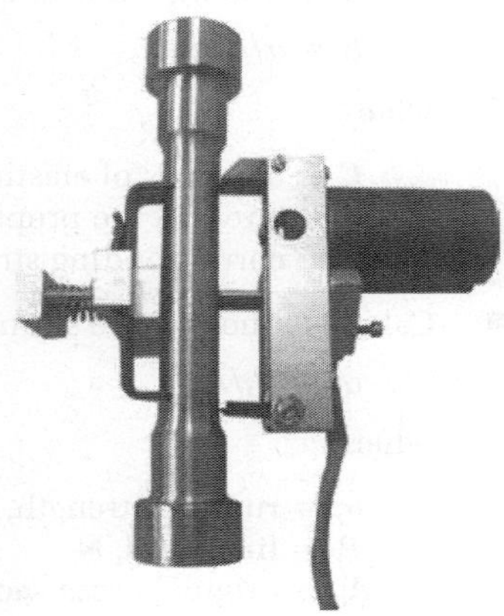

FIGURE A.7 Extensometer attached to a steel specimen. (Courtesy of Instron®)

5. Continue applying the load until the specimen breaks.
6. Record load and deformation every 2.2 kN increment for steel and every 890 N increment for aluminum, both before and after the yield point.

Analysis and Results

- Calculate the stress and strain for each load increment until failure.
- Plot the stress versus strain curve.
- Determine the yield strength using the offset method, extension method (Figure 1.7), or by observing the sudden increase in deformation.
- Calculate the tensile strength

$$\sigma = P_{\text{max}}/A_o$$

where

σ = tensile strength, MPa
P_{max} = maximum load carried by the specimen during the tension test, N
A_o = original cross-section area of the specimen, mm^2

- Calculate the elongation

Percent elongation = $[(L_s - L_0)/L_0] \times 100$

where

L_s = gauge length after rupture, mm
L_o = original gauge length, mm

For elongation > 3.0%, fit the ends of the fractured specimen together and measure L_s as the distance between two gauge marks. For elongation ≤3.0%, fit the fractured ends together and apply an end load along the axis of the specimen sufficient to close the fractured ends together, then measure L_s as the distance between gauge marks.

- Calculate the modulus of elasticity

$$E = \sigma/\varepsilon$$

where

E = modulus of elasticity, MPa
σ = stress at the proportional limit, MPa
ε = corresponding strain, m/m

- Calculate the rupture strength

$$\sigma_r = P_f/A_o$$

where

σ_r = rupture strength, MPa
P_f = final load, N
A_o = original cross-sectional area, mm^2

- Calculate the reduction of cross-sectional area

Percent reduction in cross-sectional area = $[(A_o - A_s)/A_o] \times 100$

where

A_s = cross section after rupture, mm^2

To calculate the cross section after rupture, fit the ends of the fractured specimen together and measure the mean diameter or width and thickness at the smallest cross section.

Replacement of Specimens

The test specimen should be replaced if

- the original specimen had a poorly machined surface.
- the original specimen had wrong dimensions.
- the specimen's properties were changed because of poor machining practice.
- the test procedure was incorrect.
- the fracture was outside the gauge length.
- for elongation determination, the fracture was outside the middle half of the gauge length.

Report

- Stress–strain relationship, data table, and graph
- Yield strength and the method used, e.g., 0.2% offset method
- Tensile strength
- Elongation and original gauge length
- Modulus of elasticity
- Rupture strength
- Reduction of cross-sectional area

Experiment No. 3
Torsion Test of Steel and Aluminum

ASTM Designation

ASTM E143—Shear Modulus at Room Temperature

Purpose

To determine the shear modulus of structural materials, such as steel and aluminum.

Significance and Use

Shear modulus is a material property that is useful in calculating the compliance of structural materials in torsion, provided that they follow Hooke's law; that is, that the angle of twist is proportional to the applied torque.

Apparatus

- Torsion testing machine (Figures A.8 and 3.20)
- Grips to hold the ends of the specimen in the jaws of the testing machine
- Twist gauges to measure the angle of twist

Test Specimens

Either solid or hollow specimens with circular cross section can be used (Figure A.9). Specimens should be chosen from sound, clean material. Slight imperfections near the surface, such as fissures that would have negligible effect in determining Young's modulus, may cause appreciable errors in shear modulus. The gauge length should be at least four diameters. The length between grips should at least be equal to the gauge length plus two to four diameters.

Test Procedure

1. Measure the diameter (and the wall thickness in case of tubular specimens) (Figure A.10).
2. Place the specimen in the torsion testing machine with careful alignment and apply torque (Figure A.11).
3. Make simultaneous measurements of torque and angle of twist, and record the data.
4. Maintain the speed of testing high enough to make creep negligible.

FIGURE A.8   Torsion testing machine. (Courtesy of Instron®)

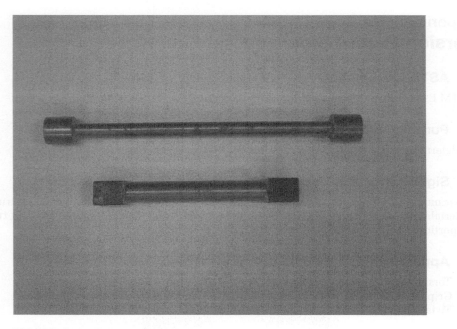

FIGURE A.9 Torsion specimens.

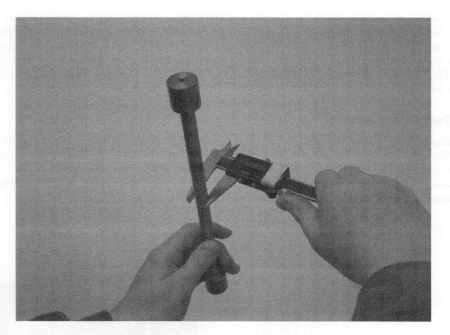

FIGURE A.10 Measuring specimen diameter with a caliper.

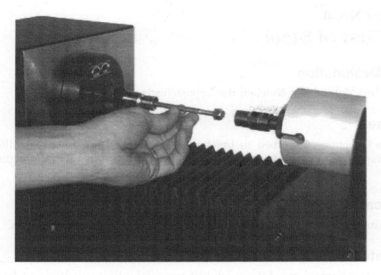

FIGURE A.11 Placing specimen in torsion machine. (Courtesy of Instron®)

Analysis and Results

- Calculate the maximum shear stress (τ_{max}) and shear strain (γ) using the formulas

$$\tau_{max} = Tr/J$$
$$\gamma = \theta r/L$$

where

T = torque,
r = radius,
J = polar moment of inertia of the specimen about its center,
θ = angle of twist in radians, and
L = gauge length.

- Plot a graph of shear stress versus shear strain, as shown in Figure 3.21.
- Determine the shear modulus G as the slope of the straight portion of the shear stress versus strain relation

$$G = \tau_{max}/\gamma$$

Report

- Shear stress versus shear strain graph
- Shear modulus

Experiment No. 4
Impact Test of Steel

ASTM Designation

ASTM E23—Test Methods for Notched Bar Impact Testing of Metallic Materials

Purpose

To determine the energy absorbed in breaking notched steel specimens at different temperatures, using the Charpy V Notch test. The energy value is a measure of toughness of the material.

Significance and Use

This test measures a specimen's change in toughness as the temperature changes.

Apparatus

Use an impact-testing machine of the pendulum type of rigid construction and capacity more than sufficient to break the specimen in one blow (Figures A.12 and 3.23).

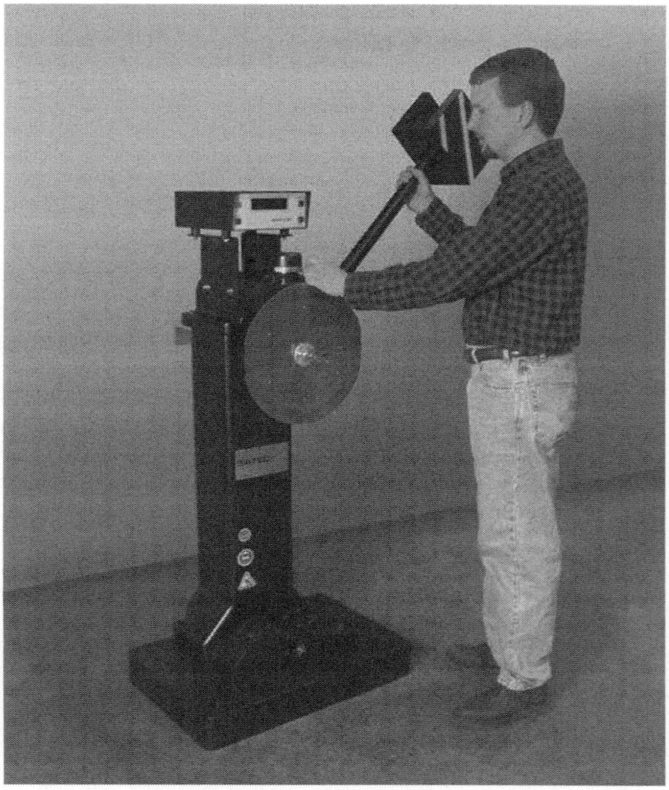

FIGURE A.12 Charpy V notch impact testing machine. (Courtesy of Instron®)

Test Specimen

Steel specimens prepared according to ASTM E23, as shown in Figure 3.22.

Test Conditions

The test can be performed at the following four temperatures:

- −40°C (dry ice + isopropyl alcohol)
- −18°C (dry ice + 30% isopropyl alcohol + 70%water)
- 4°C (ice + water)
- 40°C (oven)

Test Procedure

1. Prepare the impact-testing machine by lifting up the pendulum and adjusting the gauge reading to zero. Since the pendulum is heavy, be careful to handle it safely.
2. Remove the test specimen from the temperature medium (Figure A.13) using tongs, and immediately position it on the two anvils in the impact testing machine as shown in Figure A.14.
3. Release the pendulum without vibration by pushing the specified button. The time between removing the specimen from the temperature medium and the completion of the test should be less than 5 seconds.
4. Record the energy required to break the specimen by reading the gauge mark.

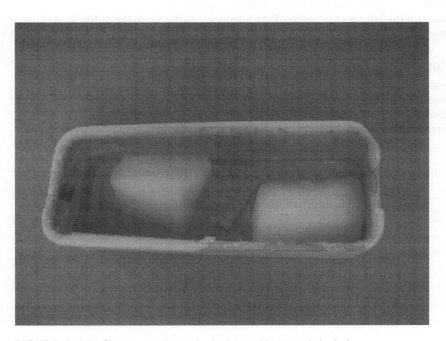

FIGURE A.13 Charpy specimens in dry ice and isopropyl alcohol.

FIGURE A.14 Placing specimen in Charpy machine.

5. Observe the fracture surface appearance (Figure 3.24).
6. Measure the lateral expansion of the specimen using a caliper or the lateral expansion gauge specified in ASTM E23.

Report

- Energy required to cause fracture versus temperature, data table, and plot
- Ductile-to-brittle transition temperature zone
- Fracture surface appearance (each specimen and temperature)
- Lateral expansion (each specimen and temperature)

Experiment No. 5
Microscopic Inspection of Materials

▨ ASTM Designation

None.

▨ Purpose

- To observe the microstructural characteristics of materials
- To compare the microstructure of different materials
- To compare metals subjected to different heat treatments
- To observe grain boundaries of metals
- To examine fractured surfaces
- To observe microcracks in concrete (Figure A.15)
- To observe fibers in fiber-reinforced concrete
- To observe the microscopic properties of asphalt concrete
- To observe grains in wood, and so on.

▨ Apparatus

Optical microscope (Figure A.16) or scanning electron microscope

▨ Material

Various materials can be used, such as metals, portland cement concrete, asphalt concrete, or wood.

▨ Report

Compare and discuss the microstructural characteristics of the materials.

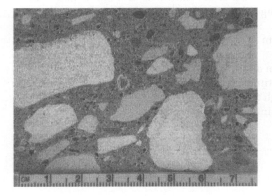

FIGURE A.15 Concrete microstructure using optical microscope.

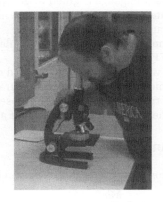

FIGURE A.16 Viewing concrete specimen with an optical microscope.

Experiment No. 6
Creep in Polymers

 ASTM Designation

None.

 Purpose

To demonstrate the concept of creep and viscoelasticity using a simple experimental tool and to predict the creep behavior of a polymer using simple equations for long term loading.

Significance and Use

The creep test results can be used to predict the long term deformation of a polymer structure subjected to sustained loading.

Apparatus

Tip loads, strain gauge and a data acquisition system that plots strains as a function of time.

Materials and Test Specimens

One or more polymers with different molecular structures can be used such as:

- Low molecular weight polyethylene
- High molecular weight polyethylene
- Polycarbonate
- Polystyrene
- Poly methyl methacrylate (PMMA)

The specimen is a beam with dimensions of 300 mm long × 50 mm wide × 15 mm thick.

Test Procedure

1. A strain gauge is glued to the specimen at the tip of the cantilever and connected to the data acquisition system. Typical strain gauge glues can be used, which are typically bought with strain gauges.
2. The specimen is fixed to a table in the form of a cantilever beam with a clamp as shown in Figure A.17(a) and (b).
3. A weight is hung from the end of the beam and the strain in the beam is monitored as time passes so that a plot of strain versus time can be obtained directly. Any frequency of recording the data can be used, although 2-5 data points per minute are good enough. The test is expected to last for about 30 minutes.

Analysis and Results

1. The stress at the gauge is given by:

$$\sigma = \frac{Mh}{2I}$$

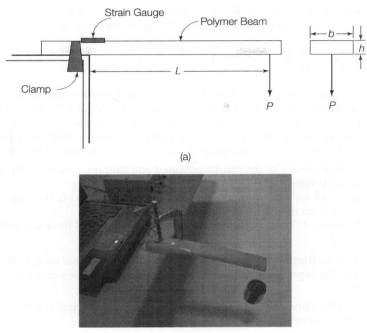

(a)

(b)

FIGURE A.17   Experimental set up: (a) schematic and (b) photo.

where

σ = applied stress, MPa

M = bending moment = $P \times L$, N · mm

h = thickness, mm

I = moment of inertia = $\dfrac{bh^3}{12}$, mm⁴

b = width, mm

2. The Burgers model (Figures A.18, 1.13) will be used to fit an equation for the creep curves obtained. In the Burgers model, each of its mechanical elements has a constant associated with it. It should be noted that polymers are not actually made up of spring and dashpots, but it is merely a convenient way to visualize the various viscoelastic responses in order to derive an equation that describes them.

Equation 1.10 shows the Burgers model equation in terms of force F and deformation δ. For small deformations, the deformation δ can be converted into strain ε using

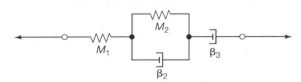

FIGURE A.18   Mechanical representation of Burgers model.

material constants and the force F can be converted into stress σ using the above-mentioned equation. Therefore, Equation 1.10 can be rewritten as:

$$\varepsilon(t) = \left[\frac{\sigma}{M_1} + \left(\frac{\sigma}{M_2} \right)\left(1 - e^{-M_2 t/\beta_2} \right) + \frac{\sigma t}{\beta_3} \right]$$

where $\varepsilon(t)$ is strain, σ stress, t time, and M_1, M_2, β_2, and β_3 are model constants. Note that these parameters do not have the same values as Equation 1.10. Note also that a particular polymer may not exhibit all three material responses (described by the three terms of the equation), in which case a term is simply dropped from the equation. The data fitting in the next step would show if a certain region is needed or not.

3. Fit the previous equation to the results obtained and find the constants M_1, M_2, β_2, and β_3. The steps to fit the equation are as follows:
   a. For the polymer tested, decide how many terms are required in the above-mentioned equation based on the observed behavior.
   b. Find the values of the constants in the equation, which will have units of a modulus or a viscosity.

The procedure for finding the values of these constants are based on the creep curve obtained from the experiment. A generalized version of the creep curve is shown in Figure A.19.

   a. Section I of the curve is the instantaneous elastic response and is governed by the M_1 term, which is given by

   $$M_1 = \frac{\sigma}{\varepsilon}$$

   b. Section III of the curve corresponds to the long-term viscous flow of the material and is associated with the β_2, which is governed by the slope of the linear section III:

   $$\beta_2 = \frac{\sigma}{\dot{\varepsilon}}$$

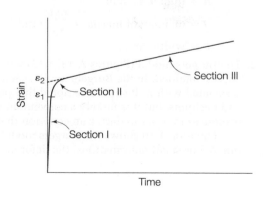

FIGURE A.19 **Generalized creep curve.**

c. The extrapolation of the section III portion of the creep curve back till it intersects with zero time provides ε_2, which is the sum of both the instantaneous and delayed elastic strains. Thus, $\varepsilon_2 - \varepsilon_1$ represents the amount of delayed elastic strain and is governed by M_2 in the general equation, given by

$$M_2 = \frac{\sigma}{\varepsilon_2 - \varepsilon_1}$$

d. The value β_3 controls the curvature of section II of the creep curve. It can be found by taking data points within section II (t_{II}, ε_{II}) and putting them into the overall equation with the previously calculated values of M_1, M_2, and β_2 and solving for β_3. It is best to do this along several points along the curved portion and using the average as β_3.

e. Use a graph to compare the measurements and the results of the equation.

4. Use the equation developed in Step 3 to calculate the expected strain at the gauge if a 1-kg mass is hung on the beam for 1 month. Also, calculate the time that will be required for a 1-kg mass to generate a strain of 0.1 m/m at the gauge. Report these values in tabular form in your results.

Report

1. Include a tabular summary of the constants for polymer(s) tested, with the moduli being reported in kPa and viscous constants in units of poise; 1 poise $= 0.1$ Pa $\cdot$ s.
2. Include the strain versus time plots used for the calculations, adequately labeling the different sections and the corresponding model constants/variables on the plot (similar to Figure A.19).
3. Comment on the following items:

   a. Why is the material "viscoelastic"? What is meant by this term?
   b. If more than one material is tested, how did the creep deformation of the different beams differ? Which ones demonstrated the most and least strains?
   c. Understanding the relationship between instantaneous and delayed elastic response, how does the strain behave at prolonged periods? What do you think would happen after the load is removed?

Experiment No. 7
Sieve Analysis of Aggregates

▩ ASTM Designation

ASTM C136—Sieve Analysis of Fine and Coarse Aggregates

▩ Purpose

To determine the particle size distribution of fine and coarse aggregate by dry sieving.

▩ Significance and Use

This test is used to determine the grading of materials that are to be used as aggregates. It ensures that particle size distribution complies with applicable requirements and provides the data necessary to control the material of various aggregate products and mixtures containing aggregates. The data may also be useful in developing relationships concerning porosity and packing.

▩ Apparatus

- Balances or scales with a minimum sensitivity of 0.5 g for coarse aggregate or 0.1 g for fine aggregate
- Sieves
- Mechanical sieve shaker (Figures A.20, A.21, and 5.12)
- Oven capable of maintaining a uniform temperature of 110 ± 5°C
- Sample splitter to reduce the quantity of the material to the size required for sieve analysis (Figure 5.21)

▩ Test Specimens

Thoroughly mix the aggregate sample and reduce it to an amount suitable for testing, using a sample splitter or by quartering. The minimum sample size should be as follows:

	Minimum Mass, kg
Fine aggregate with at least 95% passing 2.36-mm sieve	0.1
Fine aggregate with at least 85% passing 4.75-mm sieve	0.5
Coarse aggregate with a nominal maximum size of 9.5 mm	1
Coarse aggregate of a nominal maximum size of 12.5 mm	2
Coarse aggregate of a nominal maximum size of 19.0 mm	5
Coarse aggregate of a nominal maximum size of 25.0 mm	10
Coarse aggregate of a nominal maximum size of 37.5 mm	15

▩ Test Procedure

1. Dry the aggregate test sample to a constant weight at a temperature of 110 ± 5°C, then cool to room temperature.
2. Select suitable sieve sizes to furnish the information required by the specifications covering the material to be tested. Common sieves in millimeters are 37.5, 25, 19, 12.5, 9.5, 4.75, 2.36, 1.18, 0.6, 0.3, 0.15, and 0.075 mm.

FIGURE A.20   Table-top sieve shaker and sieves. (Courtesy of Humboldt Mfg. Co.)

FIGURE A.21
Hanging-type sieve shaker and sieves for small samples of aggregates.

3. Nest the sieves in order of decreasing size of opening, and place the aggregate sample on the top sieve (Figure A.22).
4. Agitate the sieves by hand or by mechanical apparatus for a sufficient period. The criterion for sieving time is that, after completion, not more than 1% of the residue on any individual sieve will pass that sieve during 1 minute of continuous hand sieving.
5. Determine the weight of each size increment (Figure A.23).
6. The total weight of the material after sieving should be compared with the original weight of the sample placed on the sieves. If the amounts differ by more than 0.3%, based on the original dry sample weight, the results should not be used for acceptance purposes.

Analysis and Results

1. Calculate percentages passing, total percentages retained, or percentages of various sizes of fractions to the nearest 0.1%, on the basis of the total weight of the initial dry sample.
2. Plot the grain size distribution on a semilog graph paper (Figure A.24).
3. Plot the grain size distribution on a 0.45 power graph paper (Figure A.25).
4. Calculate the fineness modulus.

FIGURE A.22 Placing aggregate sample in the sieves before sieving.

FIGURE A.23 Weighing aggregate retained in sieves.

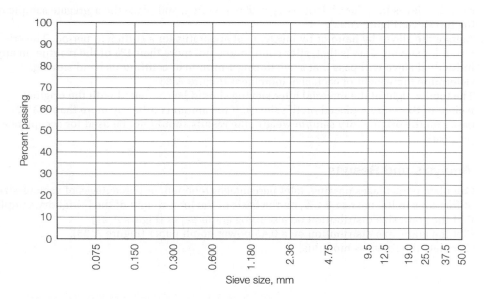

FIGURE A.24 Semi-log aggregate gradation chart.

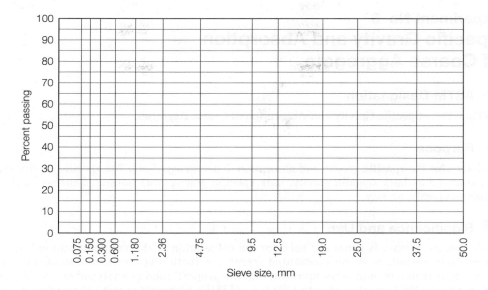

FIGURE A.25 0.45 power gradation chart.

Report

- Percentage of material retained between consecutive sieves, cumulative percentage of material retained on each sieve, or percentage of material passing each sieve. Report percentages to the nearest whole number; except if percentage passing 0.075 mm sieve is less than 10%, it should be reported to the nearest 0.1%.
- Grain size distribution plots using both semilog and 0.45 power gradation charts.
- Fineness modulus to the nearest 0.01.

Experiment No. 8
Specific Gravity and Absorption of Coarse Aggregate

ASTM Designation

ASTM C127—Specific Gravity and Absorption of Coarse Aggregate

Purpose

To determine the specific gravity and absorption of coarse aggregate. The specific gravity may be expressed as bulk specific gravity, bulk specific gravity SSD (saturated-surface dry), or apparent specific gravity.

Significance and Use

Bulk specific gravity is generally used for the calculation of the volume occupied by the aggregate in various mixtures containing aggregates, including portland cement concrete, bituminous concrete, and other mixtures that are proportioned or analyzed on an absolute volume basis. Bulk specific gravity SSD is used if the aggregate is wet. Absorption values are used to calculate the change in the weight of aggregate due to water absorbed in the pore spaces within the constituent particles, compared with the dry condition.

Apparatus

- Balance accurate to 0.05% of the sample weight or 0.5 g, whichever is greater
- Wire basket 3.35 mm or finer mesh (Figure A.26)
- Water tank
- 4.75-mm sieve or other sizes as needed

Test Specimens

- Thoroughly mix the aggregate sample and reduce it to the approximate quantity needed, using an aggregate sample splitter or by quartering.

FIGURE A.26 Wire basket and water tank used to determine bulk specific gravity and absorption of coarse aggregate.

- Reject all materials passing 4.74 mm sieve by dry sieving and thoroughly washing to remove dust or other coatings from the surface.
- The minimum mass of test specimen to be used depends on the nominal maximum size as follows:

Nominal Maximum Size, mm	Minimum Mass, kg
12.5	2
19.0	3
25.0	4
37.5	5

Test Procedure

1. Immerse the aggregate in water at room temperature for a period of 24 ± 4 hours.
2. Remove the test specimen from water and roll it in a large absorbent cloth until all visible films of water are removed. Wipe the larger particles individually.
3. Weigh the test sample in saturated surface–dry condition, and record it as B. Record this weight and all subsequent weights to the nearest 0.5 g or 0.05% of the sample weight, whichever is greater.
4. Place the specimen in the wire basket and determine its weight while it is submerged in water at a temperature of 23 ± 1.7°C, and record it as C. Take care to remove all entrapped air before weighing it by shaking the container while it is immersed.
5. Dry the test sample to a constant weight at a temperature of 110 ± 5°C, and weigh it and record this weight as A.

Analysis and Results

1. Bulk specific gravity = $A/(B - C)$
   where

   > A = mass of oven-dry sample in air, grams
   > B = mass of saturated surface–dry sample in air, grams
   > C = mass of saturated sample in water, grams

2. Bulk specific gravity (SSD) = $B/(B - C)$
3. Apparent specific gravity = $A/(A - C)$
4. Absorption, % = $[(B - A)/A] \times 100$

Report

- Bulk specific gravity
- Bulk specific gravity SSD
- Apparent specific gravity
- Absorption

Experiment No. 9
Specific Gravity and Absorption of Fine Aggregate

ASTM Designation

ASTM C128—Specific Gravity and Absorption of Fine Aggregate

Purpose

To determine the specific gravity and absorption of fine aggregate. The specific gravity may be expressed as bulk specific gravity, bulk specific gravity SSD (saturated-surface dry), or apparent specific gravity.

Significance and Use

Bulk specific gravity is the characteristic generally used for calculating the volume occupied by the aggregate in various mixtures, including portland cement concrete, bituminous concrete, and other mixtures that are proportioned or analyzed on an absolute volume basis.

Apparatus

- Balance or scale with a capacity of 1 kg or more, sensitive to 0.1 g or less, and accurate within 0.1% of the test load
- Pycnometer or other suitable container into which the fine aggregate test sample can be readily introduced. A volumetric flask of 500 cm³ capacity with a pycnometer top is satisfactory for a 500-g test sample of most fine aggregates (Figure A.27).
- Mold in the form of a frustum of a cone
- Tamper having a mass of 340 ± 15 g

Test Procedure

1. Measure the weight of the pycnometer filled with water to the calibration mark. Record the weight as *B*.

FIGURE A.27 Mold, tamper, and volumetric flask used to determine bulk specific gravity and absorption of fine aggregate.

2. Obtain approximately 1 kg of the fine aggregate sample.
3. Dry the aggregate sample in a suitable pan to constant weight at temperature of 110 ± 5°C and allow it to cool; then cover it with water, either by immersion or by the addition of at least 6% moisture to the fine aggregate, and permit it to stand for 24 ± 4 hours.
4. Decant excess water with care to avoid loss of fines, spread the sample on a flat, nonabsorbent surface exposed to a gently moving current of warm air, and stir frequently to cause homogeneous drying. If desired, mechanical aids such as tumbling or stirring may be used to help achieve the saturated surface–dry condition. Continue this operation until the test specimen approaches a free-flowing condition.
5. Hold the mold firmly on a smooth, nonabsorbent surface with the large diameter down. Place a portion of the partially dried fine aggregate loosely in the mold by filling it to overflowing and heaping additional material above the top of the mold by holding it with the cupped fingers of the hand.
6. Lightly tamp the fine aggregate into the mold with 25 light drops of the tamper. Each drop should start about 5 mm above the top of surface of the aggregate. Permit the tamper to fall freely under gravitational attraction on each drop.
7. Remove loose sand from the base and lift the mold vertically. If the surface moisture is still present, the fine aggregate will retain the molded shape. If this is the case, allow the sand to dry and repeat steps 4, 5, and 6 until the fine aggregate slumps slightly indicating that it has reached a surface-dry condition.
8. Weigh 500 ± 10 g of SSD sample and record the weight; record as S.
9. Partially fill the pycnometer with water and immediately introduce into the pycnometer the SSD aggregate weighed in step 8. Fill the pycnometer with additional water to approximately 90% of the capacity. Roll, invert, and agitate the pycnometer to eliminate all air bubbles. Fill the pycnometer with water to its calibrated capacity.
10. Determine the total weight of the pycnometer, specimen, and water, and record it as C.
11. Carefully work all of the sample into a drying pan. Place in a 110 ± 10°C oven until it dries to a constant weight. Record this weight as A.

Analysis and Results

Bulk specific gravity = $A/(B + S - C)$
where

A = mass of oven-dry specimen in air, grams
B = mass of pycnometer filled with water, grams
S = mass of the saturated surface-dry specimen, grams
C = mass of pycnometer with specimen and water to the calibration mark, grams

Bulk specific gravity (SSD) = $S/(B + S - C)$

Apparent specific gravity = $A/(B + A - C)$

Absorption, % = $[(S - A)/A] \times 100$

Report

- Bulk specific gravity
- Bulk specific gravity SSD
- Apparent specific gravity
- Absorption

Experiment No. 10
Bulk Unit Weight and Voids in Aggregate

ASTM Designation

ASTM C29—Bulk Density (Unit Weight) and Voids in Aggregate

Purpose

To determine the bulk unit weight and voids in aggregate in either a compacted or loose condition.

Significance and Use

The bulk density of aggregate is needed for the proportioning of portland cement concrete mixtures. The bulk density may also be used to determine the mass/volume relationships for conversions in purchase agreements. The percentage of voids between the aggregate particles can also be determined, on the basis of the obtained bulk density.

Apparatus

- Measure. Use a rigid metal watertight container with a known volume. A minimum volume of the measure is required for different nominal maximum sizes of coarse aggregate. For a 25-mm nominal maximum aggregate size, a minimum volume measure of 0.0093 m³ (9.3 liters) is required.
- Balance, tamping rod, shovel or scoop, and a plate glass.

Test Procedure

1. Calibrate the measure as follows:
   a. Fill the measure with water at room temperature and cover with a plate glass in such a way as to eliminate bubbles and excess water.
   b. Determine the mass of the water in the measure.
   c. Measure the temperature of the water, and determine its density as shown in the table. Interpolate as necessary.

Temperature	Density
°C	kg/m³
15.6	999.01
21.1	997.97
26.7	996.59
29.4	995.83

   d. Calculate the volume of the measure by dividing the mass of the water by its density.

2. Fill the measure with aggregate and compact it, either by rodding [for aggregates having nominal maximum size of 37.5 mm or less], or jigging

FIGURE A.28   Rodding aggregate in the container.

[for aggregates having a nominal maximum size of 37.5 to 125 mm], or shoveling (if specifically stipulated).

   a. Rodding Procedure: Fill the measure with aggregate in three layers of approximately equal volumes. Rod each layer of aggregate with 25 strokes of the tamping rod, evenly distributed over the surface (Figure A.28).
   b. Jigging Procedure: Fill the measure with aggregate in three layers of approximately equal volumes. Compact each layer by placing the measure on a firm base, raising the opposite sides alternately about 50 mm, and allowing the measure to drop 25 times on each side.
   c. Shoveling Procedure: Fill the measure to overflowing by means of a shovel or scoop, discharging the aggregate from a height not to exceed 50 mm above the top of the measure. Exercise care to avoid segregation.

3. Level the surface of the aggregate with the fingers or a straightedge. Determine the net weight of the aggregate to the nearest 0.05 kg (Figure A.29).

▓ Analysis and Results

$$M = \frac{G - T}{V}$$

$$\% \text{ Voids} = \frac{(SW) - M}{SW} \times 100$$

FIGURE A.29 Weighing the aggregate.

where

M = bulk unit mass of aggregate, kg/m³
G = mass of the aggregate plus the measure, kg
T = mass of the measure, kg
V = volume of the measure, m³
S = bulk specific gravity (dry basis) (ASTM C127 or C128)
W = unit mass of water, 998 kg/m³

Report

Report the bulk unit mass (or loose bulk unit mass in case of shoveling), void content, and method of compaction.

Experiment No. 11
Slump of Freshly Mixed Portland Cement Concrete

ASTM Designation
ASTM C143—Slump of Portland Cement Concrete

Purpose
To determine the slump of freshly mixed portland cement concrete, both in the laboratory and in the field.

Significance and Use
This method measures the consistency of freshly mixed portland concrete cement (PCC). To some extent, this test indicates how easily concrete can be placed and compacted, or the workability of concrete. This test is used both in the laboratory and in the field for quality control.

Apparatus
- Mold in the form of lateral surface of frustum with a top diameter of 102 mm, bottom diameter of 203 mm, and height of 305 mm (Figure 7.4)
- Tamping rod with a length of 0.6 m, diameter of 16 mm, and rounded ends

Test Procedure
1. Mix concrete either manually or with a mechanical mixer (Figure A.30). If a large quantity of mixed concrete exits, obtain a representative sample.
2. Dampen the mold and place it, with its larger base at the bottom, on a flat, moist, nonabsorbent rigid surface.
3. Hold the mold firmly in place by standing on the two foot pieces.
4. Immediately fill the mold in three layers, each approximately one-third of the volume of the mold. Note that one-third of the volume is equivalent to a depth of 67 mm [the top of the first layer should be 238 mm from the top of the mold], and two-thirds of the volume is equivalent to 155 mm [the top of the second layer should be 150 mm from the top of the mold].
5. Rod each layer 25 strokes, using the tamping rod (Figure A.31). Uniformly distribute the strokes over the cross section of each layer. Rod the second and top layers each throughout its depth so that the strokes penetrate the underlying layer. In filling and rodding the top layer, heap the concrete above the mold before rodding is started. If the rodding operation results in subsidence of concrete below the top edge of the mold, add additional concrete to keep an excess of concrete above the top of the mold at all times.
6. After the top layer has been rodded, strike off the surface of concrete by means of a screening and rolling motion of the tamping rod.
7. Remove the mold immediately from the concrete by raising it up carefully without lateral or torsional motion. The slump test must be completed within 2.5 minutes after taking the sample.

FIGURE A.30   Mixing concrete in a mixer.

FIGURE A.31   Rodding fresh concrete in the slump cone with a tamping rod.

FIGURE A.32   Measuring the slump of freshly mixed concrete.

8. Measure the slump by determining the vertical difference between the top of the mold and the displaced original center of the top of the specimen, as shown in Figure A.32. If two consecutive tests on a sample of concrete show a falling away or a shearing off of a portion of concrete from the mass of the specimen, the concrete probably lacks the necessary plasticity and cohesiveness for the slump test to be applicable and the test results will not be valid.

Report

- Report the slump value to the nearest 5 mm

Experiment No. 12
Unit Weight and Yield of Freshly Mixed Concrete

ASTM Designation

ASTM C138—Unit Weight, Yield, and Air Content (Gravimetric) of Concrete

Purpose

To determine the unit weight, yield, cement content, and air content of freshly mixed portland cement concrete. Yield is defined as the volume of concrete produced from a mixture of known quantities of the component materials.

Significance and Use

The unit weight value is used to calculate the volume of portland cement concrete produced from a mixture of known quantity.

Apparatus

- Measure. Use a rigid metal watertight container with a known volume. A minimum volume of the measure is required for different nominal maximum sizes of coarse aggregate. For a 25-mm nominal maximum aggregate size, a minimum volume measure of 6 liters is required.
- Balance, tamping rod, internal vibrator (optional), strike-off plate, and mallet.

Test Procedure

1. Place the freshly mixed concrete in the measure in three layers of approximately equal volume.
2. Rod each layer with 25 strokes of the tamping rod when a 0.014 m^3 or smaller measure is used, otherwise use 50 strokes per layer. Distribute the strokes uniformly over the cross section of the measure. Rod the bottom layer throughout its depth, but do not forcibly strike the bottom of the measure. For the top two layers, penetrate about 25 mm into the underlying layer.
3. Tap the sides of the measure smartly 10 to 15 times with the mallet to release trapped air bubbles.
4. An internal vibrator can be used instead of the tamping rod. In this case, the concrete is placed and vibrated in the measure in two approximately equal layers.
5. When the consolidation is complete, the measure must not contain a substantial excess or deficiency of concrete. A small quantity of concrete may be added to correct a deficiency.
6. After consolidation, strike off the top surface of the concrete, and finish it smoothly with the flat strike-off plate, using great care to leave the measure just level full.
7. After strike-off, clean all excess concrete from the exterior of the measure and determine the net weight of the concrete in the measure (Figure A.33).

Analysis and Results

- W = (Net weight of concrete)/(Volume of the measure)

FIGURE A.33 Weighing concrete inside the container.

- $Y \, (\text{m}^3) = W_1/W$
- $R_y = Y/Y_d$
- $N = N_t/Y$
- $A = [(T - W)/T] \times 100$

where

W = unit mass of concrete, kg/m^3
Y = yield = volume of concrete produced per batch, m^3
W_1 = total mass of all materials batched, kg
R_y = relative yield
Y_d = volume of concrete that the batch was designed to produce, m^3
N = actual cement content, kg/m^3
N_t = mass of cement in the batch, kg
A = air content (percentage of voids) in the concrete
T = theoretical unit mass of concrete computed on an air-free basis, kg/m^3

Report

- Report the unit mass, yield, relative yield, actual cement content, and air content.

Experiment No. 13
Air Content of Freshly Mixed Concrete by Pressure Method

ASTM Designation

ASTM C231—Air Content of Freshly Mixed Concrete by Pressure Method

Purpose

To determine the air content of freshly mixed portland cement concrete by the pressure method.

Significance and Use

Air content plays an important role in workability of freshly mixed concrete and the strength and durability of hardened concrete. The air content of freshly mixed concrete is needed for the proper proportioning of the concrete mix.

Apparatus

- Air meter Type B, consisting of a measuring bowl of a capacity at least 0.006 m^3 and cover assembly fitted with air valves, air bleeder valves, petcocks, and suitable hand pump as shown in Figure 7.14. The air meter must be frequently calibrated according to ASTM C231 procedure to ensure proper measurements.
- Miscellaneous items, including trowel, tamping rod, mallet, and strike-off bar.

Test Procedure

1. Place a representative sample of the plastic concrete in the measuring bowl in three equal layers.
2. Consolidate each layer of concrete by 25 strokes of the tamping rod, evenly distributed over the cross section.
3. After rodding each layer, tap the sides of the measuring bowl 10 to 15 times with the mallet to remove any voids.
4. Strike off the top surface by sliding the bar across the top rim with a sawing motion.
5. Thoroughly clean the flanges and cover the assembly to obtain a pressure-tight seal.
6. Using a rubber syringe, inject water through one petcock until water emerges from the opposite petcock.
7. Jar the meter gently until all air is expelled from the same petcock.
8. Pump air into the air chamber until the gauge indicator is on the initial line.
9. Open the air valve between the air chamber and the measuring bowl.
10. Tap the sides of the measuring bowl sharply, and lightly tap the pressure gauge.
11. Read the percentage of air content on the dial gauge (Figure A.34).
12. Determine the aggregate correction factor according to ASTM C231, and subtract it from the reading obtained in step 11.

FIGURE A.34   Reading pressure meter.
(Shutterstock/TFoxFoto)

Report

- Report the air content and the method used (pressure method).

Experiment No. 14
Air Content of Freshly Mixed Concrete by Volumetric Method

ASTM Designation

ASTM C173—Air Content of Freshly Mixed Concrete by Volumetric Method

Purpose

To determine the air content of freshly mixed portland cement concrete by the volumetric method.

Significance and Use

Air content plays an important role in the workability of freshly mixed concrete and the strength and durability of hardened concrete. The air content of freshly mixed concrete is needed for the proper proportioning of the concrete mix.

Apparatus

- Air meter, consisting of a bowl and a top section, as shown in Figure 7.15
- The bowl has a diameter of 1 to 1.25 the height and a minimum capacity of 0.002 m³. The capacity of the top section is 1.2 times the capacity of the bowl.
- Miscellaneous items including funnel, tamping rod, strike-off bar, measuring cup, syringe, pouring vessel, trowel, scoop, isopropyl alcohol, and mallet.

Calibration

1. The volume of the bowl must be calibrated by accurately weighing the amount of water required to fill it at room temperature and dividing this weight by the unit weight of water at the same temperature.
2. The accuracy of the graduation on the neck of the top section and the volume of the measuring cup must be calibrated according to ASTM C173.

Test Procedure

1. Fill the bowl with freshly mixed concrete in three layers of equal depth.
2. Rod each layer 25 times with the tamping rod.
3. After each layer is rodded 10 to 15 times, tap the sides of the measuring bowl with the mallet to release air bubbles.
4. After placement of the third layer of concrete, strike off the excess concrete with the strike-off bar until the surface is flush with the top of the bowl. Wipe the flanges of the bowl clean.
5. Clamp the top section on the bowl, insert the funnel, and add water until it appears in the neck. Remove the funnel and adjust the water level, using the rubber syringe, until the bottom of the meniscus is leveled with the zero mark. Attach and tighten the screw cap.
6. Invert and agitate the unit many times until the concrete settles free from the base.

FIGURE A.35 Reading air meter.

7. When all the air rises to the top of the apparatus, remove the screw cap. Add, in 1-cup increments using the syringe, sufficient isopropyl alcohol to dispel the foamy mass on the surface of the water. Note that the capacity of the cup is equivalent to 1.0% of the volume of the bowl.
8. Make a direct reading of the liquid in the neck to the bottom of the meniscus to the nearest 0.1% (Figure A.35).
9. The percent air content is calculated as the reading in step 8 plus the amount of alcohol used.

Report

■ Report the air content and the method used (volumetric method).

Experiment No. 15
Making and Curing Concrete Cylinders and Beams

■ ASTM Designation

ASTM C31—Making and Curing Concrete Test Specimens

■ Purpose

To determine how to make and cure concrete cylindrical and beam specimens.

■ Significance and Use

This practice provides standardized requirements for making and curing portland cement concrete test specimens. Specimens can be used to determine strength for mix design, quality control, and quality assurance.

■ Apparatus

- Cylindrical molds made of steel or another nonabsorbent and nonreactive material. The standard specimen size used to determine the compressive strength of concrete is 152 mm diameter by 304 mm high for a maximum aggregate size up to 50 mm. Smaller specimens, such as 102 mm diameter by 203 mm high, are sometimes used, but they are not ASTM standards.
- Beam molds made of steel or another nonabsorbent, nonreactive material. Several mold dimensions can be used to make beam specimens with a square cross section and a span three times the depth. The standard ASTM inside mold dimensions are 152 × 152 mm in cross section and a length of not less than 508 mm, for a maximum aggregate size up to 50 mm.
- Tamping rod with a length of 0.6 m, diameter of 16 mm, and rounded ends.
- Moist cabinet or room with not less than 95% relative humidity and 23 ± 1.7°C temperature or a large container filled with lime-saturated water for curing.
- Miscellaneous items including vibrator (optional), scoop, and trowel.

■ Test Procedure

1. Weigh the required amount of coarse aggregate, fine aggregate, portland cement, and water.
2. Mix the materials in the mixer for 3 to 5 minutes. If an admixture is used, it should be mixed with water before being added to the other materials.
3. Check slump, air content, and temperature of concrete.
4. For cylindrical specimens, place concrete into the mold using a scoop or trowel. Fill the cylinder in three equal layers, and rod each layer 25 times. Tap the outside of the cylinder 10 to 15 times after each layer is rodded. Strike off the top and smooth the surface. Vibrators can also be used to consolidate the concrete instead of rodding. Vibration is optional if the slump is between 25 and 75 mm and is required if the slump is less than 25 mm (Figures A.36–A.38).

FIGURE A.36 Making concrete cylinders and beams.

FIGURE A.37 Filling and rodding concrete in beams.

FIGURE A.38 Consolidating concrete in a beam mold using an external vibrator.

5. For beam specimens, grease the sides of the mold and fill the molds with concrete in two layers. Consolidate the concrete by either tamping each layer 60 times until uniformly distributed throughout or vibrating. After consolidation, finish the surface by striking off the surface and smoothing.
6. Cover the mold with wet cloth to prevent evaporation.
7. Remove the molds after 16 to 32 hours.
8. Cure the specimen in a moist cabinet or room at a relative humidity of not less than 95% and a temperature of 23 ± 1.7°C or by submersion in lime-saturated water at the same temperature (Figure A.39).

Precautions

1. Segregation must be avoided. Over vibration may cause segregation.
2. In placing the final layer, the operator should attempt to add an amount of concrete that will exactly fill the mold after compaction. Do not add nonrepresentative concrete to an under-filled mold.
3. Avoid overfilling by more than 6 mm.

Report

- Record mix design weights, slump, temperature of the mix, and air content
- Specimen type, number of specimens, dimensions, and any deviations from the standard preparation procedure
- Include this information with the report on the strength of the concrete

FIGURE A.39   Curing concrete cylinders in lime-saturated water.

Experiment No. 16
Capping Cylindrical Concrete Specimens with Sulfur or Capping Compound

ASTM Designation

ASTM C617—Capping Cylindrical Concrete Specimens

Purpose

To cap hardened portland cement concrete cylinders and drilled concrete cores with sulfur mortar or other capping compounds in order to prepare the specimen for compressive strength testing.

Significance and Use

This procedure provides plane surfaces perpendicular to the specimen axis on the ends of concrete cylinders before performing the compression test.

Apparatus

- Alignment device consisting of a frame with guide bars and a cup, as shown in Figure A.40. The size of the alignment device should match the specimen size.

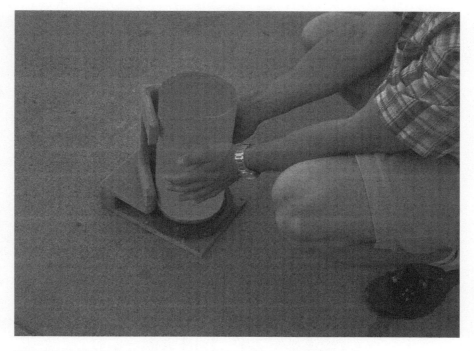

FIGURE A.40   Capping a concrete cylinder using an alignment device.

■   Melting pot, used for melting sulfur mortars or capping compound, equipped with automatic temperature control. The melting pot should be used either outdoors or under an exhaust hood. Heating over an open flame is dangerous because the mixture may ignite if overheated.

Capping Procedure

1. Prepare the sulfur mortar or capping compound by heating to about 130°C. Use a metal thermometer to check the temperature. Make sure to empty any old mortar and to use fresh mortar to avoid the loss of strength due to successive heating. The fresh sulfur mortar must be dry when it is placed in the pot because dampness may cause foaming.
2. Warm the capping cup or device slightly before use to slow the rate of hardening and to permit the production of thin caps.
3. Oil the capping cup lightly and stir the molten sulfur mortar or the capping compound immediately prior to pouring into the cup. Make sure that the ends of moist-cured specimens are dry enough at the time of capping, so there will be no steam or foam pockets.
4. Hold the concrete cylinder with two hands and push it against the guide bars of the capping device. Carefully lower the specimen until it rests in the cup (Figure A.40). This step must be completed quickly before the sulfur or capping compound solidifies. The thickness of the cap should be about 3 mm and not more than 8 mm in any part.
5. Before the cylinder is tested for compressive strength, the cap should be cured, in order to have strength comparable to that of the concrete.

Experiment No. 17
Compressive Strength of Cylindrical Concrete Specimens

ASTM Designation

ASTM C39—Compressive Strength of Cylindrical Concrete Specimens

Purpose

To determine the compressive strength of cylindrical PCC specimens, such as molded cylinders and drilled cores.

Significance and Use

This test provides the compressive strength of concrete, which is used universally as a measure of concrete quality.

Apparatus

Loading machine with two hardened steel breaking blocks. The upper block is spherically seated, and the bottom block is solid surface (Figure A.41).

FIGURE A.41 Concrete cylindrical specimen being tested for compressive strength.

Test Specimens

- The standard specimen size used to determine the compressive strength of concrete is 152 mm diameter by 304 mm high for a maximum aggregate size up to 50 mm. Smaller specimens, such as 102 mm diameter by 203 mm high, are sometimes used, but they are not ASTM standardized.
- Conduct the compression test on the moist-cured specimens directly after removing them from the curing room. Test specimens must be moist when tested.
- Neither end of compressive test specimen shall depart from perpendicularity by more than 0.5°, approximately 3 mm in 0.3 m
- If the ends of the specimen are not plane within 0.05 mm, they should be capped with sulfur or capping compound. The use of neoprene caps (Figure A.42) is covered in ASTM C1231.
- Specimen age, at time of testing, should be 24 hours $\pm$ 0.5 hours, 3 days $\pm$ 2 hours, 7 days $\pm$ 6 hours, 28 days $\pm$ 20 hours, or 90 days $\pm$ 2 days.

Test Procedure

1. Measure the diameter of the test specimen to the nearest 0.25 mm by averaging two diameters measured at right angles to each other at the middle height of the specimen.
2. Adjust the bearing blocks into position.
3. Clean the faces of the bearing blocks and the specimen.
4. Carefully align the axis of the specimen with the center of the thrust of the spherically-seated block.
5. Apply the load continuously and without shock. For screw-type machines, use a rate of loading of 1.25 mm/min. For hydraulically operated machines, apply the load at a constant rate within the range of 138 to 335 kPa/s. During the first half of the anticipated loading phase, a higher rate of loading is permitted. No adjustment in the control of the testing machine should be made while the specimen is yielding rapidly, immediately before failure.

FIGURE A.42 Neoprene caps used for capping a concrete cylinder.

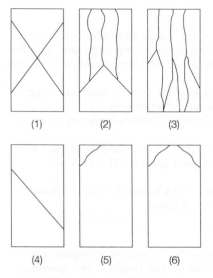

FIGURE A.43 Typical fracture patterns of concrete cylinders: (1) cones on both ends, (2) cone on one end and vertical cracks, (3) vertical cracks, (4) diagonal fracture, (5) side fracture, and (6) side fractures with pointed ends.

FIGURE A.44 Example of fracture with vertical cracks. (Courtesy of FHWA)

6. Continue applying the load until the specimen fails.
7. Record the maximum load carried by the specimen during the test.
8. Note the type of failure and the appearance of concrete (Figures A.43 and A.44).

Analysis and Results

■ Calculate the compressive strength as

$$f'_c = P_{max}/A$$

where

$$f'_c = \text{compressive strength, MPa}$$
$$P_{max} = \text{maximum applied load, N}$$
$$A = \text{cross-sectional area, mm}^2$$

Report

■ Specimen identification number
■ Diameter (and length, if outside the range of 1.8 to 2.2 times the diameter)
■ Cross-sectional area
■ Maximum load
■ Compressive strength, calculated to the nearest 0.07 MPa
■ Type of failure, if other than the usual one
■ Defects in either specimen or caps
■ Age of specimen

Experiment No. 18
Flexural Strength of Concrete

ASTM Designation

ASTM C78—Flexural Strength of Concrete (Using Simple Beam with Third-Point Loading)

Purpose

To determine the flexural strength of portland cement concrete by using a simple beam with third-point loading.

Significance and Use

The flexural strength of concrete is a measure of concrete quality.

Apparatus

- Loading machine capable of applying loads at a uniform rate
- Loading device capable of applying load configuration as shown in Figure 7.38. Forces applied to the beam shall be perpendicular to the face of the specimen and applied without eccentricity.

Test Specimens

- The standard ASTM specimen dimensions are 152 mm × 152 mm in cross section and a length of not less than 508 mm for a maximum aggregate size up to 50 mm.
- Sides of the specimen should be at right angles to its top and bottom. All surfaces in contact with load-applying and support blocks should be smooth and free of scars, indentations, holes, or inscribed identifications.

Test Procedure

1. Turn the test specimen on its side, with respect to its position as molded, and center it on the bearing blocks.
2. Center the loading system in relation to the applied force. Bring the load-applying blocks in contact with the surface of the specimen at the third points between the supports (Figure A.45).
3. If full contact is not obtained at no load between the specimen and the load-applying blocks and the supports so that there is a 25 mm or larger gap in excess of 0.1 mm, grind or cap the contact surfaces of the specimen, or shim with leather strips.
4. Apply the load rapidly up to approximately 50% of the breaking load. Thereafter, apply the load continuously at a rate that constantly increases the extreme fiber stress between 860 and 1210 kPa/min until rupture occurs.

Analysis and Results

- Take three measurements across each dimension (one at each edge and at the center) to the nearest 1.3 mm to determine the average width, average depth, and line of fracture location of the specimens at the section of fracture.

FIGURE A.45 Testing concrete beam in the flexure testing machine.

■ If the fracture initiates in the tension surface within the middle third of the span length, calculate the modulus of rupture as follows:

$$R = \frac{Mc}{I} = \frac{PL}{bd^2}$$

where

R = modulus of rupture, MPa
M = maximum bending moment, N·mm

$$c = \frac{d}{2}, \text{mm}$$

I = moment of inertia

$$= \frac{bd^3}{12}, \text{mm}^4,$$

P = maximum load, N
L = span length, mm
b = average width, mm
d = average depth, mm

■ If the fracture occurs in the tension surface outside the middle third of the span length, by not more than 5% of the span length, calculate the modulus of rupture as follows:

$$R = \frac{3Pa}{bd^2}$$

where

a = average distance between line of fracture and the nearest support on the tension surface of the beam in millimeters

- If the fracture occurs in the tension surface outside the middle third of the span length, by more than 5% of the span length, discard the results of the test.

Report

- Specimen identification number
- Average width
- Average depth
- Span length
- Maximum applied load
- Modulus of rupture to the nearest 0.03 MPa
- Curing history and apparent moisture condition at time of testing
- If specimens were capped, ground, or if leather shims were used
- Defects in specimens
- Age of specimens

Experiment No. 19
Rebound Number of Hardened Concrete

 ASTM Designation

ASTM C805—Rebound Number of Hardened Concrete

Purpose

To determine the rebound number of hardened portland cement concrete.

Significance and Use

The rebound number may be used to assess the uniformity and strength of concrete. It can also be used to determine when forms and shoring may be removed.

Apparatus

Rebound hammer (Figures 7.39, A.46, and A.47)

FIGURE A.46 Testing concrete slab with a rebound hammer.

FIGURE A.47 Testing concrete with a rebound hammer. (Courtesy of Proceq)

Test Procedure

1. Grind and clean the concrete surface by rubbing with the abrasive stone that accompanies the rebound device.
2. Firmly hold the instrument in a position that allows the plunger to strike perpendicularly to the test surface.
3. Gradually increase the pressure on the plunger until the hammer impacts.
4. After impact, record the rebound number to two significant digits.
5. Take 10 readings from each test area.

Test Conditions

1. No two impact tests shall be closer together than 25 mm.
2. Discard the reading if the impact crushes or breaks through a near-surface air void.
3. Discard readings differing from the average of 10 readings by more than 7 units, and determine the average of the remaining readings.
4. If more than two readings differ from the average by 7 units or more, discard all readings.
5. The rebound hammer should be periodically serviced and verified using metal anvils, according to manufacturer recommendations.

Report

- Structure identification, location, and curing condition.
- Average rebound number.
- Position of the hammer during the test, such as downward, upward, or at a specific angle.
- Estimated compressive strength using available correlations, such as those obtained from the manufacturer. It should be noted, however, that the rebound hammer test is not intended as an alternative for strength determination of concrete.

Experiment No. 20
Penetration Resistance of Hardened Concrete

▨ ASTM Designation

ASTM C803—Penetration Resistance of Hardened Concrete

▨ Purpose

To assess the uniformity of hardened portland cement concrete and indicate its in-place strength.

▨ Significance and Use

The penetration resistance may be used to assess the uniformity and strength of concrete.

▨ Apparatus

The penetration resistance (Windsor Probe) apparatus (Figure 7.40) consists of the following:

- Driver unit
- Probe made of hardened steel alloy
- Measuring instrument
- Positioning device

▨ Test Procedure

1. Place the positioning device on the concrete surface at the location to be tested. Two positioning devices are available: a single positioning device and a triangular device.
2. Mount a probe in the driver unit.
3. Position the driver and probe in the positioning device.
4. Fire and drive the probe into the concrete. In case of the triangular device, repeat for each position (Figures A.48 and A.49).
5. In the case of the single positioning device, place the measuring base plate over the probe, and measure the length of the probe not embedded in the concrete with the measuring instrument. In the case of a triangular device, place the triangular plate on the three probes, position the measuring instrument in the hole at the center of the triangular plate, and read the average of the exposed length. Alternatively, the lengths of individual probes not embedded in the concrete can be measured and the average of the three lengths is obtained (Figure A.50).

▨ Report

- Structure identification, location, and curing condition.
- The average exposed length of the probe.
- Estimated compressive strength, using available correlations such as those obtained from the manufacturer. (Note, however, that the penetration resistance test is not intended as an alternative for strength determination of concrete.)

FIGURE A.48   Firing and driving Windsor probe into concrete using triangular positioning device.

FIGURE A.49   Three probes driven into concrete.

FIGURE A.50   Measuring penetration of one probe.

Experiment No. 21
Testing of Concrete Masonry Units

▨ ASTM Designation
ASTM C140—Sampling and Testing Concrete Masonry Units and Related Units

▨ Purpose
To test concrete masonry units for compressive strength, absorption, moisture content, and density.

▨ Significance and Use
The test produces compressive strength, absorption, moisture content, and density data for the control and specification of concrete masonry units. These data are important for the safety and proper performance of masonry structures.

▨ Apparatus
- Testing machine
- Steel-bearing blocks and plates
- Balance

▨ Test Procedure for Compressive Strength
1. Three representative units are needed for testing within 72 hours after delivery to the laboratory, during which time they are stored continuously in air at a temperature of $24 \pm 8°C$ and a relative humidity of less than 80%.
2. The test is performed on either a full-sized unit or a part of a unit prepared by saw cutting. A part of a unit is used if the capacity or size of the testing machine does not allow the testing of a full-sized unit.
3. Measure the length, width, and height of the specimen.
4. Cap the bearing surfaces of the unit, using either sulfur and granular materials or gypsum plaster.
5. Position the test specimen with its centroid aligned vertically with the center of thrust of the spherically seated steel-bearing block of the testing machine (Figure A.51).
6. Apply the load up to one-half the expected maximum load at any convenient rate, after which apply the load at a uniform rate of travel of the moving head, so that the test is completed between 1 and 2 minutes.
7. Record the maximum compressive load in newtons as P_{max}.

▨ Test Procedure for Absorption
1. Three representative full-sized units are needed for absorption testing.
2. Weigh the specimen immediately after sampling and record the weight as the received weight (W_r).
3. Immerse the specimen in water at a temperature of 15 to 26°C for 24 hours.
4. Weigh the specimen while it is suspended by a metal wire and completely submerged in water; record the weight as the immersed weight (W_i).

FIGURE A.51 Compression testing of a concrete masonry unit.

5. Remove the specimen from the water and allow it to drain for 1 minute by placing it on a 9.5-mm or coarse wire mesh and removing visible surface water with a damp cloth. Weigh the specimen and record the weight as the saturated weight (W_s).
6. Dry the specimen in a ventilated oven at 100 to 115°C for not less than 24 hours and until two successive weights at intervals of 2 hours show a difference of not greater than 0.2%. Record the weight as the oven-dried weight (W_d).

Analysis and Results

Gross area compressive strength, MPa $= P_{max}/A_g$

where

P_{max} = maximum compressive load, N
A_g = gross area, mm^2 = $L \times W$
L = average length, mm
W = average width, mm
Net area compressive strength, MPa $= P_{max}/A_n$

where

A_n = average net area, mm^2 = V_n/H
V_n = net volume, mm^3 = $(W_s - W_i) \times 10^6$

H = average height, mm
W_s = saturated mass of unit, kg
W_i = immersed mass of unit, kg

Absorption, kg/m^3 = $[(W_s - W_d)/(W_s - W_i)] \times 1000$

Absorption, % = $[(W_s - W_d)/W_d] \times 100$

Moisture content, % of total absorption = $[(W_r - W_d)/(W_s - W_d)] \times 100$

Density, kg/m^3 = $[W_d/(W_s - W_i)] \times 1000$

where

W_d = oven dry mass, kg

W_r = received mass, kg

Report

- Gross area and net area compressive strengths to the nearest 70 kPa for each specimen and the average for three specimens
- Absorption, moisture content as a percent of total absorption, and density

Experiment No. 22
Viscosity of Asphalt Binder by Rotational Viscometer

ASTM Designation

ASTM D4402—Viscosity Determinations of Asphalt at Elevated Temperatures Using a Rotational Viscometer

Purpose

To determine the apparent viscosity of asphalt binder from 38 to 260°C using a rotational viscometer and a temperature-controlled thermal chamber for maintaining the test temperature.

Significance and Use

The viscosity is needed to ensure proper handling of the asphalt binder and for quality control and quality assurance. It is also used to determine the mixing and compaction temperatures of asphalt concrete. This test is used for the Superpave performance grading of asphalt binders.

Apparatus

- Rotational viscometer (Figure 9.19)
- Spindles
- Thermosel system, consisting of thermo-container and sample chamber, SCR controller and probe, and graph-plotting equipment

Test Procedure

1. Turn on the Thermosel power and set the proportional temperature controller to the desired test temperature.
2. Wait 1-1/2 hours (or until equilibrium temperature is obtained) with the spindle in the chamber (check control lamp).
3. Remove sample holder and add the volume of sample specified for the spindle, approximately 8 to 10 mL. Exercise caution to avoid sample overheating and to avoid ignition of samples with a low flash point. Do not overfill the sample container. The sample volume is critical to meeting the system calibration standard.
4. Thoroughly stir the filled asphalt container to obtain a representative sample.
5. The liquid level should intersect the spindle shaft at a point approximately 3 mm above the upper conical body–spindle shaft interface.
6. Using the extracting tool, put the loaded chamber back into the thermo-container.
7. Lower the viscometer and align the thermo-container.
8. Insert the selected spindle into the liquid in chamber, and couple it to the viscometer (Figure A.52). Either spindle number 27 or spindle number 20 is used for asphalt binders.
9. Allow the asphalt to come to the equilibrium temperature (about 15 minutes).
10. Start the viscometer at a 20 rpm setting.
11. Record three readings 60 seconds apart at each test temperature.
12. Multiply the viscosity factor by the rotational viscometer readings to obtain viscosity in centipoises.

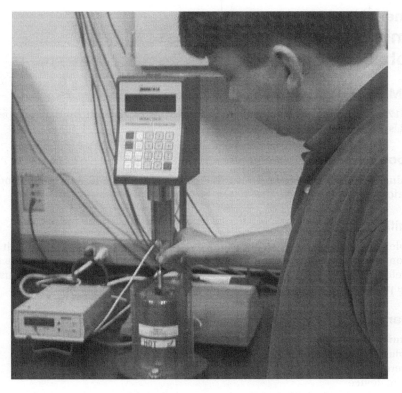

FIGURE A.52 Inserting the spindle into chamber for the rotational viscometer test.

Report

- Viscosity at each test temperature
- Spindle number and rotational speed

Experiment No. 23
Dynamic Shear Rheometer Test of Asphalt Binder

ASTM Designation

ASTM D7175—Determining the Rheological Properties of Asphalt Binder for Specification Purposes Using a Dynamic Shear Rheometer (DSR)

Purpose

To determine the complex shear modulus (G^*) and phase angle (δ) of asphalt binders using the dynamic shear rheometer.

Significance and Use

The complex shear modulus is an indicator of the stiffness resistance of asphalt binder to deformation under load. The complex shear modulus and phase angle define the resistance to shear deformation of the asphalt binder in the viscoelastic region. This test is used for the Superpave performance grading of asphalt binders.

Apparatus

- Dynamic shear rheometer (DSR) (Figures 9.20 and A.53)
- Test plates
- Temperature controller
- Loading device
- Control and data acquisition system
- Miscellaneous items, such as specimen mold, specimen trimmer, environmental chamber, reference thermal detector, and calibrated temperature detector

Test Procedure

1. Heat the asphalt binder until fluid enough to pour the required specimens.
2. Carefully clean and dry the surfaces of the test plates so that the specimen uniformly adheres to both plates. Bring the chamber to approximately 45°C, so that the plates are preheated prior to the mounting of the test specimen.
3. Place the asphalt binder sample in the DSR, using one of the following methods:

   - Remove the removable plate and, while holding the sample container approximately 15 mm above the test plate surface, pour the asphalt binder at the center of the upper test plate continuously until it covers the entire plate, except for an approximate 2-mm-wide strip at the perimeter. Wait several minutes for the specimen to stiffen, and then mount the test plate in the rheometer for testing.

   - Pour the hot asphalt into a silicon rubber mold that will form a pellet with a diameter approximately equal to the diameter of the upper test plate and a height approximately equal to 1.5 times the width of the test gap. Allow the silicon rubber to cool to room temperature. Remove the specimen from the mold and center the pellet on the lower plate of the DSR.

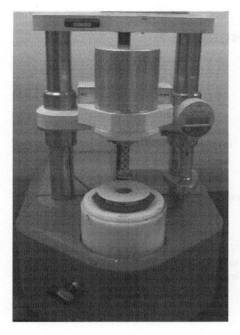

FIGURE A.53 Dynamic shear rheometer apparatus.

FIGURE A.54 Water bath for controlling specimen temperature during testing.

4. Move the test plates together to squeeze the asphalt mass between the two plates. Move the plates until the gap between the plates equals the testing gap plus 0.05 mm. Trim the specimen by moving a heated trimming tool around the upper and lower plate perimeters while trimming excess asphalt.

5. After trimming is completed, decrease the gap by 0.05 mm. The sample thickness should now equal the desired test gap.

6. Bring the specimen to the test temperature $\pm 0.1°C$ (Figure A.54). Start the test after the temperature has remained at the desired temperature, $\pm 0.1°C$ for at least 10 minutes. The test temperature can be selected from the specifications in Table 9.3.

7. When the temperature has been reached, condition the specimen by applying the required shear strain for 10 cycles at a radial frequency of 10 rad/s. Shear strain values vary from 1% to 12%, depending on the stiffness of the binder being tested. High strain values are used for relatively soft binders tested at high temperatures, whereas low strain values are used for hard binders. The rheometer measures the torque required to achieve the set shear strain and maintains this as the maximum torque during the test.

8. The data acquisition system automatically calculates G^* and δ from test data acquired when properly activated.

Report

- Identification and description of the material tested
- Test temperature and sample dimensions, including thickness
- Stress level
- G^* and δ values

Experiment No. 24
Penetration Test of Asphalt Cement

 ASTM Designation

ASTM D5—Penetration of Bituminous Materials

 Purpose

To determine the penetration of semisolid and solid bituminous materials.

 Significance and Use

The penetration test is used as a measure of consistency. High values of penetration indicate soft consistency.

 Apparatus

- Penetration apparatus and needle (Figures 9.23 and A.55)
- Sample container, water bath, transfer dish, timing device for hand-operated penetrometers, and thermometers

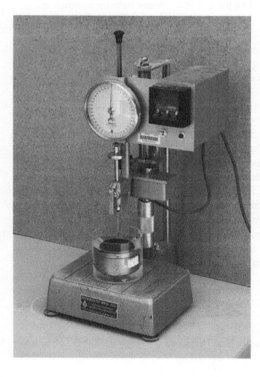

FIGURE A.55 Apparatus for penetration test of asphalt binder.

Test Procedure

1. Heat the asphalt binder sample until it has become fluid enough to pour.
2. Pour the sample into the sample container and let it cool for at least 1 hour.
3. Place the sample, together with the transfer dish in the water bath at a temperature of 25°C for 1 to 2 hours.
4. Clean and dry the needle with a clean cloth, and insert the needle into the penetrometer. Unless otherwise specified, place the 50-g mass above the needle, making the total moving load 100 g.
5. Place the sample container in the transfer dish, cover the container completely with water from the constant temperature bath, and place the transfer dish on the stand of the penetrometer.
6. Position the needle by slowly lowering it until its tip just makes contact with the surface of the sample. This is accomplished by bringing the actual needle tip into contact with its image reflected by the surface of the sample from a properly placed source of light.
7. Quickly release the needle holder for the specified period of time (5 seconds) and adjust the instrument to measure the distance penetrated in tenths of a millimeter.
8. Make at least three determinations at points on the surface of the sample not less than 10 mm from the side of the container and not less than 10 mm apart.

Report

- Report the average of the three penetration values to the nearest whole unit.

Experiment No. 25
Absolute Viscosity Test of Asphalt

 ASTM Designation

ASTM D2171—Viscosity of Asphalts by Vacuum Capillary Viscometer

 Purpose

To determine the absolute viscosity of asphalt by vacuum capillary viscometer at 60°C.

 Significance and Use

The viscosity at 60°C characterizes flow behavior and may be used for specification requirements for cutbacks and asphalt cements.

 Apparatus

- Viscometers such as the Cannon–Manning vacuum viscometer (Figure 9.24) or modified Koppers vacuum viscometer
- Bath, with provisions for visibility of the viscometer and the thermometer
- Thermometers, vacuum system, and timing device

 Test Procedure

1. Maintain the bath at a temperature of 60°C.
2. Select a clean, dry viscometer that will give a flow time greater than 60 seconds, and preheat to 135°C.
3. Charge the viscometer by pouring the prepared asphalt sample to the fill line.
4. Place the charged viscometer in an oven or bath maintained at 135°C for a period of 10 minutes, to allow large air bubbles to escape.
5. Remove the viscometer from the oven or bath, and within 5 minutes, insert the viscometer in a holder, and position the viscometer vertically in the bath so that the uppermost timing mark is at least 20 mm below the surface of the bath liquid.
6. Establish a 300-mm Hg vacuum in the vacuum system, and connect the vacuum system to the viscometer.
7. After the viscometer has been in the bath for 30 minutes, start the flow of asphalt in the viscometer by opening the toggle valve or stopcock in the line leading to the vacuum system.
8. Measure to within 0.1 second the time required for the leading edge of the meniscus to pass between successive pairs of timing marks. Report the first flow time that exceeds 60 seconds between a pair of timing marks, and identify the pair of timing marks (Figure A.56).

 Analysis and Results

Select the calibration factor that corresponds to the pair of timing marks. Calculate and report the viscosity to three significant figures formula

$$P = Kt$$

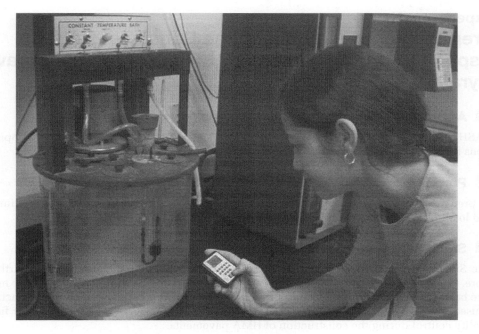

FIGURE A.56 Measuring flow time using a stop watch.

where

P = absolute viscosity, Pa · s
K = selected calibration factor, Pa · s/seconds
t = flow time, s

▨ Report

- Absolute viscosity
- Test temperature and vacuum

Experiment No. 26

Preparing and Determining the Density of Hot-Mix Asphalt (HMA) Specimens by Means of the Superpave Gyratory Compactor

▓ AASHTO Designation

AASHTO T 312—Preparing and Determining the Density of Hot-Mix Asphalt (HMA) Specimens by Means of the Superpave Gyratory Compactor

▓ Purpose

To prepare asphalt concrete specimens to densities achieved under actual pavement climate and loading conditions, or as needed for laboratory testing.

▓ Significance and Use

The Superpave gyratory compactor is capable of accommodating large aggregates. Furthermore, this device affords a measure of compaction ability that permits potential tender mixture behavior and similar compaction problems to be identified. This method of compaction is used for the Superpave volumetric mix design of asphalt concrete mixture and for field quality control during the construction of HMA pavements.

▓ Apparatus

- Reaction frame, rotating base, and motor (Figure 9.26)
- Loading system, loading ram, and pressure gauge
- Height measuring and recording system
- Mold and base plate

▓ Procedure

1. Specimens must be mixed and compacted under equiviscous temperature conditions of 0.170 ± 0.02 Pa·s and 0.280 ± 0.03 Pa·s, respectively. Mixing is accomplished by a mechanical mixer.
2. After mixing, loose test specimens are subjected to 2 hours of short-term aging in a forced draft oven at the compaction temperature which corresponds to a binder viscosity of 0.28 ± 0.03 Pa·s. During this period, loose mix specimens are required to be spread into a thickness resulting in 21 to 22 kg per cubic meter. The sample is stirred after 1 hour to ensure uniform aging.
3. Place the compaction molds and base plates in an oven at the compaction temperature for at least 30 to 45 minutes prior to use.
4. If specimens are to be used for volumetric determinations only, use sufficient mix to arrive at a specimen 150 mm in diameter by approximately 115 mm high. This requires approximately 4500 g of aggregates. If needed, specimens with other heights can also be prepared for further performance testing (Figure A.57).
5. Turn on the power to the compactor. Set the vertical pressure to 600 kPa.
6. Set the gyration counter to zero, and set it to stop when the desired number of gyrations is achieved. Three gyrations are of interest: design number of gyrations (N_d), initial

FIGURE A.57   Placing HMA in mold.

number of gyrations (N_i), and maximum number of gyrations (N_m). The design number of gyrations is a function of the climate in which the mix will be placed and the traffic level it will withstand.

7. After the base plate is placed, place a paper disk on top of the plate and charge the mold with the short-term aged mix in a single lift. The top of the uncompacted specimen should be slightly rounded. Place a paper disk on top of the mixture. Depending on the model of the compactor, it may be necessary to place a top plate on the mix.

8. Place the mold in the compactor and center it under the ram (Figure A.58). Lower the ram until it contacts the mixture and the resisting pressure is 600 kPa.

9. Apply the angle of gyration and begin compaction.

10. When the desired number of gyrations has been reached, the compactor should automatically cease. After the angle and pressure are released, remove the mold containing the compacted specimen.

11. Print the results of specimen height versus number of gyrations.

12. After a suitable cooling period, extrude the specimen from the mold and mark it (Figure A.59).

13. Measure the bulk specific gravity of the test specimens according to ASTM D2726 procedure.

FIGURE A.59 Specimen prepared with gyratory shear compactor.

FIGURE A.58 Placing the mold in the gyratory compactor.

Analysis and Results

- Draw a graph of the logarithm of the number of gyrations versus the height (densification curve) for each specimen.

Report

- Mixture ingredients, source, and relevant information
- Densification table
- Densification curve

Experiment No. 27
Preparation of Asphalt Concrete Specimens Using the Marshall Compactor

▊ ASTM Designation

ASTM D1559—Resistance to Plastic Flow of Bituminous Mixtures Using Marshall Apparatus

▊ Purpose

To prepare asphaltic concrete specimens using the Marshall hammer.

▊ Significance and Use

This method is used to design the mix using the Marshall procedure and measure its properties.

▊ Apparatus

- Either mechanical or manual compaction hammer with 4.5-kg mass and 0.48-m drop height can be used (Figure 9.27)
- Molds with 102-mm inside diameter and 75 mm high, base plates, and collars
- Compaction pedestal, specimen extruder, and miscellaneous items, such as mold holder, spatula, pans, and oven.

▊ Procedure

1. Determine the mixing and compaction temperatures so that the kinematic viscosities of the binder are 170 ± 20 cSt and 280 ± 30 cSt, respectively.
2. Separate all the required sizes of aggregates and oven dry.
3. Weigh 1200 g batches so that gradation matches the target gradation for the mix design. Either a state specification or ASTM D3515 can be followed.
4. Place both asphalt binder and aggregate in the oven until they reach the mixing temperature (approximately 150°C).
5. Add the asphalt to the aggregate in the specified amount.
6. Using the mechanical mixer, or manually, mix the aggregates and asphalt thoroughly (Figure A.60). Some agencies require curing the mix as specified in AASHTO R30.
7. Place a release paper inside the mold. Then, place the entire batch of asphaltic concrete in the mold heated to the compaction temperature, and spade with a heated spatula 15 times around the perimeter and 10 times in the middle.
8. Put in place a collar and a release paper, and put the mold on the pedestal. Clamp the mold with the mold holder, and apply the required number of blows using either a manual hammer or a mechanical hammer (Figures A.61 and A.62). The typical number of blows on each side is either 50 or 75, depending on the expected traffic volume on the road where the mix is intended to be used. Invert the mold and apply the same number of blows on the other face.

FIGURE A.60 Mixing HMA using a mechanical mixer.

FIGURE A.61 Compacting an HMA specimen using a manual Marshall hammer.

9. After cooling to room temperature, extrude the specimen using an extruding device. Cooling can be accelerated by placing the mold and the specimen in a plastic bag and subjecting them to cold water. Remove the paper disk, as shown in Figure A.63.

Report

- Mixture ingredients, source, and relevant information
- Number of blows on each side of the specimen

FIGURE A.62 Compacting an HMA specimen using a mechanical Marshall hammer.

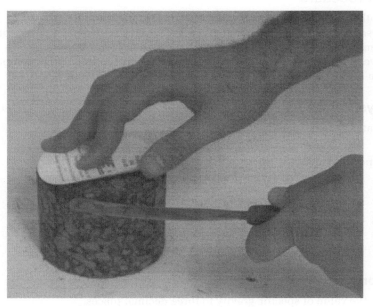

FIGURE A.63 Removing paper disk from a Marshall plug.

Experiment No. 28
Bulk Specific Gravity of Compacted Bituminous Mixtures

ASTM Designation

ASTM D2726—Bulk Specific Gravity of Compacted Bituminous Mixtures

Purpose

To determine the bulk specific gravity of compacted asphalt mixture specimens.

Significance and Use

The results of this test are used for voids analysis of the compacted asphalt mix.

Test Specimens

Laboratory-molded bituminous mixtures or cores drilled from bituminous pavements can be used.

Apparatus

- Balance equipped with suitable suspension and holder to permit weighing the specimen while it is suspended from the balance
- Water bath

Test Procedure

1. Weigh the specimen in air and record it as A (Figure A.64).
2. Immerse the specimen in water at 25 $\pm$ 1°C while it is suspended from the balance for 3 to 5 minutes, and record the immersed mass as C (Figure A.65).
3. Remove the specimen from the water, surface dry by blotting with a damp towel (Figure A.66), determine the surface dry mass, and record it as B.

Analysis and Results

- Calculate the bulk specific gravity as

$$\text{Bulk specific gravity} = \frac{A}{(B - C)}$$

- where

    A = mass of specimen in air, grams
    B = mass of surface-dry specimen, grams
    C = mass of specimens in water, grams

Report

- Report the value of specific gravity to three decimal places.

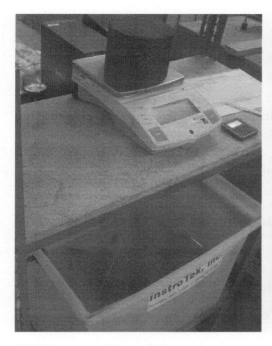

FIGURE A.64   Weighing HMA specimen in air.

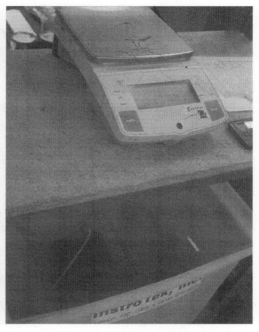

FIGURE A.65   Weighing HMA specimen in water.

FIGURE A.66   Drying HMA specimen with a damp towel.

Experiment No. 29
Marshall Stability and Flow of Asphalt Concrete

ASTM Designation

ASTM D1559—Resistance to Plastic Flow of Bituminous Mixtures Using Marshall Apparatus

Purpose

To determine the Marshall stability and flow values of asphalt concrete.

Significance and Use

This test method is used in the laboratory mix design of bituminous mixtures according to the Marshall procedure. The test results are also used to characterize asphalt mixtures.

Apparatus

- Testing machine producing a uniform vertical movement of 50.8 mm/min, as shown in Figures 9.30, A.67
- Breaking heads having an inside radius of curvature of 50.8 mm
- Load cell or ring dynamometer, strip chart recorder or flow meter, water bath, and rubber gloves

Proving Ring

Flowmeter

Breaking Head

FIGURE A.67 Marshall stability machine. (Courtesy of FHWA)

▦ Test Procedure

1. Bring the specimen prepared in experiment number 25 to a temperature of 60°C by immersing it in a water bath 30 to 40 minutes or by placing it in the oven for 2 hours.
2. Remove the specimen from the water bath then "lightly" dry (or oven) and place it in the lower segment of the breaking head. Place the upper segment of the breaking head on the specimen, and place the complete assembly in position on the testing machine.
3. Prepare the strip chart recorder or place the flowmeter (where used) in position over one of the guide rods, and adjust the flowmeter to zero while holding the sleeve firmly against the upper segment of the breaking head.
4. Apply the load to the specimen by means of the constant rate of movement of 50.8 mm/min until the maximum load is reached and the load decreases (Figure A.68). The elapsed time for the test from removal of the test specimen from the water bath (or oven) to the maximum load determination should not exceed 30 seconds.
5. From the chart recorder, record the Marshall stability (maximum load) and the Marshall flow (deformation when the maximum load begins to decrease in units of 0.25 mm). In some machines, the maximum load and the flow values are read from the ring dynamometer and the flowmeter, respectively.
6. If the specimen height is other than 63.5 mm, multiply the stability value by a correction factor (Table 9.13) (ASTM D1559).

▦ Report

- Specimen identification and type (laboratory prepared or core)
- Average Marshall stability of at least three replicate specimens, corrected when required, kN
- Average Marshall flow of at least three replicate specimens

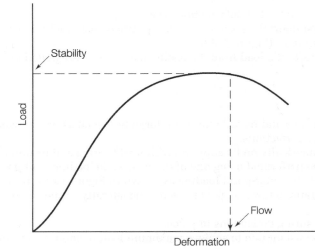

FIGURE A.68 Typical plot of load versus deformation during the Marshall stability test showing the stability and flow.

Experiment No. 30
Bending (Flexure) Test of Wood

ASTM Designations

ASTM D 198—Testing Structural Sizes Lumber
ASTM D143—Testing Small, Clear Specimens

Purpose

To determine the flexure properties of wood such as modulus of rupture, modulus of elasticity, and failure conditions.

Testing Structural Size Lumber

Significance and Use

The structural size flexural test provides extensive data including flexural properties, grading rules and specifications, working stresses, influence of imperfections on mechanical properties, strength properties, effect of chemical or environmental conditions, effect of fabrication variables, and relationships between mechanical and physical properties.

Test Specimens

The test specimen may be a solid wood beam such as a 50 mm × 100 mm, laminated wood, or a composite construction of wood or of wood combined with plastics or metals in sizes that are commonly used in structural applications.

Apparatus

- Testing machine of the controlled deformation type
- Support apparatus to support the specimen at the specified span, including reaction supports and lateral support (Figure A.69)
- Load apparatus to transfer the load from the testing machine at designated points to the specimen

Test Procedure

1. Measure beam dimensions and record relevant information such as type of wood and type and location of imperfections.
2. Locate the beam symmetrically on its supports with load bearing and reaction bearing blocks. The test can be performed using one of three possible loading configurations: two-point, third-point, or center-point loading as shown in Figure 10.15. Beams having a depth-to-width ratio of three or greater need to be laterally supported with lateral supports.
3. Set apparatus for measuring deflections in place.
4. Conduct the test at a constant rate to achieve maximum load in about 10 minutes (or 6–20 minutes).

(a) (b)

FIGURE A.69 Center-point loading on 4 × 4 wood beam: (a) before loading and (b) after failure.

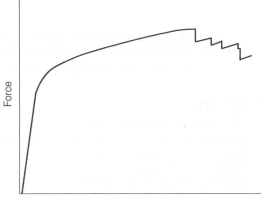

FIGURE A.70 Typical load–deflection plot.

5. Record load–deflection data (Figure A.70). Note the load and deflection at first failure, at the maximum load, and at the points of sudden change. Continue loading until complete failure or an arbitrary terminal load has been reached.
6. Describe failures in detail as to type, manner, order of occurrence, and position in beam. Drawings and photographs may be used to record failures. Hold the section of the beam containing the failure for examination and reference until analysis of the data has been completed.

Analysis and Results

■ Draw a schematic diagram of loading method (see Figure 10.15). Note: The calculations below assume a beam with solid rectangular homogeneous cross-section subjected to a center-point loading. For two-point and third-point loadings and for more detailed calculations, please refer to ASTM D198.

■ Calculate the fiber stress at proportional limit, S_f

$$S_f = \frac{3P'L}{2bh^2}$$

where

P' = applied load at proportion limit, N
L = span length, mm
b = average width, mm
h = average depth, mm

■ Calculate the modulus of rupture, S_R

$$S_R = \frac{3P_{\text{max}}L}{2bh^2}$$

where

P_{max} = maximum load, N

■ Calculate the apparent modulus of elasticity, E_f

$$E_f = \frac{PL^3}{4bh^3\Delta}$$

where

P = increment of applied load below proportional limit, N
Δ = increment of deflection of beam's neutral axis measured at midspan over distance L and corresponding load P, mm

■ **Report**

■ Specimen identification, dimensions, span length, loading conditions, rate of load application, lateral supports if used, deflection apparatus, and other relevant information, such as moisture content
■ Load–deflection plot
■ Fiber stress at proportional limit, S_f
■ Modulus of rupture, S_R
■ Apparent modulus of elasticity, E_f
■ Failure condition
■ Any deviations from the prescribed methods

Testing Small, Clear Specimens

Significance and Use

The test provides data for comparing the mechanical properties of various species and data for the establishment of strength functions. The test also provides data to determine the influence of such factors as density, locality of growth, change of properties with seasoning or treatment with chemicals, and changes from sapwood to heartwood on the mechanical properties.

Test Specimens

Specimens 50 mm × 50 mm × 760 mm are used for the primary method, and 25 mm × 25 mm × 410 mm for the secondary method. For the loading span and supports, use center loading and a span length of 710 mm for the primary method and 360 mm for the secondary method.

Apparatus

- Testing machine of the controlled deformation type
- Bearing blocks for applying the load

Test Procedure

1. Place the specimen so that the load will be applied at the center of the span. The load is applied through the bearing block to the tangential surface nearest the pith (Figure A.71).
2. Apply the load continuously throughout the test at a rate of motion of 2.5 mm/min for primary method specimens and at a rate of 1.3 mm/min for secondary method specimens.
3. Record the load–deflection curve up to or beyond the maximum load. Continue recording up to a 150-mm deflection, or until the specimen fails to support a load of 890 N for the primary method specimens, and up to a 76 mm deflection or until the specimen fails to support a load of 220 N for secondary method specimens.
4. Within the proportional limit, take deflection readings to the nearest 0.02 mm. After the proportional limit is reached, the deflection is read by means of the dial gauge until it reaches the limit of its capacity, normally 25 mm. Where deflections beyond 25 mm are encountered, the deflections may be read by means of the scale mounted on the moving head.
5. Read the load and deflection of the first failure, the maximum load, and points of sudden change, and plot the load–deformation data.

Analysis and Results

- Calculate the modulus of rupture as

$$R = \frac{Mc}{I}$$

where

R = modulus of rupture, MPa
M = bending moment = PL/4
P = maximum load, N

FIGURE A.71 Wood specimen during static bending test. (Courtesy of Instron®)

L = span length, mm
c = distance from neutral axis to edge of sample = $h/2$
I = moment of inertia = $bh^3/12$
b = average width, mm
h = average depth, mm

■ Static bending (flexural) failures are classified in accordance with the appearance of the fractured surface and the manner in which the failure develops, as shown in Figure A.72. The fracture failure may be roughly divided into *brash* and *fibrous,* the term *brash* indicating abrupt failure and *fibrous* indicating a fracture showing splinters.

Report

■ Specimen identification, dimensions, span length, and other relevant information, such as moisture content
■ Load–deflection plot
■ Modulus of rupture
■ Failure condition.

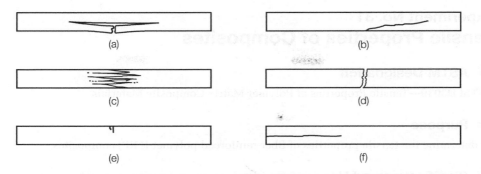

FIGURE A.72 Types of failure in static bending: (a) simple tension (side view), (b) cross-grain tension (side view), (c) splintering tension (view of tension surface), (d) brash tension (view of tension surface), (e) compression (side view), and (f) horizontal shear (side view).

Experiment No. 31
Tensile Properties of Composites

■ ASTM Designation

ASTM D3039—Tensile Properties of Polymer Matrix Composite Materials

■ Purpose

To determine the tensile properties of fiber-reinforced polymer (FRP) composites.

■ Significance and Use

This test method is designed to produce tensile property data for material specifications, research and development, quality assurance, and structural design and analysis.

■ Apparatus

Testing machine, grips, load indicator, and extension indicator

■ Test Specimens

Test specimens are obtained by cutting beam pieces from a continuous and aligned FRP composite lamina at different angles, such as 0, 22.5, 45, 67.5, and 90 degrees as shown in Figure A.73. FRP composite laminas are commercially available at a reasonable cost. There is no standard specimen size, but the most common specimen for ASTM D3039 is a constant rectangular cross section, 25 mm wide, 250 mm long, and 3 mm.

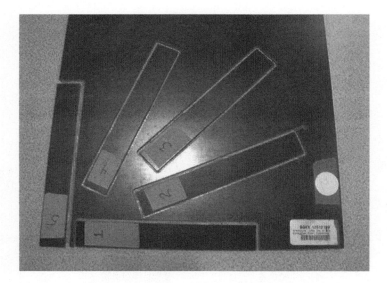

FIGURE A.73   Test specimens cut from FRP composite lamina at different angles.

Test Procedure

1. Measure the width and thickness of specimens with a suitable micrometer to the nearest 0.02 mm at several points.
2. Condition the test specimens at 23 $\pm$ 3°C and 50 $\pm$ 10% relative humidity for several hours prior to the test.
3. Place the specimen in the grips of the testing machine, taking care to align the long axis of the specimen and the grips with an imaginary line joining the points of attachment of the grips to the machine. Optional tabs may be bonded to the ends of the specimen to prevent gripping damage.
4. Attach a strain transducer or an extensometer to the specimen (Figure A.74).
5. Set the speed of testing at a rate of travel of the moving head of 1 mm/min and start the machine (Figure A.75).
6. Record the load versus strain (or transducer displacement) continuously or at frequent regular intervals. If a transition region or initial ply failures are noted, record the load, strain, and mode of damage at such points. If the specimen is to be failed, record the maximum load, failure load, and strain (or transducer displacement) at, or as near as possible to, the moment of rupture.
7. Record the mode and location of failure of the specimen.

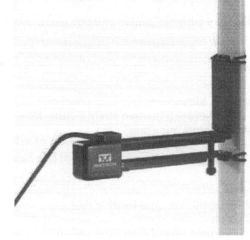

FIGURE A.74 Extensometer on an FRP composite specimen. (Courtesy of Instron®)

FIGURE A.75 Tension test on an FRP composite specimen. (Courtesy of Intertek Plastics Technology Laboratories)

Analysis and Results

- Tensile stress/tensile strength—Calculate the tensile stress at each required data point and the tensile strength as:

$$\sigma_i = \frac{P_i}{A}$$

$$\sigma_{\text{ult}} = \frac{P_{\max}}{A}$$

where

σ_i = tensile stress at ith data point, MPa
P_i = load at ith data point, N
σ_{ult} = tensile strength, MPa
$P_{\max}$ = maximum load carried by the specimen during the tension test, N
A = average cross-sectional area of the specimen, mm^2

- Tensile strain—Calculate the tensile strain as:

$$\varepsilon_i = \frac{\delta_i}{L}$$

where

ε_i = tensile strain at the ith data point, mm/mm
δ_i = extensometer displacement at the ith data point, mm
L = extensometer gauge length, mm

- Tensile chord modulus of elasticity
Calculate the tensile chord modulus of elasticity between strains of 0.001 and 0.003 as:

$$E_{\text{chord}} = \frac{\Delta_\sigma}{\Delta_\varepsilon}$$

where

E_{chord} = tensile chord modulus of elasticity, MPa
Δ_σ = difference in applied tensile stress between the two strain points, MPa
Δ_ε = difference between the two strain points, mm/mm

 Report

- Complete identification of the material tested
- Type of test specimen and dimensions, fiber orientation, and loading rate
- Stress–strain relationship, data table, and graph
- Tensile strength, strain at failure, tensile chord modulus of elasticity, and the transition strain, which is the strain at which any significant change in the slope of the stress–strain curve occurs
- Failure mode and location of failure

Experiment No. 32
Effect of Fiber Orientation on the Elastic Modulus of Fiber Reinforced Composites

Purpose

To measure the variation of the elastic modulus of continuous and aligned fiber-reinforced polymer (FRP) composite laminas with respect to the orientation of the fibers. The experiment also demonstrates another technique for measuring the elastic modulus of a material.

Significance and Use

This test provides information on the elastic modulus of FRP composite laminas at different directions of fibers. This information may be useful in material characterization, research and development, quality control, and design under certain circumstances.

Apparatus

- Simply supported set up with a span of 150 mm as shown in Figure A.76
- Dial gauge for deformation measurements
- Masses that are incremental by 250 g

Materials and Test Specimens

Test specimens are obtained by cutting beam pieces from a continuous and aligned FRP composite lamina at different angles, such as 0, 22.5, 45, 67.5, and 90 degrees as shown in Figure A.73 in Experiment 31. FRP composite laminas are commercially available at a reasonable cost. The beam size is 175 mm long × 25 mm wide × 3 mm deep.

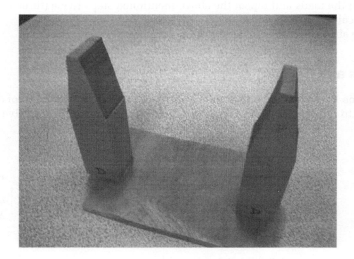

FIGURE A.76 Frame set up for supporting the beam.

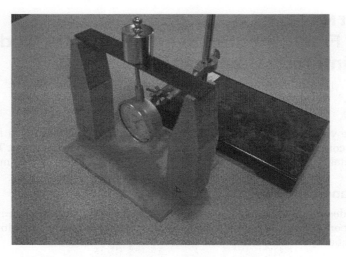

FIGURE A.77 Simply supported beam loaded at the center with a dial gauge for deflection reading.

Test Procedure

1. Examine the dial gauge; note its sensitivity and range. Record the specimen number being tested.
2. Center the specimen on the supports and set up the dial gauge at mid-span as shown in Figure A.77. Zero the dial gauge and ensure that there is good contact between the gauge head and the specimen.
3. Carefully stack the masses at mid-span in 250 g increments, up to 1250 g. Gently tap the dial indicator to ensure that the needle is not stuck and read and record the deflection readings at each load increment. In a table record the load (P) and the corresponding deformation (Δ).
4. Remove all the loads and repeat the above-mentioned steps twice (three readings in total) and calculate the average.
5. Repeat the above steps for the other four specimens.

Analysis and Results

- Using the load–deflection data, plot a graph of load (P) as a function of deflection (Δ), similar to the one shown in Figure A.78. Use the same graph for all five beams and

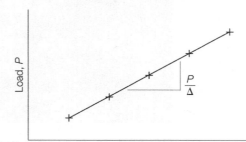

FIGURE A.78 Load versus deflection plot.

label them appropriately. Use a linear regression to find the slope P/Δ of each plot (show on plot).

- The mid-span deflection for a simply supported beam is well established and can be obtained from engineering references. This equation can be rearranged to calculate the Young's Modulus of each beam.

$$E = \frac{P}{\Delta} \times f(b,h)$$

where

$E =$ Young's modulus, MPa
$\Delta =$ mid-span deflection, mm
$P =$ applied load, N
$f =$ a function of the beam's width (b) and height (h) (consult the references to determine this function). The ratio P/Δ is the slope of the straight line formed when plotting load against deflection.

- Plot of the Young's modulus, on the ordinate, versus fiber direction, on the abscissa. Using the plot, develop a simple equation that relates E to the fiber direction.

Report

- Complete identification of the material tested
- Fiber orientation that displayed the largest/smallest deformation and the reason
- Relationship between fiber orientation and Young's modulus
- Your suggestion about the structural applications of this fiber-reinforced composite material
- Comment on any discrepancy in your data and the reasons

INDEX